UMTS NETWORKS
AND BEYOND

UMTS NETWORKS AND BEYOND

Cornelia Kappler

deZem GmbH, Germany

A John Wiley and Sons, Ltd, Publication

This edition first published 2009
© 2009 John Wiley & Sons, Ltd

Registered office

John Wiley & Sons Ltd, The Atrium, Southern Gate, Chichester, West Sussex, PO19 8SQ, United Kingdom

For details of our global editorial offices, for customer services and for information about how to apply for
permission to reuse the copyright material in this book please see our website at www.wiley.com.

Library of Congress Cataloging-in-Publication Data

Kappler, Cornelia.
  UMTS networks and beyond / Cornelia Kappler.
    p. cm.
  Includes bibliographical references and index.
  ISBN 978-0-470-03190-2 (cloth)
  1. Universal Mobile Telecommunications System. I. Title.
  TK5103.4883.K36 2009
  621.3845'6–dc22   2008041818

A catalogue record for this book is available from the British Library.

ISBN 9780470031902 (H/B)

Typeset in 10/12pt Times by Thomson Digital, Noida, India.

Contents

Preface

What is this book about?

This is a book on **Mobile Communication Networks**, in particular the **Universal Mobile Telecommunication System** (UMTS) and its successor technologies. UMTS is the successor of GSM, and is expected to become one of the world's most common mobile telecommunication technologies.

This is also a book on the **evolution** of Mobile Communication Networks—towards something called **4G**—and the **design principles** guiding both this evolution and individual technical choices.

As we will show, the design principles of a **Telecommunication Network** such as UMTS differ somewhat from those of traditional IP Networks, called **Computer Networks** in this book. In fact, this book is written especially for readers with a background in Computer Networks and aims at introducing them to the world of Telecommunication Networks.

This book therefore offers a systematic overview of UMTS and a thorough explanation of the technical details, comparing consistently the "Telecommunication solution" for solving a particular problem with the "Computer Network solution" for solving the analogous problem. We also discuss why particular technical solutions are favoured, why particular choices were made and how Communication Networks will develop in the future.

Today's Telecommunication Networks—e.g. UMTS—employ the IP protocol, as do Computer Networks. Telecommunication Networks, as do Computer Networks, offer data services, voice service and many other services. Therefore, it is often said that the evolution of Communication Networks consists of an overall technical **convergence**. The book's ongoing comparisons of "Telecommunication solutions" and "Computer Networks solutions" allows for a more detailed investigation of the phenomenon.

How is this book structured?

The first part of the book provides an in-depth description of UMTS network technology as specified in 2008. We cover architecture, protocols and overall functionality. We also discuss how UMTS evolved from the earlier mobile telephony system **Global System for Mobile Communications** (GSM).

The second part of the book discusses the successor technologies to today's mobile Telecommunication Networks and Computer Networks, e.g. **Long Term Evolution** (LTE)/ **System Architecture Evolution** (SAE)—also known as **Evolved Packet System** (EPS), **Ultra Mobile Broadband** (UMB), **Mobile WiMAX**, and **Next Generation Networks** (NGN). We give an overview of what these technologies are likely to offer, and describe in detail the ongoing research and development, in particular regarding the evolution of UMTS.

A typical chapter is structured as follows:
• An introduction to the technical problems covered in the chapter.
• Two or more subsections detailing the solutions for these problems employed by the different communities—typically the UMTS solution and the Computer Networks Solution.
• A discussion of the motivation for technical choices, and a comparison of the different solutions.
• A brief summary of the main results of the chapter.

A note about the discussion sections: it is in the nature of discussion to move beyond technical fact and to offer interpretations. While the author substantiates these interpretations with solid arguments, they are obviously inspired by her own—possibly somewhat opinionated—view of things. Readers are encouraged to review the discussions critically and to construct their own interpretation.

This book was completed in mid-2008 and accordingly reflects the status-quo at that time.

Conventions—what should the reader know before starting to read?

As the reader will soon become aware, *terminology* plays a crucial role in Communication Networks: there is a veritable abundance of technical terms and acronyms. Furthermore, basic terms may be interpreted differently in the various networking communities, or, alternatively, different terms mean the same thing! For example, what is called a **Mobile Node** in a Computer Network is conceptually equivalent to the **User Equipment** in UMTS, which in turn is not quite identical to what is called a User Equipment in a fixed Telecommunication Network. A very interesting example is **IP Protocol** for which various interpretations exist.

Correspondingly, special care is taken with the terminology in this book:
• Since this book covers parallel networks from different communities, generic terms are introduced for typical network components and functions. For example, "Mobile Station" is used as a technology-independent term for referring to the user's mobile device, known as the Mobile Node or User Equipment in specific technologies.
• The meaning of some terms is clarified and/or restricted (e.g. "IP Protocol") for the purpose of this book. The book-specific definition of these terms can be looked up in Appendix A. Obviously, the terms are used more loosely outside this book.
• Technical terms carrying a special meaning—either technology-specific or book-specific—are put in **bold** when they are introduced or explained. Terms referring to concepts defined in a narrow technical context are additionally capitalized in order to alert the reader to that fact, e.g. "**GPRS Attach**". Since capitalizing is somewhat cumbersome and also inhibits the reading flow, it is not applied to well-known terms such as "IP address".
• The most important terms and acronyms discussed in each chapter are summarized in "terminology boxes". Since some concepts are discussed more than once, the corresponding terms appear in more than one "terminology box". Vice versa, very special terms appearing only once in a chapter do not make it into the box. Of course, the index at the back of the book will also help in finding the corresponding reference. Book-specific terms are put in Arial in the terminology box in order to simplify their identification.

- Appendix A summarizes the key terminology, also indicating which terms are book-specific, and which terms valid generally.

What is this book not about?

The scope of a book becomes clearer by defining that which it does not include. This book focuses on the *network aspects* of UMTS and other Communication Networks. It is not a book on radio technology or services—although an overview of the UMTS radio technology is in fact given since it is one of UMTS's most important distinguishing features. UMTS's **Virtual Home Environment** (VHE) or **Personalized Service Environment** (PSE) are not the subject of this book.

Who is the envisaged reader?

This book is a textbook for advanced students and professionals.

As a textbook, its focus is on explaining the relevant concepts and on enabling the reader to study the original technical specifications.

As a book for advanced students, it assumes basic knowledge of communication networks, in particular IP networks, at about the level provided by [Tanenbaum 2002]. For example, the reader should be familiar with the **OSI Reference Model** and the protocols commonly used in IP Networks, such as IP, the **Dynamic Host Configuration Protocol** (DHCP), etc.

Finally, this book is aimed at professionals who wish to extend their knowledge both within and beyond the UMTS area. As such, the book goes beyond basic concepts and examines some aspects in detail. Special emphasis is given to providing the relevant references that will enable the reader to deepen his or her own expertise on the subject.

Storyline overview

This preface closes with a brief overview of the book's "storyline" in order to help readers orient themselves.

The book begins with Part I on UMTS. It provides three introductory chapters with background information that is vital in order to understand the hard-core technical chapters that follow:

- Chapter 1
  Introduces the main characters: Mobile **Telecommunication Networks**, in particular **UMTS**, and **Computer Networks**. We derive different **design principles** which guide their technical development. We also show how the difference in design principles is indeed rooted in the difference of the **business models** of the operators of these networks. We ask ourselves what **convergence** could mean in the light of the different approaches to network design.
- Chapter 2
  Provides an overview of both the technical and the non-technical sides of the evolution of mobile Communication Networks, in particular UMTS.
- Chapter 3
  Introduces **standardization** as a key activity in specifying mobile Telecommunication Networks. We cover a number of standardization bodies, in particular the **3rd Generation Partnership Project** (3GPP) responsible for UMTS.

The book now becomes more technical. We introduce the central concept of network **architectures**. We then provide a concise overview of the UMTS radio interface as one of UMTS's characteristic features. In subsequent chapters we present the individual architectural components of UMTS in more detail: what is their role, what is their substructure, and which protocols do they employ?

- Chapter 4
  Explains the concept of network architectures and presents the basic architectural components of UMTS—**Packet-switched Domain**, **Circuit-switched Domain**, **UMTS Terrestrial Radio Access Network**, **User Equipment** and **IP Multimedia System**—and other mobile Communication Networks such as WLAN.
- Chapter 5
  Covers the physical layer of the UMTS radio interface.
- Chapter 6
  Covers the Packet-switched Domain.
- Chapter 7
  Is about the Circuit-switched Domain.
- Chapter 8
  Describes the UMTS Terrestrial Radio Access Network.
- Chapter 9
  Deals with the User Equipment.
- Chapter 10
  Presents the IP Multimedia Subsystem.

The following chapters are function-oriented, and thus in some sense orthogonal to the previous architecture-oriented chapters. Each chapter deals with a key functionality provided by UMTS and shows how the individual architectural components presented above collaborate in order to provide this functionality. Concurrently, the Computer Network approach for providing the same functionality is presented. We also discuss how the design principles influenced the technical solutions.

- Chapter 11
  Presents *basic functionality* such as establishing connectivity between User Equipment and network, and setting up user sessions.
- Chapter 12
  Is concerned with *mobility support*.
- Chapter 13
  Explains the *security* concepts.
- Chapter 14
  Discusses *Quality of Service*.
- Chapter 15
  Deals with *session control*.
- Chapter 16
  Covers *charging*.
- Chapter 17
  Is on *policy control*.

- Chapter 18
  Moves on to an advanced functionality of UMTS, *the support of alternative access technologies*, e.g. WLAN. This feature is an indication of what to expect from successor technologies.
- Chapter 19
  Is the last chapter of Part I. It establishes the time line in which the functionalities discussed in previous chapters were introduced into UMTS and by doing so also reviews the previous chapters.

An Epilogue summarizes the first part of the book, and in particular revisits the original question about convergence of UMTS and Computer Networks.

In Part II we discuss the ongoing, highly active evolution towards the next generation of Mobile Networks, called the **4th Generation** (4G).

- Chapter 20
  Reflects Chapter 2 and sets the stage by considering the evolution of Mobile Communication Networks since UMTS was conceived. What are the business models of future Communication Networks, what functionality will they provide, how are they different from networks of previous generations?
- Chapter 21
  Presents EPS—sometimes called LTE/SAE—the evolution of UMTS towards 4G. We cover architecture, protocols and functionalities.
- Chapter 22
  Introduces other technologies whose immediate evolution may be candidates for 4G: **Ultra Mobile Broadband** (UMB), **Mobile WiMAX**, **ETSI's Next Generation Networks**—which in fact evolved from a fixed telephony network, and **PacketCable 2.0**—which originally evolved from cable TV.
- Chapter 23
  Gives an overview of the technology and ideas that are under discussion for 4G which, however, are at this point not included.

In an Epilogue to Part II we ask ourselves to what extent convergence between Telecommunication Networks and Computer Networks will be achieved in 4G.

Acknowledgements

The author would like to sincerely thank a number of colleagues for providing support in writing this book—unless otherwise stated, all of them are with Nokia Siemens Networks: Ralph Kühne (Univ. Tübingen, Germany), Frank-Uwe Andersen, Nadeem Akhtar (IIT Madras, India) and Ulrike Meyer (Univ. Aachen, Germany) reviewed critically several chapters, gave constructive as well as most welcome advice, and were invaluable discussion partners.

Mirko Schramm helped me keep up-to-date with the latest developments in 3GPP SA2, Günther Horn and Ulrike Meyer clarified the 3GPP work on security, Ralph Kühne and Uwe Föll did the same with 3GPP charging issues, as did Ulrich Thomas with the UTRAN. Jörg Swetina (NEC) provided me with insight into the ongoing work on 3GPP SA1, Max Riegel answered my numerous questions on WiMAX, Hannes Tschofenig shared his expertise on security issues in the IETF, Andreas Köpsel supported me regarding IETF mobility protocols and Pierre Lescuyer (Nortel Networks) explained radio interface issues.

A number of key ideas in this book were developed together with Robert Hancock and Eleanor Hepworth (Roke Manor Research, UK) in the course of our joint work on future Mobile Communication Networks.

Thanks go also to Georg Carle (Univ. Tübingen, Germany) who was crucial in the initial process of formulating what this book would be about.

This book is based on a course which I taught at the Technical University of Berlin, Germany. Thanks therefore go also to my students for their critical and curious questions which helped to improve both the book's concept and its technical accuracy.

Finally, I'd like to thank my editors at Wiley—Birgit Gruber, Richard Davies, Sarah Hinton and Sarah Tilley who supported the writing process, answered all of my queries and managed the production process seamlessly.

About the Author

Cornelia Kappler studied physics at the Ludwigs-Maximilians University in Munich, Harvard University and the University of Toronto where she earned a Ph.D. in 1995. Later, she moved into the area of future communication networks. She has worked for NEC, Siemens and Nokia Siemens Networks, managing international research projects, contributing to standards in the IETF and 3GPP, publishing scientific articles, writing patents and teaching courses at universities. Since April 2008 she has been responsible for the technology concepts of a Berlin start-up company, deZem, which is developing sensor networks for supporting energy efficiency.

Acronyms

1G	1st Generation
2G	2nd Generation
3G	3rd Generation
3GPP	3rd Generation Partnership Project
3GPP AN	3GPP Access Network
3GPP2	3rd Generation Partnership Project 2
4G	4th Generation
AAA	Authentication, Authorization, Accouting
AAL2	ATM Adaptation Layer 2
AF	Application Function
AGW	Access Gateway
AIPN	All-IP Network (in 3GPP System)
AKA	Authentication and key agreement
ALG	Application Level Gateway
AN	Access Network
AP	Access Point
APN	Access Point Name
AS	Application Server
ASN	Access Service Network (in WiMAX)
AT	Access Terminal (in UMB)
ATM	Asynchronous Transfer Mode
B3G	Beyond 3G
BB	Bandwidth Broker
BCCH	Broadcast Control Channel
BGCF	Breakout Gateway Control Function
BICC	Bearer Independent Call Control
BMC	Broadcast and Multicast Control
CAN	Converged Access Network
CARD	Candidate Access Router Discovery
CC	Call Control protocol
CCCH	Common Control Channel
CDF	Charging Data Function
CDMA	Code Division Multiple Access
cdma2000	Code Division Multiple Access 2000
cdmaOne	Code Division Multiple Access One

CDR	Charging Data Record
cell ID	Cell Identifier
CGF	Charging Gateway Function
ChC	Channelization code
CK	Cryptographic Key
CMTS	Cable Modem Termination System (in PacketCable)
CN	Core Network
CN	Correspondent Node
COPS	Common Open Policy Service Protocol
CPICH	Common Pilot Channel
CS Domain	Circuit-switched Domain
CSCF	Call State Control Function
CSN	Connectivity Service Network (in WiMAX)
CTF	Charging Trigger Function
DCCH	Dedicated Control Channel
DECT	Digital Enhanced Cordless Telecommunications
DiffServ	Differentiated Service
DL-SCH	Downlink Shared Channel (in EPS)
DOCSIS	Data Over Cable Service Interface Spec. (in PacketCable)
DRNC	Drift RNC
DSCH	Downlink Shared Channel (in GPRS)
DSCP	DiffServ Code Point
DSL	Digital Subscriber Line
DSMIPv6	Dual-Stack Mobile IPv6
EAP	Extensible authentication protocol
EAPOL	EAP over LAN
EDGE	Enhanced Data rates for GSM Evolution
EGAN	Enhanced GAN
EGANC	EGAN Controller
eNB	Evolved Node B
EPC	Evolved Packet Core
ePDG	Evolved PDG
ePDIF	Evolved Packet Data Interworking Fct.
EPS	Evolved Packet System
ESS	Extended Service Set (in WLAN)
ETSI	European Telecommunication Standards Institute
E-UTRA	Evolved UMTS Radio Access Network
E-UTRAN	Evolved UTRAN
FDD	Frequency Division Duplex
FDMA	Frequency Division Multiple Access
FMC	Fixed-Mobile Convergence
FMIP	Fast Handoff for Mobile IP
GAN	Generic Access Network
GANC	GAN Controller
GBR	Guaranteed Bitrate
GCID	GPRS Charging Identifier

GERAN	GSM/EDGE RAN
GGSN	Gateway GPRS Support Node
GMM	GPRS Mobility Management
GMSC	Gateway Mobile Switching Center
GPRS	General Packet Radio Service
GPS	Global Positioning System
GRE	Generic Routing Encapsulation
GRX	GPRS Roaming Network
GSM	Global System for Mobile Communication
GSMA	GSM Asscociation
GSN	SGSN or GGSN
GTP	GPRS Tunnelling Protocol B132
GTP-C	GPRS Tunneling Protocol - Control Plane
GTP-U	GPRS Tunneling Protocol - User Plane
Heterogeneous AN	Heterogeneous Access Network
HIP	Host Identity Protocol
HLR	Home Location Register
HMIP	Hierarchical Mobile IP
HPLMN	Home PLMN
HSDPA	High Speed Downlink Packet Access
HSPA	High Speed Packet Access
HSPA	High Speed Packet Access
HSS	Home Subscriber Server
HSUPA	High Speed Uplink Packet Access
IAPP	Inter Access Point Protocol (in WLAN)
I-CSCF	Interrogating CSCF
ID	Internet Draft
IEEE	Institute of Electrical and Electronics Engineers
IETF	Internet Engineering Task Force
IK	Integrity Key
IKE	Internet Key Exchange
IMS	IP Multimedia Subsystem
IMS-GWF	IMS Gateway Function
IMSI	International Mobile Subscriber Identity
IMT-2000	International Mobile Telecommunications at 2000 MHz
IntServ	Integrated Service
IP	Internet Protocol
IP-CAN	IP Connectivity Access Network
IPX	IP Roaming Exchange
ISDN	Integrated Services Digital Network
ISIM	IP Multimedia Services Identity Module
ISUP	ISDN User Part
ITU	International Telecommunications Union
I-WLAN	Interworking WLAN
LA	Location Area
LMA	Local Mobility Anchor

LSP	Label Switched Path
LTE	Long Term Evolution
M3UA	MTP3-User Adaptation Layer
MAC	Message Authentication Code
MAG	Mobile Access Gateway
MANET	Mobile Ad-hoc Network
MAP	Mobile Application Part
MAP	Mobility Anchor Point
MBMS	Multimedia Broadcast Multicast Service
MCC	Mobile Country Code
ME	Mobile Equipment
MEGACO	Media Gateway Control Protocol
MGCF	Media Gateway Control Function
MGW	Media Gateway
MIH	Media Independent Handover (in IEEE)
MIMO	Multiple-Input Multiple-Output
MIPv4	Mobile IPv4
MIPv6	Mobile IPv6
MM	Mobility Management (protocol)
MM Context	Mobility Management Context
MME	Mobility Management Entity
MMS	Multimedia Message Service
MNC	Mobile Network Code
MN-HoA	Mobile Node Home Address
MOBIKE	Mobile IKE
MPLS	Multiprotocol Label Switching
MRFC	Media Resource Function Controller
MRFP	Media Resource Function Processor
MSC	Mobile Switching Center
MSIN	Mobile Subscription Identification Number
MSISDN	Mobile Station International ISDN number
MT	Mobile Terminal
MT	Mobile Termination
MTP	Message Transfer Part
NAI	Network Access Identifier
NAS	Network Access Server
NASS	Network Attachment Subsystem
NAT	Network Address Translater
NEMO	Network Mobility
NGN	Next Generation Networks (in ITU)
non-3GPP AN	Non-3GPP Access Network
NSIS	Next Steps in Signalling
OCF	Online Charging Function
OCS	Online Charging System
OFDMA	Orthogonal Frequency Division Multiple Access
OMA	Open Mobile Alliance

PAN	Personal Area Network
PANA	Protocol for Carrying Authentication for Network Access
PCC	Policy and Charging Control
PCEF	Policy and Charging Rules Enforcement Function
PCRF	Policy and Charging Rules Function
P-CSCF	Proxy CSCF
PDCP	Packet Data Control Protocol
PDF	Policy Decision Function
PDG	Packet Data Gateway
PDP	Policy Decision Point
PEP	Policy Enforcement Point
PGW	Packet Data Network Gateway
PHB	Per-Hop Behaviour
PLMN	Public Land Mobile Network
PMIP	Proxy Mobile IP
PN	Personal Network
PNM	Personal Network Management
PPP	Point-to-Point Protocol
PS Domain	Packet-switched Domain
P-SCH	Primary Synchronization Channel
PSTN	Public Switched Telephone Network
P-TMSI	Packet Temporary Mobile Subscriber Identity
QAM	Quadrature Amplitude Modulation
QoS	Quality of Service in 3GPP Systems
QPSK	Quaternary Phase Shift Keying
R99	Release 99
RA	Routing Area
RACS	Resource and Admission Ctrl. Subsystem (in ETSI NGN)
RADIUS	Remote Authentication Dial-In User Service
RAI	Routing Area Identifier
RAN	Radio Access Network
RANAP	RAN Application Protocol
RAO	Router Alert Option
Rel-4	Release 4
Rel-5	Release 5
Rel-6	Release 6
Rel-7	Release 7
Rel-8	Release 8
RFC	Request for Comment
RFID	Radio Frequency Identification
RLC	Radio Link Control
RNC	Radio Network Controller
RRC	Radio Resource Control Protocol
RSVP	Resource Reservation Protocol
RSVP-TE	RSVP for Traffic Engineering
RTC	Real Time Protocol

RTPC	Real Time Control Protocol
SA	IPsec Security Association
SAAL	Signalling ATM Adaption Layer
SAE	System Architecture Evolution (see also EPS)
SBLC	Service-based Local Policy
SC	Scrambling code
SCCP	Signalling Connection and Control Part
SC-FDMA	Single-Carrier Frequency Division Multiplex Access
S-CSCF	Serving CSCF
SCTP	Stream Control Transport Protocol
SDP	Session Description Protocol
SDR	Software Defined Radio
SEG	Security Gateway (general)
SEGW	Security Gateway (in GAN)
SFT	Service Flow Template
SGSN	Serving GPRS Support Node
SGW	Serving Gateway (in EPS)
SGW	Signalling Gateway
SIM	Subscriber Identity Module
SIP	Session Initiation Protocol
SLA	Service Level Agreement
SM	Session Management
SMS	Short Message Service
SPR	Subscriber Profile Repository
SRNC	Serving RNC
SS7	Signalling System Number 7
S-SCH	Secondary Synchronization Channel
SSID	Service Set Identity (in WLAN)
STUN	Session Traversal Untilites for NAT
TAF	Terminal Adaptation Function
TCAP	Transaction Capabilities Application Part
TDD	Time Division Duplex
TDM	Time Division Multiplex
TDMA	Time Division Multiple Access
TE	Terminal Equipment
TFT	Traffic Flow Template
TLS	Transport Layer Security
TMSI	Temporary Mobile Subscriber Identity
TR	Technical Report
TS	Technical Specifications
UE	User Equipment
UICC	Universal Integrated Circuit Card
UIM	User Identity Module
UL-SCH	Uplink Shared Channel
UMA	Unlicensed Mobile Access
UMB	Ultra Mobile Broadband

UMTS	Universal Mobile Telecomm. System
URA	UTRAN Registration Area
URI	Uniform Resource Identifier
USIM	Universal Subscriber Identity Module
UTRA	UMTS Terrestrial Radio Access
UTRAN	UMTS Terrestrial Radio Access Network
UWB	Ultra Wideband
VANET	Vehicular Ad-hoc Network
VLR	Visited Location Register
VoIP	Voice over IP
VPLMN	Visited PLMN
WAG	WLAN Access Gateway
WAVE	Wireless Access in Vehicular Environments
WCDMA	Wideband CDMA
WEP	Wired Equivalent Privacy (in WLAN)
WiBro	Wireless Broadband
WiMAX	Worldwide Interoperability for Microwave Access
WLAN	Wireless Local Area Network
WSN	Wireless Sensor Network

Part I
UMTS Networks

1

Introduction

This is a book on mobile **Telecommunication Networks**, in particular the **Universal Mobile Telecommunication System** (UMTS) and its successors. UMTS is an evolution of the popular mobile Telecommunication Network **Global System for Mobile Communication** (GSM). Compared to GSM, UMTS offers a much higher throughput — in the Megabit range as compared to kilobit — and a much larger variety of services, e.g. web surfing, information services, mobile TV applications, etc.

UMTS has a long history. The first concrete ideas date from the 1990s; commercial deployment started in 2002. In mid-2008, UMTS had 250 million subscribers worldwide. Compared to GSM with over 2.5 billion subscribers, it is fair to say UMTS is not yet a mass-market phenomenon. In the medium term however, UMTS is expected to supersede GSM and become one of the world's predominant mobile telecommunication technologies.

This book offers a technical introduction to the networking aspects of UMTS. Additionally, UMTS is placed in the context of the **evolution** of mobile Telecommunication Networks — from GSM, to UMTS, to the technologies succeeding UMTS and finally to the so-called **4th Generation**.

The evolution of mobile Telecommunication Networks is strongly influenced by the evolution of **Computer Networks**, particularly the Internet: Mobile Telecommunication Networks are increasingly based on IP, the protocol used by Computer Networks. And just as Computer Networks, they offer whatever service is possible over IP: telephony, web-surfing, video-downloading etc. It is therefore often said that these technologies **converge**. However, mobile Telecommunication Networks and Computer Networks are designed under fundamentally different approaches, which makes this overall convergence somewhat surprising.

This chapter sets the stage for the book that follows. We start with a characterization of mobile Telecommunication Networks and Computer Networks, respectively. We then consider a seemingly unrelated, non-technical question, which however will turn out to be vital: what are the business models of the operators of these networks? On the basis of these business models we then derive the fundamentally different design principles which result in a **Telecommunication approach** and a **Computer Networks approach** to network design, respectively. Finally, we ask ourselves what convergence could mean in the light of these different approaches.

UMTS Networks and Beyond   Cornelia Kappler
© 2009 John Wiley & Sons, Ltd

Terminology discussed in Chapter 1:	
Bazaar-style	
Business Model	
Cathedral-style	
Computer Network	
Computer Network approach	
Convergence	
Design principle	
Global System for Mobile Communication	GSM
Integrated Services Digital Network	ISDN
IP Network	ISDN
(Network) architecture	
Operator control	
Protocol	
Telecommunication approach	
(mobile) Telecommunication Network	
Universal Mobile Telecomm. System	UMTS
User control	
Voice over IP	VoIP
Wireless Local Area Network	WLAN
Worldwide Interperability for	
Microwave Access	WiMAX

1.1 Mobile Telecommunication Networks and Computer Networks

The family of **Communication Networks** includes two very different members:
- Mobile Telecommunication Networks are an offspring of the plain old circuit-switched telephone networks and support services with devices moving at medium or high speed. UMTS is an evolution of GSM, which in turn could be called a mobile **Integrated Services Digital Network** (ISDN).
- Computer Networks comprise small home networks or office networks, Access Networks, as well as, of course, the Internet itself. Also, Computer Networks have—to some extent—become mobile. While access from truly moving devices is rarely supported, access from wireless, nomadic devices is common today, e.g. employing **Wireless Local Area Networks** (WLAN). The increasingly popular **Mobile Worldwide Interoperability for Microwave Access** (Mobile WiMAX) even supports mobility.

Telecommunication Networks and Computer Networks converge, both in the fixed and the mobile realm: they are increasingly offering the same services, and increasingly employ the same networking technology.

However, Telecommunication Networks and Computer Networks are also very different: a Telecommunication Network is a monolithic black box, offering a spectrum of high-quality services to a pampered user. A Computer Network, by contrast, resembles an open toolbox offering connectivity. More services are possible; however this requires technical sophistication on the user's side.

Despite these differences, we still observe the above mentioned technological convergence. This book examines both the differences and the convergence in detail.

1.2 Network Design Principles and Business Models

Telecommunication Networks and Computer Networks are developed on the basis of different design principles. This difference in design principles is rooted in the different business models of the respective operators. We examine both business models and design principles in the following subsections.

1.2.1 Business Models

What is a business model? While formal definitions exist, in this book on technology we use the term in a somewhat loose sense to denote the following: what products and services does a company offer? What is the competence area of the company? Who are the target customers? How does the company generate revenue?

The business model of operators of mobile Telecommunication Networks consists in selling *high-quality user services*, e.g. telephony, building on *carefully controlled connectivity*. That is connectivity which hides complicating facts such as that the user is moving at a high speed and that several other users are trying to run high-bandwidth applications and compete for the same radio resources. Ideally, the user just sees a button for "service X", presses it, and the service is delivered. The operator serves users that do not have the time, the ability or the inclination to deal with technical details.

It is noteworthy that telecommunication operators do not sell services only, nor do they just sell connectivity! Rather, they sell an end-to-end solution. Consequently, they pursue an integrated approach in running their business that is also called **vertical silo**: they run the network, they provide carefully-controlled connectivity, they deliver services on top of this connectivity, they manage subscriber credentials and preferences, they bill their subscribers, etc.

The business model of the operator of a Computer Network—e.g. a Internet Service Provider (ISP), a WLAN hotspot operator, or the manager of a university network—is different. Such operators just provide *connectivity* with other Computer Networks, particularly the Internet, and possibly a web-hosting and email-hosting service. Usually, the user is responsible for hosting and running anything on top of this, e.g. Voice over IP.

Consequently, there is no need for operators or owners of Computer Networks to own the entire business chain. They can outsource, e.g. network administration, subscriber handling and billing. Service provisioning is not their problem.

Based on these different business models, we now identify a number of design principles followed in Telecommunication Networks and Computer Networks, respectively. To bring the point across, these design principles are deliberately simplified, and black-and-white; of course, real-life networks are much more complicated!

1.2.2 The Cathedral and the Bazaar

In his well-known and thought-provoking essay "The cathedral and the bazaar", Eric S. Raymond contrasts two fundamentally different approaches to software engineering

[Raymond 2000]. He calls them **cathedral style** and **bazaar style**. Whereas cathedral-style is the typical closed-group development model of proprietary software (the example at the time the essay was written was GNU Emacs), bazaar-style is the public development model of open-source software, exemplified by Linux. We take the liberty of reusing the metaphor in the context of mobile networks.

A mobile Telecommunication Network is very much like a cathedral. A cathedral is carefully crafted as a whole system, and will only ever be built as a whole system. A master plan is drawn up in order to guide constructions. A selected group of engineers work according to the master plan. There is a clear end point at which the cathedral is finished. Each building block fulfils a particular role. If a building block is missing, faulty, or in the wrong place, the entire work can become endangered.

Note how the cathedral-style design reflects the integrated business model of operators: a Telecommunication Network is designed as an integrated whole that enables carefully controlled connectivity, services, subscriber management, etc.

An important property of cathedral-like Telecommunication Networks is that the coordinated action of all parties involved makes it feasible to achieve major updates and a directed evolution of the technology. For example, we are currently witnessing the eventual transition from GSM to UMTS.

A Computer Network is natively rather like a bazaar.[1] A bazaar consists of a random number of stalls, and there is only a rough plan and limited coordination. People set up their stalls in the bazaar independently. The bazaar develops, and there is no end to its evolution. If the bazaar is large enough it is of minor importance whether a particular stall is present, the bazaar as a whole continues to work.

An important property of the bazaar-like Computer Network is that the evolution of technology can only happen locally, in small steps. However, a substantial update of the technology, e.g. in the entire Internet, is rather difficult because it involves changes in a large number of network elements in a large number of subnets. For example, the transition from IPv4 to IPv6 has not yet happened although there are many good reasons for it.

1.2.3 Operator Control and User Control

Operators of Telecommunication Networks sell an end-to-end solution to users who are not expected to engage in technicalities. Consequently, the operator needs to have control of the network and services as well as a detailed understanding of the network conditions. We thus conclude that a Telecommunication Network must offer comprehensive operator control. The control exerted by a telecommunications operator is traditionally realized by locating powerful, centralized control nodes at the core of the network.

Operators of Computer Networks, by contrast, provide just connectivity. Their interest in controlling the network is limited to ensuring this connectivity. Any additional features are the responsibility and under the control of the user. Consequently, the Computer Network design approach has always been to locate functionality and intelligence at the edge of the network, e.g. in the mobile devices and in Access Routers, and to keep the network as simple as possible.

---

[1] It is also possible set-up a Computer Network without many of its bazaar-like features, e.g. in company environments.

We conclude that in a Computer Network, there is less control, it is decentralized and to considerable degree in the hands of the user.

In Chapter 4 we will take up the central term "control" and provide the reader with a more detailed discussion.

1.2.4 In the Beginning is the Architecture and In the Beginning is the Protocol

When the engineers of a Telecommunication Network start designing a new system, they think about the functionality they would like to achieve—and design an **architecture**. They break down the functionality—e.g. mobility support and charging—group it into network elements and determine the relation between the network elements. Once the architecture is fixed, **protocols** are defined that enable the network elements to talk to each other. Typically, there is one protocol per pair of network elements, and the protocol carries everything that these two network elements need to tell each other. The advantage of this approach is that a coordinated network operation is easy to achieve. For example, when a mobile terminal moves and changes its point of attachment to the core network, it is simple to trigger the appropriate security functions. This is because the network element that is in control of mobility is acquainted with the network element in control of security.

The problem with this approach is that when the architecture is redesigned, the functionality associated with each network element changes, and hence the protocols need to be redesigned, also.

When engineers design new features for Computer Networks, e.g. the Internet, they think about the functionality they would like to achieve—and design one or more protocols. There is one (or more) protocol(s) per functionality, for example there is one protocol to support mobility and another protocol for metering. Network elements in a Computer Network, particularly routers, are comparatively generic. Often, they just need to be fitted with a specific piece of software in order to react meaningfully to the protocols they are supposed to process. Thus, architectural considerations are minimal. The advantage of this approach is that the resulting network is rather flexible with regard to the functionality which it provides. If more functionality is desired, another protocol is added. The problem with this approach is that each protocol operates independently from the others. This way it is more difficult to achieve a coordinated network operation.

Figure 1.1 illustrates the concept by providing concrete examples. We see a number of communicating network elements and their protocol stacks. *At this point the reader should not worry about the acronyms!* Rather, the focus should be on the *topology* of the protocol stacks: in (a) the telecommunications-style protocol stacks of three UMTS network entities (called UTRAN, SGSN and GGSN) are illustrated. Each pair of network elements has its own, specific, protocol stack with an IP layer in between. The top-most protocol carries the actual functionality: the middle network element, the SGSN, communicates with the GGSN on the right via GTP-C over UDP. GTP-C carries information about, e.g. mobility, charging, session control, etc. To the left, the SGSN communicates with the UTRAN via RANAP over SCCP/M3UA/SCTP. RANAP carries information about mobility and charging as well as about other functionality. By contrast, the wireless Access Network in (b) exhibits network elements featuring a toolbox of ellipses on top of the IP protocol, each ellipse standing for one protocol providing one functionality, in this case, e.g. mobility, admission control and metering. In order to achieve a different functionality, the ellipses are exchanged.

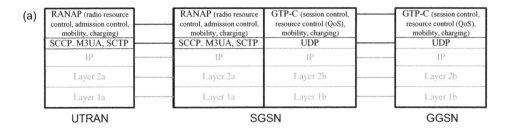

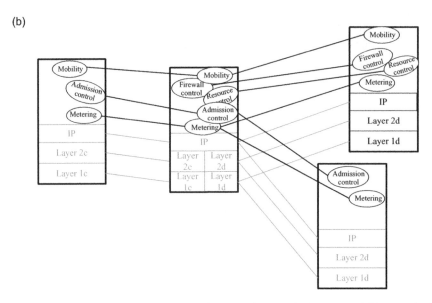

Figure 1.1   Examples of protocol stacks of (a) UMTS network elements and (b) network elements in a wireless Access Network to the Internet

1.2.5 Convergence

In the light of the comparison above, what does it mean that Telecommunication Networks and Computer Networks converge? In the strict sense of the word, **convergence** means that the networks evolve towards becoming indeed technically identical, indistinguishable and exchangeable at some point in the future. Is this the case here? Or is the convergence rather superficial and indeed stops at both being IP Networks and offering the same services?

We will discuss this question in two tranches. In Part I of this book we look at the convergence that is realized by UMTS and its contemporary Computer Network technologies. In Part II we will investigate the same question for the mobile Telecommunications Networks that come after UMTS.

1.3 Summary

This chapter introduced concepts that form the basis for the entire book. A mobile Telecommunication Network was originally a mobile, circuit-switched telephone network. Today, however, Mobile Telecommunication Networks are IP Networks in addition to circuit-switched networks, and offer data services such as web-surfing as well as telephony, video streaming, etc.

The business model of the operator of a mobile Telecommunication Network provides an end-to-end solution including carefully-controlled connectivity and high-quality user services. Correspondingly, the network design principles are typically cathedral-style development, support of operator-control and a central role for the network architecture.

A Computer Network is an IP Network, and can be found in many orders of magnitude, from home offices to the Internet. As IP Networks, Computer Networks may potentially support any communication service. Access to Computer Networks can be nomadic. Truly mobile support can currently only be offered on the basis of Mobile WiMAX.

The business model of the operator of a Computer Network provides connectivity with other Computer Networks. The user is responsible for any additional service. Correspondingly, the network design principles are typically bazaar-style development, user-control and a central role for protocols.

Since mobile Telecommunication Networks and Computer Networks are based on the same IP technology and offer the same services it is often observed that Communication Network technology converges.

2

UMTS Motivation and Context

UMTS and future networks evolving from UMTS are commercial mobile Telecommunication Networks. The attribute "commercial" means that these networks come into existence because somebody hopes to engage in commerce with them, and—put simply—to make money. One makes money by selling the network, or services offered by the network, to customers. Customers, however, only buy UMTS if it satisfies a need which they have. This need must first of all exist, and, additionally, there must not be a more convenient or cheaper way to satisfy this need.

This simple economic statement is crucial for understanding UMTS. UMTS was developed in order to create and satisfy a future market that could not be satisfied with the mobile Telecommunication Networks existing at the time. We therefore start this chapter with a section reviewing the development of the mobile telecommunications market. We also highlight the regional differences that are the important drivers in the market. We then provide some general background on mobile telecommunication technology.

In the next section, we begin thinking about UMTS in much the same way people did in the mid-1990s: Given the booming mobile voice market, on the one hand, and the exploding Internet, on the other—what applications and services would customers want ten years on? And, once this is settled, what requirements can be derived for the system-to-be?

In the last section of this chapter we change the angle of observation. From the enlightened perspective of a decade later we analyse how UMTS compares to other mobile technologies—WLAN, WiMAX, etc.—which have since come into existence. WLAN in particular was first perceived as a competitor for UMTS. Today, however, this and other mobile technologies are mostly seen as complementing UMTS. In fact, the latest release of UMTS describes how a WLAN access can be integrated into a UMTS network. Integration of other alternative access technologies is being worked on. This changed viewpoint also sets the scene for what comes beyond UMTS, as will be described in Part II of this book.

UMTS Networks and Beyond   Cornelia Kappler
© 2009 John Wiley & Sons, Ltd

Terminology discussed in Chapter 2:	
Bluetooth	
cdmaOne	
Cell	
Charging	cdma2000
Code Division Multiple Access 2000	
Data service	
Enhanced Data rates for GSM Evolution	EDGE
Evolved Packet System	EPS
1st Generation	1G
2nd Generation	2G
3rd Generation	3G
4th Generation	4G
General Packet Radio Service	GPRS
Handover	
High Speed Packet Access	HSPA
HSPA +	
International Mobile Telecommunications at 2000 MHz	IMT-2000
International Telecommunication Union	ITU
IP Multimedia Subsystem	IMS
Location-based service	
Long Term Evolution	LTE
Multimedia Message Service	MMS
Multimedia service	
Personalized service	
Quality of Service	QoS
Roaming	
Security	
Short Message Service	SMS
Standardization	
Ultra Mobile Broadband	UMB
Ultra Wideband	UWB

2.1 The Evolution of the Mobile Telecommunication Market

2.1.1 Overall Market Evolution

Both mobile Telecommunication Networks and the fixed Computer Networks, in particular the Internet, experienced a breakthrough in development starting in the 1990s. The same happened later again, with regard to wireless access to the Internet. Figure 2.1 illustrates the growth in the number of subscribers to mobile Telecommunication Networks since 1995 broken down into the most important regions today in terms of influence on the technology development.

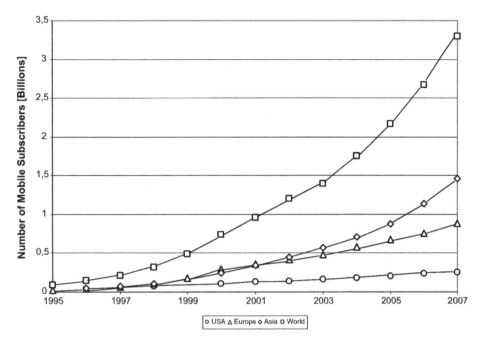

Figure 2.1 Number of mobile cellular subscribers in Mio since 1995. "Europe" includes all countries of the European continent; "Asia" likewise. Data from [ITU-D]

Subscriber numbers continue to increase. Observe the slight dip when the "new economy" broke down—the reader may recall that the highly successful stock market for communication technology crumbled in early 2001. The communication market took a severe downturn from which it only slowly recovered. As Figure 2.1 shows, however, that overall there is still a quasi-exponential growth of mobile subscriber numbers. Forecasts on market development—even today—keep underestimating the yearly increase in the number of mobile subscribers. In 2002, the number of mobile subscribers overtook the number of fixed line subscribers. For the record, it is expected that the number of newly sold mobile laptops will soon surpass that of fixed desktop computers.

Figure 2.1 also shows that Europe was initially the leading region (relative to the number of inhabitants) but has long been overtaken by Asia. Indeed, growth in Asia is continuing, whereas the markets in Europe and the US are saturating. Whereas in most Asian countries (e.g. China, India), mobile phones are not yet common place, in some European countries the mobile penetration rate is now over 100%. Asia is therefore seen currently as the most important future market for mobile telecommunications.

Mobile telecommunications is not a single, uniform technology. In fact, it is many different technologies. For example, different regions developed different technologies which are usually incompatible. On the time axis, we distinguish different **generations** of mobile Telecommunication Networks. The breakthrough came with the **2nd Generation** (2G), particularly with the GSM, which was developed in Europe. GSM is the most popular 2G Network, with a share of about three quarters of the market. UMTS is a **3rd Generation**

Network (3G) which descended from GSM. Another third generation technology is **Code Division Multiple Access 2000** (cdma2000) [3GPP2], backed mainly in the US.

In comparison with 3G Networks offering similar features, UMTS also has a share of three quarters of the market. However, market penetration is certainly behind the forecasts and has only recently picked up speed: In August 2005, there were 33 million 3G subscriptions worldwide. In the summer of 2008, this number had climbed to 265 million (including the recent update **High Speed Packet Access** (HSPA)), i.e. not even a 10th of the 3.5 billion worldwide mobile subscribers are using UMTS/HSPA. The reader may remember the high expectations and prices with which UMTS licences were procured around the year 2000.

Regional distribution is also of interest: In 2005, over 50% of subscriptions for UMTS and similar technologies were in Japan where the first commercial 3G Networks were launched in 2002, with around 40% in Europe. In 2007, 50% of subscriptions are found in the Asia-pacific region—while countries like South Korea and China have also become important players, and 50% are made in Europe. One can here also observe that the Asian market is the most dynamic with the keenest interest in new technology and applications. This is also interesting when one wonders what comes after UMTS. It is certain that the viewpoint of Asian consumers as well as the interests of Asian manufacturers and operators will play an important role.

2.1.2 Service Evolution

The **1st Generation** (1G) of mobile Telecommunication Networks offered voice service, exclusively. In 2G, support for fax and some **data services** such as a mobile-adapted web browsing functionality (WAP), **Short Message Service** (SMS) and **Multimedia Message Service** (MMS) were added. Of course it is also possible to use a 2G phone as a modem for data services or fax.

In the mid-1990s, when UMTS development started, it was already obvious that the prospects for the mobile telecommunications market prospects were very promising. However, it was also expected that the market for simple telephony would saturate at the turn of the century and that its profit margin would become extremely low. At the time, mobile telecommunications offered telephony—almost exclusively. Consequently support for new services had to be developed. But what would those services would be?

Given the explosive growth of the Internet, mobile Internet and Intranet access as well as support for data services were identified as important areas. Intranet access includes remote access to company Intranets, e.g. mobile banking or access from the mobile office, as well as remote monitoring and control, e.g. of the home. Additionally, support for **multimedia services** was identified. Multimedia services are services that literally include more than one medium. Examples are streaming video, video conferencing, music on demand, interactive gaming and joint working on documents while making a voice call (also called "rich voice"). These services were singled out for the future market that UMTS would satisfy.

Figure 2.2 shows a forecast from 2001 of what operator revenue will be composed of in 2010. It was predicted that revenue from simple voice will be barely over a quarter of total revenue. The majority of revenue will come from different multimedia services. It must be noted that in 2005, 80% of revenue was still coming from simple voice service. Most of the remaining 20% in 2005 came from SMS, so there is still some way to go in order to satisfy the forecast. On the other hand, the market is notoriously difficult to predict. For example, in Japan, watching TV on mobile devices and mobile downloading of music titles experienced a breakthrough in 2005. We will analyse today's market in more detail in Chapter 21 when we study the future evolution of mobile telecommunications.

2.2 The Evolution of Mobile Telecommunication Technology

It is now time to look at the development of mobile telecommunications from a technical perspective. Table 2.1 shows the evolution from the 1st generation (1G) up to what is expected to come after 3G, the **4th Generation** (4G). The first mobile user devices became available in the middle of the last century. They were the size of a car boot and were mobile because they indeed travelled in a car boot. Mobile user devices have since decreased in size dramatically. Today, researchers work on mobile sensors in the range of millimetres. Some of the items in Table 2.1 are singled out for more detailed discussion below. The "4G" column will be discussed in Part II of this book.

Bandwidth. An important characteristic of mobile networks is the bandwidth available to a single user on the radio interface. The radio interface is naturally the bottleneck of the network. Note that ever more bandwidth per user is only of interest when services other than voice are being utilized. We observe a steady increase of bandwidth over time. In the end of 2006, 5 Gb/s were demonstrated in a field trial.

Cell radius. Another interesting aspect of evolution is the size of a single **cell**, i.e. the region covered by a single antenna. In early 1G Networks, it was not possible to perform a **handover**, i.e. to move out of the coverage of one antenna into the coverage of another, while maintaining an ongoing phone call. Furthermore, it was not possible to initiate a call to mobile users without knowing in which cell they were currently located. Interestingly, in today's WLANs support for handover and paging is only slightly better. In early 1G Networks the missing handover support

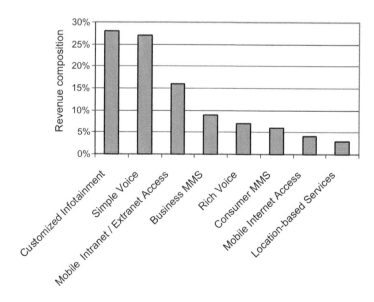

Figure 2.2   Forecast from 2001 for the composition of European UMTS revenue in 2010 [UMTS Forum 2001]

Table 2.1 Comparison of generations of mobile Telecommunication Networks

	1st Generation (1G)	2nd Generation (2G)	3rd Generation (3G)	Expectation for 4th Generation (4G)
Timeframe	1950s – mid 1990s	1990s – 2020?	2001 – ...?	2010 onwards?
Technology	NMT, AMPS,...	GSM (worldwide) (Americas, Asia), D-AMPS (America), PDC (Japan),...	IMT-2000, e.g. UMTS, CDMA2000, mobile WiMAX	IMT-Advanced
Standards	Mostly proprietary, domestic	A number of closed standards	A number of open standards	Integration of heterogeneous standards
Bandwidth (theoretical)		Initially <10 kbps, evolves to >50 kbps	Originally up to 2 Mbps, evolves to >40 Mb/s	100 Mbps mobile and 1 Gbps nomadic
A/D	Analogue radio, analogue/digital network	Digital	Digital	Digital
CS/PS	Circuit switched	Circuit switched	Circuit and packet switched	"All-IP"
Cell radius	in the 100 km range	Kilometers	Meters to kms	cm to kms
Roaming	Basic (national scope)	Advanced (continental scope)	Global (within same technology)	Global, intertechnology
Services	Speech	Speech, some data (MMS, SMS, WAP)	Speech, data, web-portals, multimedia, location-based services	Any IP-based service

was alleviated by making cell sizes as large as possible, for example in Germany with a radius up to 150 kilometers. This, however, decreased the overall capacitiy of the network because the overall capacity of a cell is (to some degree) independent of its size; in a given area, many small cells can serve more users than one big cell. In 2G, paging and handover no longer posed a technical challenge and consequently cell radii shrank down to a kilometer range. Cell radii in UMTS can be even smaller, down to meters if need be.

Analogue/Digital. The early 1G Networks were of course completely analogue. The fixed part of the network later became digital. However the radio interface remained analogue. The transition to 2G is defined by the radio interface also becoming digital.

Standards and Technology. The table also shows the growing impact of **standardization**. 1G Networks were barely standardized. At the time, telecommunication operators were usually national monopolists, buying their equipment from a single manufacturer. This implied that all subscribers of this operator were delivered mobile phones made by the same manufacturer. What is more, for each subscription, mobile communication was restricted to a single country: Usually, neither equipment nor networks could interoperate between countries. Markets, however, became more open. It became desirable for operators to allow their users to attach to and use the networks of other operators, a feature known as **roaming**. For manufacturers it became desirable to be able to sell their equipment to all operators, even at the cost of increased competition from other manufacturers. Therefore, starting with 2G, mobile Telecommunication Networks became standardized on a large scale. With 3G, the standard even became open, i.e. accessible to everybody. Standardization, however, does not imply that it is possible to agree on a single world-wide standard. Rather, a number of incompatible regional standards exist, hampering global roaming. In 2G, Europe developed GSM, the US developed **cdmaOne** and D-AMPS and Japan developed PDC. For 3G, an attempt was made to develop a single world-wide standard on the basis of the **International Mobile Telecommunications at 2000 MHz** (IMT-2000) family concept. This endeavour, however, did not succeed and in 3G we have, among others, UMTS, cdma2000 and—as the most recent addition—**Mobile WiMAX**. For 4G, a new approach to the problem promises worldwide roaming: the integration of heterogeneous standards such that inter-technology roaming is possible. We will hear more about standardization in Chapter 3.

Circuit Switched/Packet Switched. Mobile Telecommunication Networks are direct descendants of circuit-switched fixed-line telephony. Therefore, GSM is circuit-switched. This means that each call is reserved the one-size-fits-all identical bandwidth. This is perfect for voice with its constant bandwidth needs. However, with the broadening of supported services the picture changes: Data traffic tends to be bursty, and, moreover, it does not have real-time requirements: It is not crucial for an email to arrive within, say, 100 ms. Booking a constant bandwidth for data is therefore quite wasteful. Rather, data traffic is the perfect background traffic for filling up currently unutilized bandwidth when need be. Hence, for example, in "2.5G", **General Packet Radio Service** (GPRS), **packet-switching** based on the IP protocol, and consequently bandwidth-sharing were introduced. This was the first, irreversible, step towards convergence of the telecommunications world and the Internet.

2.3 The Genesis of UMTS

The history of UMTS, or, more precisely, of 3G technology, began in the mid-1980s with work by the **International Telecommunication Union** (ITU). By the mid-1990s, the ITU defined a framework and requirements for 3G Networks, called IMT-2000. It was—rightfully—considered unfeasible to expect there to be a single 3G technology worldwide: players having stakes in a particular 2G technology would of course try to protect their investment and consequently promote a 3G technology that required as little modification to their existing network as possible. Therefore, IMT-2000 became an umbrella for a family of 3G standards, among them the aforementioned UMTS, cdma2000 and Mobile WiMAX. All IMT-2000 family members must satisfy the same high-level requirements, particularly regarding radio-interface bandwidth, must have a particular network structure and must allow roaming to subscribers of other IMT-2000 family members. It must be noted that the roaming requirement is largely of theoretical value. Terminals capable of communicating with more than one IMT-2000 member technology are not exactly commonplace.

Also, in the mid-1990s, the European Union funded 3G research projects on the 4th Framework programme—just as 4G research projects are funded in the European Union's 6th and 7th Framework programmes today. In the late 1990s, the specification and, simultaneously, standardization of UMTS started. Since many vendors and operators are involved in the deployment of UMTS, it is essential to standardize the technology to a considerable extent. We will hear more on standardization in Chapter 3. The further steps taken until commercial operation are illustrated in Figure 2.3.

When one embarks on the specification and standardization of a new technology, it is important to have a clear idea of the services it will offer, and of the requirements it must satisfy in order to realize these services. These are the subject of the following sections.

2.3.1 UMTS Services

One important goal of UMTS was to make it a universal networking technology that offers communication anytime, anywhere. GSM is so widespread that it already almost satisfies this goal, at least for voice-communication. UMTS has yet to live up to it.

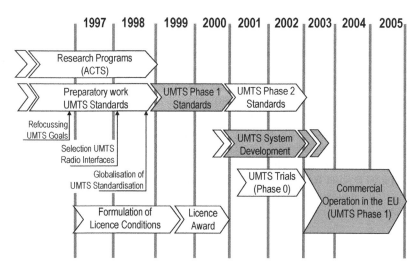

Figure 2.3 UMTS Time schedule. Reproduced by permission of John Wiley & Sons, Ltd, from *UMTS The Fundamentals* by B. Walker, P. Seidenberg, M.P. Althoff (2003)

As described in Section 2.1.2, UMTS was developed in order to support data services as well as mobile Internet and Intranet access, e.g. for telemedicine and remote monitoring and control of the home. Multimedia services were also targeted, in addition to the voice service already supported by GSM.

Furthermore, UMTS will support information services, particularly so-called **location-based services**. These services build on the fact that the location of an active mobile node is known to the network with the accuracy of a single cell. Picture a traveler abroad in London in need of some food. She orders a pizza using a location-based service via her UMTS terminal: because the network knows her location, it can direct the order to a nearby pizza-place in London rather than to a pizza-place in her home town of Berlin. This way, the pizza will still be hot when it is delivered, resulting in improved customer satisfaction. Another location-based service could provide the user with the local movie programme.

Finally, UMTS will also support **personalized services**. The idea is that users can create their own personal service environment which is accessible on all kinds of end devices and in all networks.

Location-based services and personalized services are beyond the scope of this book.

2.3.2 UMTS Technical Requirements

Many of the high-level technical requirements for UMTS are in fact requirements for IMT-2000 Networks in general [ITU Q.1701]. We group the first set of requirements around the services UMTS that will support.

Anytime anywhere communication.

- The obvious requirement is a single, global standard which, however, was not feasible. Instead we have the IMT-2000 requirement that roaming must be possible between all 3G technologies (see the comment on the practical value of this requirement earlier in this section).
- It is also helpful to have a unified spectrum, i.e. the same frequency bands being allocated to the technology world-wide. GSM, for example, does not have a unified spectrum, and dual-band or tri-band terminals are necessary for intercontinental trips. Unfortunately, a unified spectrum is rather difficult to achieve as spectrum is a scarce resource that has been assigned to the various technologies in different countries in different ways. In Chapter 5 we will hear more about this.
- Finally, a high coverage is required and is often enforced by the regulatory authorities.

Data Transport. As discussed in Section 2.2, this results in a requirement for the support of packet-switching. In Chapter 4.4 it is explained how this is achieved.

Mobile Internet and Intranet access. This mandates a gateway to the Internet, also see Chapter 4, Section 4.4 for more details.

Multimedia services. This results in a large number of requirements for the future network:

- High bandwidth must be provided, particularly on the radio interface. IMT-2000 requires 144 kb/s in vehicular environments and 2 Mb/s in indoor environments.
- Both symmetrical and asymmetrical bit-rates must be supported, i.e. equal, resp. unequal, bit-rates uplink and downlink. Voice causes, e.g. symmetrical bit-rates. Asymmetrical bit-rates must be supported because for some services, such as web surfing, downlink traffic has more volume than uplink traffic.

- Services with fixed and with variable bandwidths, both in time and between services, must be supported. As for data services, the one-size-fits-all concept for bandwidth from GSM is no longer suitable. Rather, something more flexible is needed, particularly packet-switching rather than circuit switching.
- The quality of voice must be as least as good as that in GSM in order to give users switching from GSM to UMTS the same experience as before.
  - If voice is transported over IP this may create difficulties because IP Networks serve packets on a first-come-first-serve basis. Therefore we need support for **Quality of Service** (QoS), i.e. a means of treating some packets with greater priority than others, or to explicitly reserve network resources for specific packets. This becomes even more important when real-time with higher bandwidth than voice, e.g. video conferencing, will also be supported. We will come back to this point in Chapter 14 on QoS.
  - Users should hardly notice a **handover**, i.e. when they move out of the coverage of one antenna into the coverage of another antenna. In GSM, a handover is only perceptible as a background click. IP Networks however, currently do not support **seamless handover**. We will come back to this point in Chapter 12 on Mobility.
- It must be possible to deploy new services, particularly new multimedia services, even after the network has been installed, because services are seen as a rapidly evolving field. That is, we need an extendable Service Network. Observe that in previous Telecommunication Networks, the services offered, e.g. voice, SMS, etc., have been cathedral-style, hard-coded into the network; one might even say the networks have been designed for this specific service. The telecommunications world was now inspired by the Internet world where services and applications are independent from the infrastructure and can be deployed at any time. This topic is covered in more detail in Chapter 10 on the **IP Multimedia Subsystem** (IMS).

The second set of requirements deals with the operator's interest in the network. Operators have a obvious interest in controlling their networks. They like to know who is using their resources, and they like to make sure that only authorized, paying users are using them. Resource usage should be as efficient as possible, and, generally, the costs for setting-up and operating the network as low as possible. From this we can formulate the following requirements for UMTS:

- The network must be highly secure, including authentication and authorization functionality. This, of course, is also in the user's interest. We will hear more of this in Chapter 13 on Security.
- The network must provide a sophisticated charging functionality, which is covered in Chapter 16 on Charging.
- The network must be backwards compatible to GSM. The typical operator active in UMTS standardization and specification already owns a GSM Network. It is unfeasible, both financially and technically, to invest into a completely new network, and to switch over all users to UMTS on a Day Zero. Rather, it must be possible to shift the existing network gradually to UMTS, and to first deploy UMTS in an "island fashion". Furthermore, users of UMTS must be able to hand over to GSM where UMTS coverage is not yet available.

The requirements listed in this section are of course not exhaustive; they are merely intended to provide an overview and set the stage for the next chapters. The reader interested in more requirements may study [ITU Q.1701].

2.4 Comparison of UMTS with Other Mobile Technologies

2.4.1 WLAN

When consideration was first given to UMTS, there was only fixed-line access to the Internet. UMTS—or more precisely GPRS—was intended to fill a market niche and supply mobile access. From the same line of thought, another technology branched off, cdma2000, which consequently addresses the same niche.

WLAN [IEEE 802.11], of course, also addresses this niche. Its first specification was finalized in the year 1997, somewhat earlier than UMTS. The deployment of the first UMTS Networks in 2001 then coincided with both the breakdown of the new economy and the successful introduction of WLAN Hotspots. Wireless Internet access via WLAN Hotspots at the time was much cheaper than via UMTS, and moreover, WLAN cards for laptops were much simpler to come by than UMTS cards. At this point, WLAN was identified as a serious competitor for UMTS.

Interestingly, this perception changed upon closer inspection. As already discussed in Chapter 1, WLAN and UMTS in fact target different markets. While WLAN equipment is indeed cheaper, it is also much simpler and provides less sophisticated functionality: WLAN provides wireless *nomadic* access to the Internet. That is, WLAN provides wireless access to users at home and in other places. While they are online, however, users are quasi stationary, much like in the first generation of mobile Telecommunication Networks (cf. Section 2.2). UMTS, in contrast, provides truly *mobile* access to the Internet: it also works in high-speed trains. Furthermore, UMTS provides QoS and sophisticated charging support, all of which WLAN still does not. Finally, UMTS is an integrated system that offers not only connectivity, but it also offers service support. That is, users can be "technology agnostic" and still use services such as Voice over IP. It should be added that while the UMTS specifications certainly contain these features, UMTS as rolled out today still remains somewhat beyond these possibilities.

Having performed this analysis, it was found that WLAN and UMTS in fact complement each other. WLAN is the ideal access technology for nomadic Hot Spot access, whereas UMTS is ideal for truly mobile access, and for providing high-quality services with operator support. The current trend is towards integrating WLAN and other mobile Access Networks as alternative accesses into the UMTS Network; this will be described in greater detail in Chapter 18 and Part II.

2.4.2 Other Mobile Technologies

WLAN did not remain alone. In the last couple of years, a wide variety of new mobile technologies, e.g. **Bluetooth** [Bluetooth] have been introduced successfully. More technologies, e.g. WiMAX [IEEE 802.16] are currently being deployed. Figure 2.4 compares these technologies with UMTS regarding their *theoretical* bit-rate and range. Thereby, "range" denotes—somewhat unscientifically—a combination of contiguous coverage of the technology and speed of the mobile user.

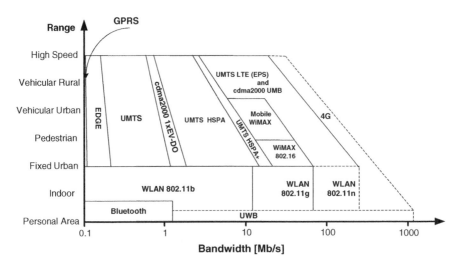

Figure 2.4 Coarse-grained comparison of UMTS with other mobile technologies. Dashed lines denote technologies still under specification. Note the logarithmic X-axis! Also note that technologies are shown in an overlapping way: of course, UMTS is also suitable in the "fixed urban" range

Figure 2.4 illustrates that the bandwidth of GPRS is about 40 kb/s, which is indeed rather negligible compared to the newer technologies. GSM cannot even be depicted in this Figure. The new radio technology for GSM, **Enhanced Data rates for GSM Evolution** (EDGE) with up to 473.6 kb/s can, however, be shown; it is classified as bordering on a 3G technology.

We see in Figure 2.4 that the original UMTS offers up to ideally 2 Mb/s. A new update of the UMTS radio interface, HSPA, however, yields downlink speeds of up to 14.4 Mb/s and an uplink of up to 5.7 Mb/s. HSPA is currently (in 2008) being rolled out and we can discern that it serves a unique market by providing comparatively high bandwidth up to high speeds. An enhancement of HSPA, called **HSPA +**, is, however, already underway which will increase the bandwidth even further. cdma2000 with its radio interface 1xEV-DO currently offers a slightly lower bit-rate. It is, however, expected to catch up soon.

With 63 Mb/s WiMAX has a higher maximum bandwidth than even HSPA +. Originally, WiMAX was a nomadic technology just like WLAN. A new WiMAX specification, Mobile WiMAX, however, also supports mobile access. If mobility support for WiMAX lives up to its promise it can indeed develop as a serious competitor for UMTS and other 3G technologies.

Figure 2.4 also depicts two short-range technologies, Bluetooth and Ultra Wideband (UWB) [UWB] which are meant for indoor usage, e.g. for communication between personal devices. UWB is still in the specification process. It is expected to support bandwidths of 500 to 1000 Mb/s.

Also shown is the outlook for two upcoming technologies that have recently finalized their specification: an augmentation of UMTS called **Long Term Evolution** (LTE) or **Evolved Packet System** (EPS), and the corresponding evolution of cdma2000 called **Ultra Mobile Broadband** (UMB). The dashed lines indicate possible future developments, such as in the field of 4G.

It must be noted that Figure 2.4 provides a superficial comparison only. Apart from the fuzzy definition of the y-axis, it is difficult to give a single, precise number for the bandwidth of each technology: Real-life values usually are significantly lower than the theoretical values, up-link

Table 2.2 Detailed comparison of UMTS with other mobile access technologies

	UMTS	WiMAX	WLAN	Bluetooth	UWB
Theoretical Bandwidth	384 kb/s–2 Mb/s 14.4 Mb/s downlink, 5.7 Mb/s uplink (HSPA) 47 Mb/s downlink, 11 Mb/s uplink (HSPA +)	63 Mb/s downlink, 28 Mb/s uplink (802.16a)	11 Mb/s (802.11b) 54 Mb/s (802.11 g) [248 Mb/s (802.11n) still being specified]	2.1 Mb/s (Bluetooth 2.0)	500–1000 Mb/s
Cell Radius	30 m–20 km	up to 50 km	50–300 m	0.1–100 m	10 m
Mobility	High	Medium	Nomadic	Nomadic	Nomadic
First Standard Availability	1999	2004	1997	1999	?
Frequency Band	2 GHz	2–66 GHz	2.4 or 5.2 GHz	2.4 GHz	3.1–10.6 GHz
Frequency Licence	Yes	Depends	No	No	No

and down-link bandwidth may be different and sustained bit-rates may be much lower than the peak bit-rates. Furthermore, the maximal real-life sustainable bandwidth is often only obtained when a single user is accessing the Access Point.

Table 2.2 provides more technical detail on a subset of the above technologies. Of particular interest is the last item, on whether a license is necessary for using the targeted radio frequency. Acquiring a license is usually costly and adds bureaucratic overheads. On the other hand, unlicensed frequencies can become rather crowded leading to serious problems with interference. While UMTS is always operated within a licensed band, WLAN is not, and WiMAX can operate both in licensed and unlicensed frequency bands.

2.5 Summary

The aim of this chapter is to provide background information on why UMTS was developed, and how it compares to other mobile technologies.

Mobile Telecommunication Networks and the fixed Internet have experienced a tremendous growth, starting in the 1990s. In the framework of IMT-2000, thought was invested early on as to what the next, the 3rd generation of mobile Telecommunication Networks would look like. Based on IMT-2000, a number of 3rd generation technologies were developed, among them UMTS, cdma2000 and Mobile WiMAX.

UMTS is a 3rd generation mobile Telecommunication Network in the tradition of GSM. It goes beyond GSM in that it offers packet-based services and, additionally, support for multimedia and location based services. It offers—within the same technology—world-wide roaming and is backwards compatible with GSM.

Compared to other 3G technologies such as cdma2000, UMTS offers similar features.

Compared to other mobile technologies such as WLAN and WiMAX, UMTS today offers a higher quality service to the user such as mobility support, also at high velocities. Furthermore, a UMTS operator offers to the user an "integrated" packet, including the services themselves. WLAN and WiMAX, as compared to UMTS, offer higher bandwidth. Furthermore, WiMAX has recently been augmented to become a full mobile network offering control features such as mobility support.

While alternative mobile technologies such as WLAN and WiMAX were first perceived as competition for UMTS, today they are seen as complementing technologies. The latest release of the UMTS standard describes how WLAN and other technologies can be employed in order to access the UMTS Network.

3

Standardization

Standards play an important role in Communication Networks. Contrary to what one might think, standardization is a very creative process. Communication technology is often conceived and standardized simultaneously. In this chapter, we explain in more detail why standardization is performed, and how standards are written. In separate subsections, we introduce the different standardization bodies that are of importance for 3G, in particular UMTS and 4G, explain how they work and how they relate to UMTS and to each other.

Terminology discussed in Chapter 3:	
802.11	
802.16	
European Telecommunication Standards Institute	ETSI
3rd Generation Partnership Project	3GPP
3rd Generation Partnership Project 2	3GPP2
IETF Protocol	
Internet Draft	ID
Institute of Electrical and Electronics Engineers	IEEE
Internet Engineering Task Force	IETF
Mobile Terminals	MT
Release 4, . . ., Release 8	Rel-4, . . ., Rel-8
Release 99	R99
Request for Comment	RFC
Technical Report	TR
Technical Specifications	TS
WiMAX Forum	

UMTS Networks and Beyond   Cornelia Kappler
© 2009 John Wiley & Sons, Ltd

3.1 The Importance of Standardization

As already explained in Chapter 2, Section 2.2, the first generation of mobile Telecommunication Networks was not standardized, or was only standardized on a national level. Manufacturers and operators had their own proprietary way of implementing the desired functionality. With the 2nd Generation, markets opened up and mobile Telecommunication Networks were standardized. Only then was it possible for users to *roam*, i.e. to access networks owned by a "visited" operator based on the subscription which they had with the "home" operator. It also became possible for operators to own heterogeneous networks consisting of equipment produced by different manufacturers. For users, it became possible to buy handsets independently produced by any—standard-compliant—manufacturer, and still access the operator's network.

Standardization can be performed in different ways. One can first develop the technology, e.g. the operating system of a computer, without standardizing it. If the technology is successful and reaches a certain market share it becomes a "de-facto" standard to which everybody had better adapt. This strategy is risky. In mobile telecommunications, a different route is usually pursued. The technology is developed and standardized simultaneously. To this end, network experts from different companies, and sometimes from academia, come together on a regular basis in order to discuss and align, in sometimes heated debates, the latest standards contributions.

It is important to realize that a standard is not a construction manual. The standard for, e.g., UMTS, while certainly lengthy, does not contain all of the information necessary to build network elements or entire networks. The standard only describes whatever is essential for **interworking** between networks and network elements. Everything else is not subject to standardization.

What are the essentials that need to be standardized?

- The basic network architecture, including the core functionality of Network Elements and the interfaces between the Network Elements.
- Protocols and protocol stacks.
- Information elements and where they are stored.

And what does not need to be standardized? Everything that is not necessary for interworking. This includes, e.g. not only the internal structure of Network Elements, but also intra-network solutions. A good example of the latter is intra-domain QoS provisioning in UMTS. The standard describes exactly what QoS levels must be achieved, e.g. the "conversational level" for voice calls, which implies a maximum transfer delay of 100ms. The standard does, however, not describe how Network Elements within the network should collaborate to achieve this quality level. We therefore see that, even when a standard exists, network engineers still have a lot of specification work to do until a network can finally be built.

There is also another aspect to the standard not being a construction manual. Despite collaborating on the standard, the contributing companies are competitors. It is important for them to differentiate their own product, i.e. network elements or networks, from the products of others. This is only possible if the standard is not a complete specification. Furthermore, sometimes it is even deemed wise to leave an "essential for interworking" unstandardized in order to shift competition elsewhere.

3.2 Standardization Bodies

Many different standardization bodies are important in the context of UMTS and for what comes after UMTS. We already encountered one standardization body, the ITU, which authored the IMT-2000 framework and thus the definition of 3G. For the standardization of UMTS, however, the **3rd Generation Partnership Project**, 3GPP [3GPP], was founded in December 1998. At about the same time, the **3rd Generation Partnership Project** 2 (3GPP2) [3GPP] began the standardization of cdma2000, also a member of the IMT-2000 family. Both 3GPP and 3GPP2 draw heavily on protocol standards developed for IP Networks by the **Internet Engineering Task Force**, IETF. The **Institute of Electrical and Electronics Engineers**, the IEEE, also plays an important role. The IEEE is active in standardization of wireless networking and is responsible for the lower OSI layers of WLAN and WiMAX.

3.2.1 ITU

The ITU was originally founded in 1865 as International Telegraph Union, with the goal of standardizing telegraph equipment. Later, the title was amended to **International Telecommunication Union** and its scope was broadened to all information and communication technologies. In 1947, the ITU became an agency of the United Nations. Its members consist of 191 states and 700 "sector members", i.e. operators, manufacturers and international organizations. The ITU covers three core areas:

- Radiocommunication, in the ITU-R, where the international radio-frequency spectrum and satellite orbit resources are managed. Particularly, the ITU-R holds the World Radio Conference (WRC) every four years, where the radio spectrum is (re)-assigned to particular services, e.g. satellite communication, mobile, broadcasting, amateur, space research, emergency telecommunications, meteorology, global positioning systems, etc.
- Standardization, in the ITU-T, where standards regarding all fields of information and communication are specified, such as the IMT-2000 standard defining 3G. ITU-T Standards are published as "Recommendations" and are available online at no cost.
- Development, in the ITU-D, where equitable, sustainable and affordable access to information and communication technologies is supported. For example, the ITU-D assists developing countries in enhancing their telecommunications and information infrastructure and applications.

3.2.2 3GPP

3GPP was established in December 1998. A number of regional telecommunication standardization bodies cooperate in the global organization 3GPP in order to produce the standard for an entire mobile Telecommunication Network, UMTS. These regional standardization bodies are ARIB of Japan, ATIS of the USA, CCSA of China, the **EuropeanTelecommunication Standards Institute** (ETSI), TTA of Korea and TTC of Japan. We see that the key players in today's mobile telecommunication, the Far East, Europe and the US, are all represented.

While 3GPP is formally a cooperation of standardization bodies, the actual work is carried out by the delegates from companies. The companies, in turn, are members of the constituent standardization bodies. Typically, they are large international manufacturers and operators. 3GPP meetings are closed to non-members. Therefore, academia does not play an active role in 3GPP.

The original scope of 3GPP was to produce a UMTS standard based on an evolved GSM. In particular, 3GPP is responsible for specifying the **Mobile Terminals** (MT), the **Radio Access Network** (RAN) and the **Core Network** (CN). This list reflects the basic architecture of a UMTS Network, see Chapter 4 and Figure 4.1. The scope of 3GPP was subsequently amended to also include the maintenance and development of GSM including evolved radio access technologies (e.g. EDGE) and GPRS.

As illustrated in Figure 3.1, the work in 3GPP is organized into different Technical Study Groups (TSGs). Each TSG is in turn subdivided into Working Groups (WGs). The TSGs are coordinated by a Project Coordination Group.

One TSG, namely "Services and System Aspects" (SA), is responsible for the overall architecture and service capabilities of UMTS. New ideas for the overall service evolution of UMTS are usually first discussed in SA WG1, resulting in feasibility studies and requirement documents. SA WG2 is responsible for the continued development of the 3GPP architecture, including UMTS, GSM and GPRS, thereby taking up ideas from SA WG1. SA WG2 defines the functional blocks of the architecture and the high-level information flow between the functional blocks. The specifications resulting from the work of SA WG2 are then passed to the other WGs and TSGs for detailed work on protocols and physical Network Elements. Of course, it sometimes also works the other way round, with other WGs asking SA WG2 to update the architecture as a consequence of detailed work on e.g. protocols. TSG CT is responsible for Core Network and Terminals, TSG RAN is responsible for the Radio Access Network of UMTS and TSG GERAN is responsible for the Radio Access Network of GSM/EDGE.

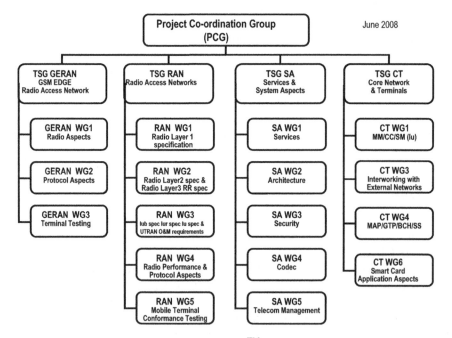

Figure 3.1 3GPP organizational charts. ©2008, 3GPP™ materials are the property of ARIB, ATIS, CCSA, ETSI, TTA and TTC who jointly own the copyright in them. They are subject to further modifications and are therefore provided to you "as is" for information purposes only. Further use is strictly prohibited

The daily work in 3GPP consists of company representatives submitting standardization contributions—new ideas or updates of existing documents—to meetings taking place about six times a year all around the globe. Outside of meetings, the discussion proceeds via mailing lists. The work of the 3GPP is all on paper. The reality-check occurs when companies implement the standard in their prototypes and products.

The work of 3GPP is documented in Technical Specifications (TS) or Technical Reports (TR). Appendix B gives an overview of the systematics for numbering 3GPP specifications. According to a very rough estimate, the Technical Specifications and Reports currently run to 30 000 pages. All of them are available to the public on the 3GPP web site [3GPP]. This openness is new as compared to GSM, whose specifications were only available to members. Besides generally becoming more open, one motivation was that bazaar-style public review is an extremely efficient method for uncovering errors and problems.

UMTS is evolving. Therefore, the UMTS standard is not finalized at any particular time. But then, UMTS design is cathedral-style. Therefore, UMTS proceeds in so-called *releases* which are finalized about every other year. With each release, new features are added, and some old ones may be abandoned. The first UMTS release was published in 1999. It was called Release 99 (R99). The numbering of subsequent releases was decoupled from the calendar years. They are now numbered consecutively: Release 4 (Rel-4), Release 5 (Rel-5), etc. Rel-4 was finalized in 2001. Currently, 3GPP is working on Rel-8. Chapter 19 on the evolution of UMTS describes the differences between the individual releases in more detail.

Work in 3GPP proceeds according to a master plan for each release, with timely completion as a guiding principle: companies need input for the next product release, or for planning updates to an already deployed network. Delays are expensive. Work on a new release is started with a list of features to be included. If a certain feature cannot be specified in time it is deferred to the next release.

3.2.3 3GPP2

Simultaneously with 3GPP, 3GPP2 was founded. While 3GPP specifies UMTS, the 3rd generation evolving from the European GSM, 3GPP2 specifies cdma2000, the 3rd generation technology evolving from the American cdmaOne. As in 3GPP, regional standardization bodies cooperate in 3GPP2, in particular from America and Asia. A European partner is missing. This does, however, not mean that there are no European companies in 3GPP2. European companies which are active in standardization are typically large and international, and are also members of non-European standardization bodies. This way, they are also eligible to work in 3GPP2. In other words, European companies do not restrict themselves to selling UMTS. They also sell cdma2000. And the same is, of course, true of companies based in other continents.

3GPP2 standards are available to the public on the 3GPP2 web site [3GPP2].

3.2.4 IETF

The IETF is a loose, open community of engineers and researchers from industry and academia. In fact, anybody can participate in IETF standardization, their success is however coupled to the technical merit of the contribution.

The IETF develops standards for the Internet, or more generally standards for use in IP Networks. Most standards relate to protocols. As explained in Chapter 1, architectures are not so important. An example of an IETF protocol is IP. More recently developed protocols include the **Session Initiation Protocol** (SIP) for session signalling and the **Diameter** protocol for Authentication, Authorization and Accounting. In this book we call IETF-developed protocols for Computer Networks **IETF Protocols**.

The technical work of the IETF is carried out in its Working Groups, which are organized by topic into several areas (e.g., routing, transport, security, etc.). Much of the work is handled via mailing lists. The IETF holds meetings three times a year. Ongoing work is documented in so-called **Internet Drafts** (IDs). Results, including the actual standards, are found in documents called politely **Request for Comment** (RFC). Of course, all IETF documents are available to the public on the IETF web site [IETF].

The IETF's working style is very different from that of other standardization bodies, e.g. 3GPP. New working groups are founded when enough people are interested in doing the work, can agree on a charter that defines a reasonable scope and convince an Area Director of the IETF that the project is worthwhile. Following a true "bazaar approach" as described in Chapter 1, there is no "grand plan" for building an entire system. One of the IETF principles is "rough consensus and running code" [RFC 3935]. This means that decisions are taken based on rough consensus of IETF participants, either in meetings or on the mailing list. It is not necessary for everybody to agree. Furthermore, a protocol can only become a standard if at least two interoperable implementations exist. It is very hard to design faultless protocol machinery on paper only.

As we saw in Chapter 2, the worlds of telecommunications and the Internet converge. The majority of the mobile networks depicted in Figure 2.5 are in fact IP-based. UMTS increasingly employs protocols developed by the IETF as this improves interoperability with other networks. Therefore the IETF gains importance in the telecommunications world, and 3GPP and IETF collaborate. For example, members of the 3GPP actively participate in IETF working groups in order to develop standards of importance to 3GPP.

3.2.5 IEEE

The IEEE is a professional association for the "advancement of technology". Its members are over 350 000 engineers and scientists from all technical areas and from all over the world—with emphasis on the US. The IEEE, among other things, publishes technical literature, organizes conferences, awards grants and designs standards.

Among the IEEE standards activities, "Project 802" is of particular interest to us. It develops Local Area Network and Metropolitan Area Network standards, mainly for the lowest two OSI layers, PHY and MAC. IEEE 802 includes a number of Working Groups, among them **802.11** for WLAN, and **802.16** for Broadband Wireless Access, also known as WiMAX. A relatively new Working Group is **802.21** for **Media Independent Handover** and interoperability between heterogeneous network types, including 802 and non-802 networks.

While the original WiMAX specification provided nomadic access, a recent update also supports mobile access. The IEEE takes care of the PHY and MAC layer, network aspects are taken of by the **WiMAX Forum** [WiMAX Forum]. The IEEE and the WiMAX Forum also collaborate with the IETF.

The IEEE standards process is again different from that in 3GPP, 3GPP2 and IETF. Anybody can join the meetings and mailing lists. Decisions may, however, require a formal vote in which not everybody can participate. Voting rights are assigned to either individuals or companies, or both, based on group-specific policies. For example, in IEEE 802, voting rights are usually per individual and earned by participation in two out of four consecutive meetings. Individuals should represent their own opinion rather than that of their employer or sponsor. Standards issued by the IEEE are available to the public. IEEE 802 standards are even available free of charge six months after they have been finalized [IEEE 802].

3.3 Summary

This chapter introduces the reader to the importance of standardization. Standardization fosters the interoperability of equipment by different manufacturers and networks by different operators. Standardization facilitates roaming, and increases the choice available to customers.

A standard for a mobile network technology is not, however, a blueprint for building Network Elements or networks. The standard only encompasses those aspects of the Network Necessary for interworking. These include the basic architecture, core functionality of Network Elements, interfaces, protocols and information elements.

This chapter also provides an overview of the different standardization bodies active in the area of 3G. 3GPP is the standardization body responsible for UMTS, 3GPP2 is responsible for cdma2000. Both organizations often draw upon work by the IETF, where protocols for IP Networks are being standardized. The IEEE, particularly IEEE 802, standardizes the WLAN and WiMAX technology. The network aspects of WiMAX are covered by the WiMAX Forum.

Most of the standard documents discussed in this book are available to the public and are obviously a good source for more in-depth information.

4

UMTS Architecture and Functionality

In this chapter we provide an overview of today's UMTS **architecture**. We start by asking ourselves what is architecture, and why it is useful. We also discuss which functionality must be reflected in a mobile network architecture, and present the frequently applied split into **user-plane**, **control-plane** and **management-plane** functionality. We then focus on concrete architectures. We approach the problem in an evolutionary fashion by starting with the IMT-2000 architecture for 3G Networks. We then look at the, relatively, simple GSM, and then advance, via GPRS, to UMTS—and even to an evolution of UMTS called 3GPP System. Finally, we compare these architectures to the frugal architecture of a 802.11 WLAN.

The set of chapters following this one will go into the detail of the different components of the UMTS architecture and give an overview of the protocols. The chapters on architecture introduce a large number of acronyms. Unfortunately, acronyms are quite popular in the Communication Networks community, so the reader will have to memorize them.

Terminology discussed in Chapter 4:	
2.5G	CN
3GPP Core Network	ESS
3GPP System	
Access Network	
Access Point	AP
Access Router	
(Network) architecture	
Circuit-switched Domain	CS Domain
Core Network	CN
Control-plane	
Extended Service Set	ESS
Functional group	
GSM/EDGE RAN	GERAN

UMTS Networks and Beyond   Cornelia Kappler
© 2009 John Wiley & Sons, Ltd

IP Multimedia Subsystem	IMS
Management-plane	
Mobile Station	
Mobile Terminal	MT
Packet-switched Domain	PS Domain
Public Switched Telephone Network	PSTN
Radio Access Network	RAN
Reference point	
Shared channel	
Subscriber Identity Module	SIM
UMTS Terrestrial Radio Access	UTRA
UMTS Terrestrial Radio Access Network	UTRAN
Universal Integrated Circuit Card	UICC
User Equipment	UE
User Identity Module	UIM
User-plane	

4.1 Overview of Telecommunication Network Architecture

We already highlighted in Chapter 1, Section 1.2.4 that Telecommunication Networks are architecture-centric. That is, the architecture is designed first. A network architecture shows what functions a network must provide, e.g. data forwarding or mobility support, and how these functions are grouped in the network. The architecture also establishes **reference points** between these **functional groups**. A reference point defines what information is exchanged between functional groups. The protocols used across the reference point are selected or designed later. Figure 4.1 illustrates the IMT-2000 architecture for 3G Networks as an exemplary network architecture. It will be discussed in detail later in this chapter.

4.1.1 Overview of Mobile Network Functionality

In order to make the theoretical considerations above more concrete, we draw up a list of functions that are necessary in a mobile Telecommunication Network, in other words a list of the functions that could be grouped in an architecture. This list is by no means comprehensive, but includes in particular those functions that are important in UMTS Networks and that will be covered in the later chapters of this book.

- **Transport**. The network is only useful if it is able to move information to and from the **Mobile Station**, i.e. the mobile device of the user.
- **Routing**. This function includes population and maintenance of routing tables.

Figure 4.1   High-level architecture for 3G Networks. RPI,.. RP4 denote reference points

- **Security**. Only users with adequate credentials may access the network. Vice-versa, a user should only access a trustworthy network. Security functions, furthermore, include privacy of user identity and user location, as well as privacy and integrity of the content of communication sessions.
- **Session control or call control**. Sessions and calls must be set up, maintained and torn down. In circuit switched networks, any session is a call, whereas in a packet-switched network, sessions include the entire range from telephony calls, via data sessions to multimedia sessions.
- **QoS**. Some services, particularly real-time services might not deliver acceptable quality to the user when sharing resources, e.g. bandwidth, evenly with everybody else. These services need Quality of Service (QoS), i.e. dedicated resources.
- **Radio resource control and provisioning**. The idea here is similar to the QoS, with focus on the radio interface because this is where resources traditionally are scarce. Radio resources are, e.g. time slots and frequencies; radio resource control and provisioning takes place below layer 3.
- **Mobility**. Obviously a crucial function in mobile networks—it describes the ability of the Mobile Station and network to maintain user sessions while the Mobile Station moves and changes its point of attachment to the network.
- **Charging**. A system for accounting and charging is vital in Telecommunication Networks. Each subscriber must be charged correctly, i.e. according to the resources he or she consumed and the tariffs to be applied. Besides, accounting data might be used for analytical post-processing by the operator.
- **User Services**. The most well known example is of course telephony. Telecommunication Networks usually include support of user services in their network design.

4.1.2 User-Plane, Control-Plane and Management Plane

The functions performed by a network are often classified into user plane functions, control plane functions and management plane functions.

- The user plane is concerned with the forwarding and routing of information, and performs functions such as error correction and flow control. User plane functionality comprises simple connectivity and information transport including routing. It also includes delivery of user services such as voice, fax, video, web access, etc.
- The control plane deals with *short-term network operation*. Originally, it was concerned with the control of connections, including call set-up, call release and maintenance. With the integration of packet-oriented networks, today's control planes are concerned with the control of communication sessions. From the list above, the functions security, QoS, mobility, charging and radio resource control belong to the control plane.
- The management plane is concerned with *long-term network operation*. The classic management functions are called FCAPS: *F*ault management, *C*onfiguration management, *A*ccounting management, *P*erformance management and *S*ecurity management.

We see that some functions appear both in the control plane and the management plane. This is because there are both short-term and long-term aspects to these functions. For example, whereas control-plane security deals, e.g. with performing the authentication of individual users, management plane security deals with the maintenance of the authentication system as a

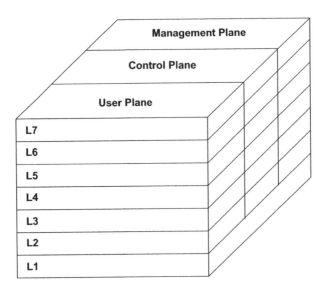

Figure 4.2   User plane, control plane and management plane with the example of an OSI protocol stack

whole. Whereas control plane charging deals with collecting and correlating charging information, management plane accounting deals with pricing and billing.

This plane structure is often reflected in the architecture of Telecommunication Networks by locating user plane, control plane and management plane functions in separate functional groups. Likewise, protocol stacks are layered into a user plane, a control plane and a management plane, see Figure 4.2.

Computer Networks such as the Internet originally only comprise the user plane. Indeed, the OSI Model (explained in [Tanenbaum 2002]) was designed for data networks, and only provides a layer model for the user plane. Later, the model was extended to also cover Telecommunication Networks and thus to include layer models for control plane and management plane [ITU I.322] along the lines depicted in Figure 4.2. With the convergence of Computer Networks and Telecommunication Networks, a large number of control protocols have also been developed for Computer Networks. Meanwhile, the IETF also works on protocols that allow a systematic architectural separation of network elements into control plane and user plane entities [RFC 3746].

Since a well-organized control plane is however one of the distinguishing features of UMTS, and since UMTS operators indeed like to control their network, cf. Chapter 1, Section 1.2.2, the control plane is treated with special emphasis in this book.

4.2 High-Level Architecture of 3G Networks

Figure 4.1 depicts the high-level architecture for IMT-2000 Networks, i.e. 3G Networks, as defined in [ITU Q.1701]—incidentally, this architecture does not show the above mentioned planes explicitly. A 3G Network is composed of the following functional groups:

- **User Identity Module (UIM)**, supporting user security and services, both in the user plane and in the control plane. The GSM **Subscriber Identity Module** (SIM, or "SIM card" or "chip") is an example of a UIM.

- **Mobile Terminal (MT)**, or more colloquially your mobile, responsible for communicating with the UIM and Radio Access Network, and supporting user services and mobility.
- **Radio Access Network (RAN)**, communicating with the MT and **Core Network**. The RAN contains the **Base Station** which sends and receives the radio signal to and from the MT. In a WLAN, the equivalent of a Base Station is called the **Access Point**. Radio Resource Control is also located in the RAN.
- **Core Network (CN)**, responsible for routing user sessions between RAN and external networks. It also supports user services and user mobility.

This architecture tells us that traffic originating from the user travels from MT via RAN to the CN and then to other networks. When user traffic is destined to another MT in the same network, it will still follow this path: from MT to CN and back to another MT. At such a high level, this structure, by the way, is not specific to 3G Networks. As we will see below, GSM has the same high-level structure.

This architecture, however, goes beyond mere structure and assigns functions to each functional group. For example, it tells us that a 3G RAN is oblivious of user services—services are located in the CN. It is at this level that GSM ceases to fit a 3G architecture.

In architecture design, the concept of functional groups is applied recursively. That is, one proceeds to a finer-granular architecture by decomposing each functional group into subgroups, thereby adding more fine-grained functionality.

Architectures such as those shown in Figure 4.1 offer the benefit that individual functional groups can be exchanged independently. If the new functional group maintains the reference point, the rest of the system does not need to be changed. For example, the original GSM was updated with a new air interface and a new RAN without changing the rest of the Core Network.

One should note that a functional group, even at the lowest level of granularity, is not necessarily a single physical network element—although it is often built as one. It is possible to merge functional groups into a single network element, e.g. in small networks. The other way round is also possible, i.e. to split a functional sub-group into several network elements. In this case, of course, the reference points and protocols between the network elements are not standardized.

4.3 GSM Architecture

A GSM Network is designed for voice service, and can also provide, e.g. fax and modem service. It is a mobile version of the fixed ISDN. Thus, it is a circuit-switched network. In other words, for all services, a dedicated connection is established between the end-points. This dedicated connection has the same bandwidth for all services and charging is performed on a per-time unit basis. It is easy to see that, when services become more varied, reserving the identical bandwidth for each service is rather wasteful, particularly on the radio interface, where resources tend to be scarce.

Figure 4.3 shows a simplified architecture of a GSM Network, with the same functional groups as in Figure 4.1.

- **SIM and MS**. In GSM, the UIM is called **Subscriber Identity Module** (SIM), and the MT is called **Mobile Station** (MS). The SIM allows a unique identification of the user. The MS communicates with the RAN and supports user services and mobility.

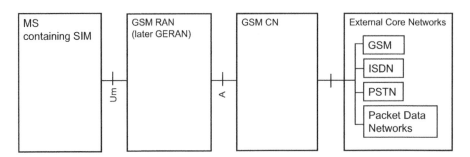

Figure 4.3   Simplified architecture of a GSM Network

- **GSM RAN**. A GSM RAN communicates with the MS via the radio interface and with the CN via a fixed line. The GSM RAN performs radio resource management and communicates with the MS about RAN issues, e.g. how to establish a channel on the radio interface. The GSM RAN also contains a functional subgroup performing a user service: voice is transcoded between different formats and rates. E.g. ISDN-coded voice arrives with 54 kb/s and must be transcoded into the 14.4 kb/s voice for the GSM radio interface. With this localization of user service functionality in the RAN, the GSM architecture is different from a 3G architecture.

  The GSM RAN, particularly the radio interface, is being evolved to higher bandwidth and called EDGE (cf. Chapter 2, Section 2.4.1) with a theoretical maximum bandwidth of 384 kb/s. The general term for referring to a RAN with either a GSM radio interface, an EDGE radio interface, or both is **GSM/EDGE RAN** (GERAN).

- **CN**. A GSM CN communicates with the GSM RAN and has a gateway to external networks, such as other GSM Networks, fixed line telephone networks such as ISDN or the analogue **Public Switched Telephone Network** (PSTN). It also connects to Packet Data Networks such as the Internet. The GSM CN is responsible for many control functions in a GSM Network. It authenticates users, routes calls, supports mobility and collects charging information.

4.4 GPRS Architecture

GPRS is a so-called 2.5G technology, between 2G and 3G. It originally updated the GSM radio interface so that a maximum of 171.2 kb/s is possible. This is better than GSM, but does not yet quite qualify as 3G (cf. Chapter 2, Section 2.3.2). Meanwhile, GPRS was augmented to operate with an EDGE radio interface and in this way improve its throughput. However, GPRS exhibits also other 3G features that clearly distinguish it from GSM:

- The GPRS architecture is 3G-conformant in that the transcoding unit, a user service, is moved into the Core Network. In GSM, the transcoding unit is located in the RAN.
- Most importantly, however, GPRS supports data traffic based on packet-switched technology. This is achieved by two changes. Firstly, a new channel is introduced on the radio interface. Secondly, the Core Network explicitly supports packet-switched services. We now describe both of these changes in more detail.

  Whereas in GSM each service is given its exclusive channel with a dedicated bandwidth, in GPRS, a **shared channel** is introduced from CN towards MS ("downlink"). This means that several services can share the same channel. Note that in other mobile technologies, e.g.

WLAN, radio interface resources are always shared unless one puts in additional effort. It is easy to see that channel sharing can be useful: data services, e.g. web-surfing, typically use their bandwidth in a bursty fashion: When the user downloads a web-page, a lot of bandwidth is needed. When the user reads a web-page, no bandwidth may be needed for a considerable time. Because users are not synchronized in their web-surfing patterns, an effect occurs called statistical multiplexing: on average the channel bandwidth is utilized evenly. Of course, this is only true on average. At times, contention might occur. In any event, radio interface resources are used much more efficiently when different web-surfing sessions share a channel. With the introduction of the shared channel it is only logical that GPRS allows charging per transmitted data volume. Charging is not necessarily per time-unit as in GSM.

4.4.1 PS Domain and CS Domain

The second change for supporting data services is architecturally more visible. As depicted in Figure 4.4, the Core Network is split into a **Circuit-switched Domain** (CS Domain) and an IP-based **Packet-switched Domain** (PS Domain). The CS Domain maintains the functionality and technology of the circuit-switched GSM Core Network. The PS Domain performs the same functions, however, on the basis of an *IP Network*.

An important note on terminology: as the book goes along we will see that "IP Network" in the context of 3GPP means that the network is packet-switched and employs the IP protocol. It does, however, not mean also other protocols employed in the network are developed by the IETF. Rather, they may be custom-built by 3GPP according to the telecommunications approach to protocols (cf. Chapter 1, Section 1.2.4 and Figure 1.1).

Since a user may simultaneously hold sessions with both CS Domain and PS Domain, e.g. a CS voice call while surfing the Internet, CS Domain and PS Domain have a reference point that allows them to coordinate, e.g. their mobility control. Access to external networks in GPRS is possible from both CS Domain and PS Domain.

4.5 UMTS Architecture

The main difference between UMTS R99 and GPRS is a new Radio Access Network with a new radio interface technology called **UMTS Terrestrial Radio Access** (UTRA). The UTRA

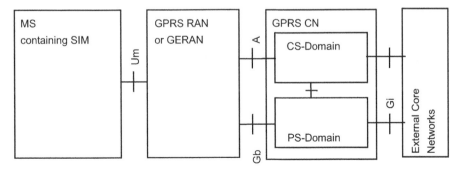

Figure 4.4 Simplified architecture of a GPRS Network

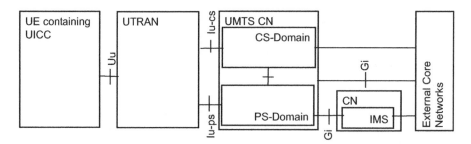

Figure 4.5   Simplified architecture of UMTS

offers considerably more bandwidth, namely (theoretically) 2 Mb/s in R99, and 10 Mb/s in Rel-6. We will hear more about this in Chapter 5. Figure 4.5 shows the UMTS architecture with the new RAN, called **UMTS Terrestrial Radio Access Network** (UTRAN). UMTS also introduces slight shifts of functionality between RAN and CN compared to GPRS, and introduces a new name for the Mobile Station, namely **User Equipment** (UE). The UIM becomes a **Universal Integrated Circuit Card** (UICC).

4.5.1 IMS

With UMTS Rel-5, an important new functional group is introduced, the **IP Multimedia Subsystem** (IMS), also shown in Figure 4.5. The IMS supports the delivery of IP-based multimedia services. The reader will remember from Chapter 2 that delivery of multimedia services is one of the important markets which UMTS is targeting. Up to Rel-5, multimedia services support is proprietary. Most UMTS Networks deployed today (2008) utilize such proprietary solutions. The first IMS's are, however, being introduced. The IMS can be accessed via the PS Domain. The PS Domain hides user mobility from the IMS; the IMS is not mobility-aware. As can be seen from Figure 4.5, the CS Domain does not interface directly with the IMS, the IMS is for IP-based services only. The IMS does, however, include gateways that allow "translating" IP-based sessions, e.g. IP telephony sessions, into circuit-switched sessions and to route them into external circuit-switched networks, e.g. GSM.

The IMS *supports* services, it does not necessarily deliver them. That is, the actual Application Server may be located outside the IMS and may be owned by a third party. IMS "service support" includes a variety of functions, such as discovering the current IP addresses of another user or service, negotiating codecs and so on. More on the IMS can be found in Chapter 10.

The IMS is accessed from the PS Domain, however it is designed so that it can also be accessed by other networks. For example, 3GPP2 standardizes its own IMS which in fact is quite similar to the UMTS IMS. Also, fixed broadband access operators are standardizing access to the 3GPP IMS. We therefore see the IMS supporting multimedia services both in mobile and in fixed networks.

4.6 3GPP System Architecture

For completeness, let us introduce yet another term, the **3GPP System**. A 3GPP System generalizes UMTS in that it is more open regarding the RAN. Whereas UMTS has, by

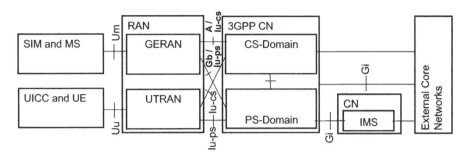

Figure 4.6 Simplified architecture of a 3GPP System

definition, a UTRAN as its Radio Access Network, a 3GPP System features both UTRAN and GERAN, connecting to both CS Domain and PS Domain. We already encountered GERAN as the evolution of the GSM RAN (cf. Section 4.3. In a 3GPP System, CS Domain and PS Domain together are called **3GPP Core Network**.

The alert reader will notice that there is a problem: GERAN originates from GSM. GSM, as 2G Network, assumes a different distribution of functionality between RAN and CN than UTRAN does. Put more formally, GERAN and, e.g., the GPRS CN interwork via the A/Gb interface rather than via Iu. 3GPP offers a twofold solution to this problem: on the one hand, the 3GPP Core Network is enabled to interwork via A/Gb—resulting in a kind of GPRS Core Network. On the other hand, GERAN is evolved such that it can also interface via Iu— effectively evolving GSM into a kind of 3G technology. A 3GPP Core Network thus can contain both 2.5G and 3G network elements. The 3GPP System architecture in Figure 4.6 shows how UTRAN and GERAN interface with both CS Domain and PS Domain.

Let us state at this point that the definition above of 3GPP System becomes broader with later releases where even non-3GPP access technologies become possible.

The content of this book often applies to 3GPP Systems as well as to UMTS. However, we will use the term "UMTS" unless referring explicitly to a 3GPP System.

4.7 WLAN Architecture

The emphasis of the 802.11 standard [IEEE 802.11] is the specification of the link layer. However, some basic architectural assumptions have been made from the beginning. While 802.11 also describes peer-to-peer or "ad-hoc" communication directly between the mobile nodes (called **Mobile Stations** (MS)), in this book we focus on so-called **infrastructure networks** involving dedicated Access Points.

The basic WLAN architecture is depicted in Figure 4.7. It shows how 802.11 concentrates on the Radio Access Network—which in the Computer Networks community would rather be called an **Access Network**.[1] The Mobile Station (beyond its radio modules) and the Core Network are not described in detail by the standard. When examining the internal structure of the RAN we see a generic router, the **Access Router**, providing access from an external network to a set of Access

---

[1] The difference between Radio Access Network and Access Network is not clear cut. In Part II we will see how the term Access Network is also introduced in Telecommunication Networks and how it includes the Radio Access Network *plus* other network entities!

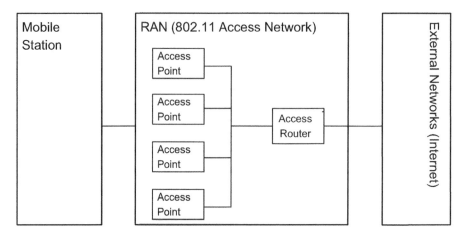

Figure 4.7   Simplified architecture of a WLAN

Points. A Mobile Station can connect to one Access Point at a time and thereby gain connectivity to the external network. The Access Points form what is called an **Extended Service Set** (ESS). The Access Points also are Layer 2 bridges, and thus are invisible from an IP perspective. The distribution medium between Access Router and Access Points usually is an Ethernet.

From a control perspective, the WLAN, particularly the Access Points, take care of authentication, radio resource control and some basic mobility support. As opposed to a telecommunication RAN, each Access Point controls its own radio resources. There is no coordination among them. The other control functions listed in Sec. 4.1.2 are not in the realm of the core 802.11 specification. They can be added by employing other specifications, both on the link-layer (e.g. link-layer QoS [IEEE 802.11e]) and above (e.g. any number of IETF-specified protocols).

4.8 Summary

This chapter introduced the concept of network architecture. We saw that a network architecture groups functionality in such a way that functional groups can be exchanged quite independently in later updates to the network. Network architectures, particularly those for Telecommunication Networks, often additionally—and orthogonally—distinguish user plane, control plane and management plane.

We also presented the high-level architecture for 3G Networks. It consists of the sequence User Identity Module, Mobile Terminal, Radio Access Network and Core Network. The Core Network provides access to external networks.

We then looked at the evolution of mobile Telecommunication Networks, starting with the circuit switched GSM. The 2.5G Network GPRS adds efficient means for data transport by introducing an IP Network in the core, the PS Domain, in parallel to the circuit-switched network, the CS Domain, adopted from GSM. We note the PS Domain being an IP Network does not mean it is designed according to the Computer Networks approach. It just means the PS Domain is a packet switched network employing the IP protocol.

UMTS is a 3G network. It differs from GPRS by a new, higher-bandwidth radio technology UTRA, a slight redistribution of functionality between RAN—called UTRAN—and CN, and a

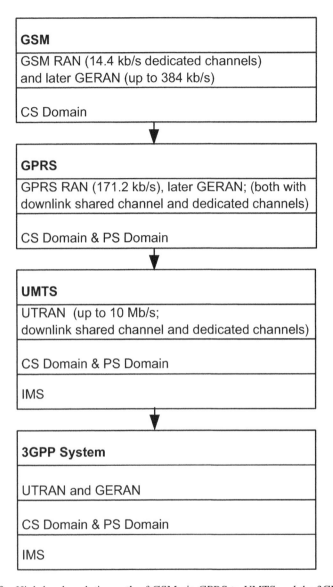

Figure 4.8   High-level evolution path of GSM via GPRS to UMTS and the 3GPP System

dedicated subsystem for service support, the IMS. We also introduced 3GPP System as a generalized UMTS with both UTRAN and GERAN. This very high-level evolution path is illustrated in Figure 4.8.

Finally, we briefly introduced the architecture of a WLAN 802.11 network. Compared to 3G network architectures, the WLAN architecture is rather simple, concentrating on the Radio Access Network and providing only limited functionality. More functionality can be added by employing additional IETF specifications.

5

UMTS Radio Interface Technology—the Physical Layer

This chapter treats a distinguishing feature of UMTS, its radio interface technology UTRA, particularly the Physical Layer.

The medium to carry information on the radio interface is electromagnetic waves. Two aspects are explored in this chapter:

- Information must be coded onto the electromagnetic waves by **modulating** them. For dealing with this problem, a variety of techniques exist. UMTS mostly employs a technique called **Quaternary Phase Shift Keying** (QPSK). HSPA employs a more advanced technique, **16 Quadrature Amplitude Modulation** (16-QAM). HSPA + and an upcoming version of the UMTS radio interface, **Evolved UTRA** (E-UTRA)—which is the result of a project known as **Long Term Evolution** (LTE)—will even go further and employ 64-QAM.
- Since many users send and receive simultaneously, we need a means of sharing the electromagnetic spectrum. A number of techniques exist in order to achieve this. UMTS today utilizes a technique known as **Code Division Multiple Access** (CDMA) which will be explained in detail. E-UTRA/LTE will use an alternative technique called **Orthogonal Frequency Division Multiplexing Access** (OFDMA), which is also introduced.

This chapter gives a brief overview of a topic that is covered in detail in other books, e.g. [Lescuyer 2004] and [Walker 2003]. HSPA is described in [3GPP 25.308, 3GPP 25.309].

Terminology discussed in Chapter 5:	
Carrier frequency	
Cell	
Cellular Network	
Channelization code	ChC
Chip	
Code division	

UMTS Networks and Beyond   Cornelia Kappler
© 2009 John Wiley & Sons, Ltd

Code Division Multiple Access	CDMA
Downlink	
Evolved UMTS Terrestrial Radio Access	E-UTRA
Frequency Division Duplex	FDD
Frequency Division Multiple Access	FDMA
High Speed Packet Access	HSPA
HSPA +	
Long Term Evolution	LTE
Macrodiversity	
Modulation	
Multiple-Input Multiple-Output	MIMO
Near-far effect	
Orthogonal Frequency Division Multiplex Access	OFDMA
Power control	
Power density spectrum	
16-QAM, 64-QAM	
Quadrature Amplitude Modulation	QAM
Quaternary Phase Shift Keying	QPSK
Scrambling code	SC
Single-Carrier Frequency Division Multiplex Access	SC-FDMA
Space division	
Spreading	
Time Division Duplex	TDD
Time Division Multiple Access	TDMA
Transmit diversity	
UMTS Terrestrial Radio Access	UTRA
Uplink	
Wideband CDMA	WCDMA

5.1 Information Coding

On a wireless interface, electromagnetic waves are used for transmitting the desired information. The amplitude P of an electromagnetic wave is a function of time t and in its simplest form is a sine of the form

$$P(t) = P_0 \sin(2\pi f t - \phi) \tag{5.1}$$

The wave thus is characterized by three parameters: the amplitude P_0, the frequency f and the phase angel ϕ. The wave in equation (5.1), however, does not yet transmit much information. Information is included by modulating the wave, i.e. by making one or more of the three parameters time dependent. One can therefore pick between amplitude modulation (AM), frequency modulation (FM) and phase modulation, or a combination of these. The rate of the modulation usually is much lower than the frequency of the modulated wave.

UMTS mostly uses a phase modulation technique for coding information, QPSK. More specifically, four different phase shifts relative to a reference wave are defined. Each phase shift

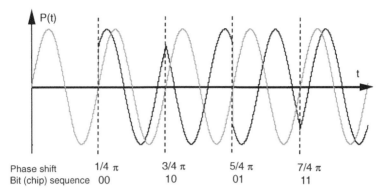

Figure 5.1   Modulation of a carrier wave with QPSK, relative to a reference wave (in grey), together with the chip sequence corresponding to each phase shift. In order to make good use of space, the rate of modulation in this figure is unrealistically high

codes one of the four possible sequences of two bits from the set $[-1,1]$. Thus, by adequately stringing phase shifts, any possible bit sequence can be constructed and transmitted over the air.

Well, more precisely one should say that in UMTS each phase shift codes two **chips** rather than two bits, as will be explained in Section 5.3. Chips, just as bits, take on the values $+1$ and -1. Since this is a peculiarity of UMTS, for generalities sake we continue talking about bits in this chapter. QPSK is illustrated in Figure 5.1.

QPSK is not the only modulation technique standardized for UMTS. In Release 5 of UMTS, HSPA was introduced as a high-speed update to the radio interface, see Chapter 2, Section 2.4. HSPA utilizes a combined phase-amplitude modulation called 16-QAM. Sixteen different phase-amplitudes shifts are defined, which allows for the coding of four bits (resp. chips) with each shift. The latest updates of the UMTS radio technology, HSPA + and E-UTRA/LTE, additionally use 64-QAM, coding six bits with each shift. An alternative means of depicting modulation techniques is used in Figure 5.2. Each phase-amplitude shift is shown as a point in complex space. The distance of the point to the origin is the relative amplitude, the angle with the real axis is the phase shift.

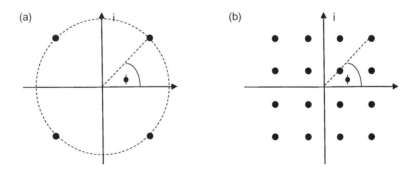

Figure 5.2   (a) QPSK and (b) 16-QAM in complex space

The wave resulting from modulations such as QPSK no longer is a pure sine wave. Rather, it is an aperiodic function (unless the information transmitted is periodic of course). As the reader knows from Fourier Analysis, any reasonably behaved aperiodic function $P(t)$ can be described as an integral of sine waves of different frequencies:

$$P(t) = \int p(f)\sin(2\pi ft - \phi(f))df \qquad (5.2)$$

Each sine of frequency f contributes to the integral with a specific amplitude $p(f)$ and a phase $\phi(f)$. The function $p(f)$ is called the Fourier Transform of $P(t)$, or the **power density spectrum**. In a case like ours, where $P(t)$ is dominated by a sine of a particular frequency f_0 as that shown in Figure 5.1, $p(f)$ has a peak at f_0, the **carrier frequency**.

Equation (5.2) states that when one transmits a modulated wave, it is not only a wave of frequency f_0 that is transmitted. Rather, it is a continuum of waves of the entire frequency band where $p(f)$ is non-zero. It can be shown that the Fourier Transform of discontinuous functions never decays to zero, i.e. the corresponding frequency bandwidth is infinite. Unfortunately, an ideal phase-modulated wave as shown in Figure 5.1 is, by definition, discontinuous—the discontinuity occurs precisely at the phase shifts. This is a problem when there is more than one sender: the waves emitted interfere with each other and the receiver cannot tell which is which. In contrast, when each sender sends on a different frequency-band of finite bandwidth, the receiver can pick out the right signal by filtering. One can solve this problem by slightly blurring the discontinuities at the phase shifts, i.e. by making the phase modulation less ideal. This translates in Fourier Space into making the corresponding frequency bandwidth finite, as illustrated in Figure 5.3.

There is, in fact, a theoretical limit to this blurring. The Nyquist criterion in equation (5.3) says the rate r of "symbols" transmitted can be at most equal to the frequency bandwidth Δf. Put differently, the higher the symbol rate, the wider the frequency band which is occupied by the waves constituting the signal.

$$r = \Delta f \qquad (5.3)$$

It is interesting to ponder the definition of symbol rate, also called baud rate. For QPSK, each phase shift constitutes a symbol. Each symbol thus transmits two bits (resp. chips). For HSPA,

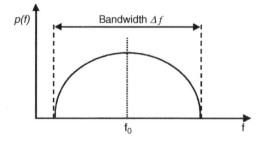

Figure 5.3   Exemplary power density spectrum of a modulated wave with a finite frequency band

each amplitude-phase shift constitutes a symbol. Each symbol in this case transmits four bits (resp. chips). Equation (5.3) by itself thus does not describe a theoretical limit to the amount of information that can be coded into electromagnetic waves.

5.2 Sharing the Electromagnetic Spectrum

The electromagnetic spectrum is a scarce resource that is shared by many technologies. In order to separate the different technologies from each other, each is assigned a particular frequency band. Mobile Telecommunication Networks and mobile Computer Networks are located typically in the frequency range between hundreds of MHz to tens of GHz. Lower frequencies are assigned to television broadcasting and AM/FM radio broadcasting. Frequencies adjacent to 300 GHz correspond to infrared and visible light.

Separating technologies, however, is not sufficient. Within one technology, the spectrum resource is shared by different companies and—for mobile Communication Networks—by different users, as well as by the two communication directions for each user, **downlink** (i.e. network towards user) and **uplink** (i.e. user towards network), see Figure 5.4. If all senders were to use the entire available frequency band indiscriminately it would be impossible for the receiver to distinguish the signal intended for her.

Many techniques exist for dividing the available frequency band. Often, they are used in combination, for example, one technique is used for separating uplink and downlink, and another technique is used for separating users.

5.2.1 Frequency Division

This technique consists straightforwardly in subdividing the available frequency band further, so that each sender is assigned its own more narrow frequency band. Senders in adjacent cells may not use the same frequency. This technique, while simple, is however somewhat inflexible, because each cell is statically assigned a frequency sub-band, and this assignment must be planned carefully when the network is set-up. When, due to unforeseen user and load distribution, the cell structure is changed, e.g. by dividing one cell into two, the frequency assignment must be redone. Another inflexibility and inefficiency inherent in this technique is that each sender is given the same bandwidth.

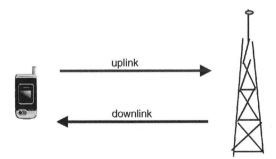

Figure 5.4 Uplink and downlink

If frequency division is used for separating uplink and downlink, it is called **Frequency Division Duplex** (FDD). If it is used to separate users, it is called **Frequency Division Multiple Access** (FDMA).

5.2.2 Time Division

With this technique, all senders utilize the same frequency band, however at different times. One possibility is that all senders are synchronized to the same clock and transmit in different time slots which are assigned by a central entity, e.g. the Access Point. A challenge for this technique is achieving synchronization of senders, particularly in the uplink direction. An advantage of this technique is that time slots can be assigned more flexibly than frequencies, simplifying network (re)planning and shifting of bandwidth between senders.

If this kind of time division is used for separating uplink and downlink, it is called **Time Division Duplex** (TDD). If it is used to separate users, it is called **Time Division Multiple Access** (TDMA).

GSM employs FDD to separate uplink and downlink, together with a combination of TDMA and FDMA: each user sends on a particular frequency band, on a particular time slot. The higher bandwidth of GPRS compared to GSM results from allowing the bundling of several time slots for a single sender.

Another, less efficient, but self-organized time division technique is for the sender to sense whether anybody else is currently transmitting before starting its own transmission. This technique, called **Carrier Sense Multiple Access** (CSMA) is known from Ethernet.

5.2.3 Space Division

A simple form of **space division** is a technique utilized by all mobile Telecommunication Networks. The antennae in the Radio Access Network tune their sending power so that they illuminate an area of a certain size, the **cell**. The antennae are carefully placed so that they fully cover the desired area. This way, the same frequency band can be used by antennae that are sufficiently distant to one another. Neighbouring antennae, however, must use different frequency bands. With the introduction of space division, the capability for handover also had to be introduced. Networks using this technique are also called **Cellular Networks**.

A more advanced, additional technique for space division is called **Multiple-Input Multiple-Output** (MIMO). Here, sender and/or receiver employ multiple antennae. Each transmission antenna results in a more or less distinct propagation path of the radio signal which can be picked up by a corresponding antenna at the receiver. These multiple antennae can be used in a variety of ways:

- Each antenna transmits the same bit sequence in a redundant way. The receiver recombines the signal and uses the redundancy to correct transmission errors. This technique is called **transmit diversity**.
- Each antenna transmits a different bit sequence. The bit sequences can belong to different users, in which case MIMO increases the overall cell capacity. Alternatively, the different bit sequences can belong to a single user who splits her high-bitrate signal into several lower-bitrate signals. In this case, MIMO increases the bit rate of a single user. This technique is called spatial multiplexing.

- The transmission antennae are operated in a coherent way so that their interference results in a "beam" directed towards the receiver. This technique allows one to reuse the same frequency band in several directions.

MIMO technique, particularly special multiplexing and transmit diversity is currently entering the standards. It is recommended in WLAN IEEE 802.11n, and 3GPP employs it in the radio technology updates HSPA + as well as E-UTRA/LTE.

5.2.4 Code Division

Since **code division** is the technique of choice for UMTS, at least originally, we will go into greater detail. This technique for dividing the resources on the radio interface is less straight-forward than the previous ones. Each sender can utilize the same frequency band, time slot and space as all the others. Each sender-receiver pair is, however, assigned a unique **CDMA code**. Codes are sequences of "one" and "minus one", so-called chips. The sender multiplies the bit sequence by the code before sending. The receiver, in turn, multiplies the received sequence of chips again with the code, thereby obtaining back the original sequence of bits. Of course, because of the non-zero travelling time between sender and receiver, the receiver must apply the code with the right time-shift, i.e. we need synchronization between sender and receiver.

Orthogonality of Codes. Code division works because codes assigned to different sender-receiver pairs have a high auto-correlation, and a very low cross-correlation. Mathematically speaking, they are orthogonal or quasi-orthogonal. Therefore, when the receiver multiplies the total of all received signals with the right code it filters out the signal destined to itself.

Spreading and Despreading. The chip rate is much higher than the bit rate. In other words, each bit is multiplied by many chips. As equation (5.3) tells us, this results in a *spreading* of the bandwidth of the signal, as illustrated in Figure 5.5. The receiver, when multiplying the received sequence of chips with the right code obtains back the original lower-rate bit sequence and simultaneously the narrow-band signal.

The entire process is illustrated in Figure 5.6. The reader is invited to test what happens when the received signal is multiplied with the wrong code: the result is not a meaningful bit-sequence.

While code division is more complex to realize than the previous techniques, it is also more flexible: because there is no shortage of codes, user and load shifts between cells can be easily

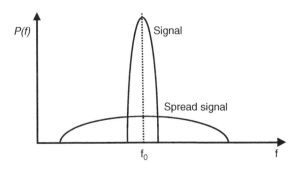

Figure 5.5   Spectrum spreading as a result of code division

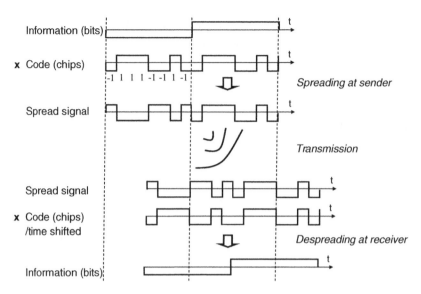

Figure 5.6   Spreading and despreading of two bits by multiplication with an example code

accommodated. Additionally, diverse bandwidth needs can be accommodated efficiently by varying the number of chips per bit: A signal with a low bit rate is spread with a long code, and a high bit rate is spread with a short code so that the resulting chip rate is always the same.

Code division is employed in order to separate users, and is thus called Code Division Multiple Access (CDMA). Traditional UMTS (excluding the upcoming E-UTRA/LTE) combines CDMA with FDD, i.e. uplink-downlink utilize different frequency bands, and users utilize different codes. The UMTS standard actually also allows TDD-CDMA, however this is rarely deployed. CDMA was already utilized by the 2G technology cdmaOne. Since UMTS works with a higher chip rate (3.84 Mega chips/s) than cdmaOne, and thus with a wider frequency band, it is also called **Wideband CDMA** (WCDMA).

The technique presented in this chapter may appear like magic at first. Why is it possible to filter out the right signal from other signals destined to other receivers, travelling simulta-neously on the same frequency band with the same signal strength? [Lescuyer 2004] illustrates the technique with a helpful analogy: Imagine a room full of people with different nationalities. They stand together in groups of fellow countrymen, each group chatting in its native language. The English are somewhere in the back. It would still be possible for an English woman entering the room to follow their conversation. She tunes into the English conversation by filtering out the "noise" created by all the other languages.

5.2.4.1 The Near-Far Effect

The example above, however, only works under one, important, condition: All groups must converse at about the same volume. If, to use a cheap cultural stereotype, the Italians in the front are especially noisy, the English lady entering will be unable to follow her countrymen's conversation at the back. The same is true for code division: the receiver is only able to despread the signal properly, if all signals arrive at about the same power level.

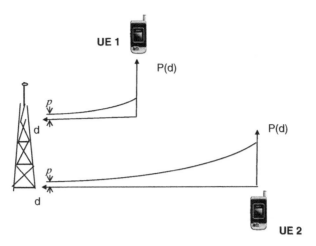

Figure 5.7   Power control to counter the near-far effect. The graphs depict the power P of the signals emitted by the UEs as a function of distance d

This requirement creates a complication for the code division technique: If all UEs were to send at the same power, the signal of UEs closer to the Base Station would be received more strongly. This is called the **near-far effect**. The problem is solved by a sophisticated **power control**: Controlled by the RAN, more distant UEs emit with more power than UEs that are closer, so that their signals arrive with the same power p at the Base Station, as illustrated in Figure 5.7.

The near-far effect is only a problem in the uplink direction. In the downlink direction, the UE of course receives all signals emitted by its nearest Base Station at the same signal strength. Note, these signals are not perturbed by downlink signals because uplink and downlink usually send on different frequency bands (FDD) or in different time slots (TDD).

5.2.4.2 Macrodiversity

With power-control alone, the near-far problem is not fully solved: the more distant a UE from the Base Station, the more power it emits. When it is currently attached to cell A and moves away in the vicinity of a neighbouring cell B, its increasingly strong signal is perturbing the signals of the UEs resident in the neighbour cell B, e.g. of UE 3 in Figure 5.8.

This effect is abated by **macrodiversity**: A UE close to a cell border can be attached to two or more cells simultaneously, as illustrated in Figure 5.8. The downlink signal is sent via all cells to which the UE is attached. The uplink signal from the UE is received and processed by all cells it is attached to. Both UE and RAN combine what was received via the different paths and this way can eliminate errors. This redundancy allows, in turn, to reduce the UE's emission power, and thereby to reduce interference. The basic idea of macrodiversity is thus the same as of MIMO transmit diversity (cf. Section 2.5.3).

In Chapter 12 on mobility we will see we how macrodiversity, which here appears as a remedy to solve the near-far problem in fact also improves handover quality.

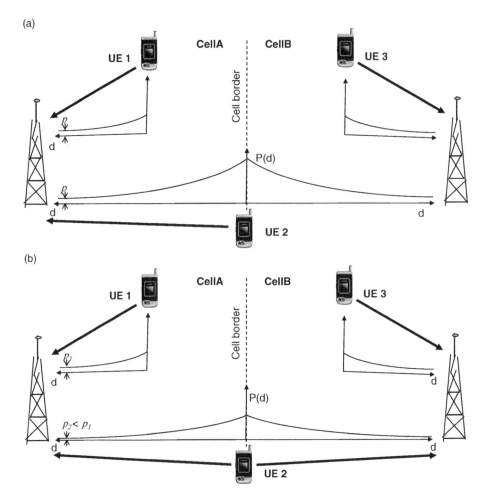

Figure 5.8 Power control (a) without and (b) with macrodiversity, allowing UE 2 to reduce emission power

5.2.4.3 Channelization Code and Scrambling Code

Looking more closely at the assignment of CDMA codes it turns out to be a two-stage process. Two different codes, the **channelization code** and the **scambling code**, each serving a different purpose, are applied consecutively to each bit sequence. The exact solution differs between FDD and TDD. Here we describe only FDD.

Channelization Code. Channelization codes are orthogonal codes of length four to 256. Each individual bit is multiplied by the full code. As explained above, the higher the bit rate, the shorter the code, so that the resulting chip rate is the same for all signals. Channelization codes thus serve to align all signals to the same chip rate.

Scrambling Code. Spreading with the orthogonal channelization codes alone is insufficient because orthogonal codes are rather sensitive to synchronization: Two orthogonal codes that are time-shifted relative to each other can have a substantial cross-correlation. This becomes a problem because, e.g. UEs are synchronized with their respective cell, however, cells are not synchronized among themselves. Therefore, with orthogonal codes the receiving cell cannot properly despread a signal which contains contributions from several UEs, possibly attached to different cells. Scrambling codes, in contrast, are quasi-orthogonal. This means that their auto-correlation is high, and their cross-correlation is almost, but not quite zero. But then, they remain quasi-orthogonal even when time-shifted relative to each other. Therefore, each sender first applies a channelization code, and then a scrambling code on top.

Scrambling codes are much longer than channelization codes, they have 384 000 chips. With this code length, the number of possible scrambling codes is very large. UMTS only utilizes 8192 different scrambling codes which is still large enough to allow the flexibility described above to serve user and load shifts between cells without too much bookkeeping.

The assignment of channelization code and scrambling code is different in uplink and downlink direction:

- **Uplink**: each of the unsynchronized UEs is assigned a different scrambling code. Furthermore, each individual UE is in command of the entire set of channelization codes, allowing the UE to manage the bandwidths of its sessions independently.

- **Downlink**: each cell uses its own scrambling code, because cells among themselves are not synchronized, either. The scrambling code is thus a cell's "finger print". Each cell has the full set of channelization codes at its disposal, assigning a different one to each UE it is serving.

As a final step we analyse the implications of the code-assignment procedure on macro-diversity. Macrodiversity allows each UE to communicate via several cells simultaneously. Since the downlink code is cell-specific, the signal from each cell is spread with a different code, and must be despread individually by the UE. The uplink code, however, is not cell-specific, and thus the UE only needs to send once. It is not necessary to send a custom-coded signal to each cell. This is good news since the paramount principle in radio interface design is saving radio resources.

The usage of channelization codes and scrambling codes is illustrated in Figure 5.9.

5.2.5 Advanced Division Techniques

Finally, we introduce two more techniques for dividing the available frequency band between users. They derive from the original frequency division technique (cf. Section 5.2.1) but are more sophisticated. These techniques are enjoying increasing popularity and are employed in the more recently developed radio technologies.

5.2.5.1 Orthogonal Frequency Division Multiple Access

The first of these techniques is called **Orthogonal Frequency Division Multiple Access** (OFDMA). With OFDMA, the available frequency band is divided into thousands of narrow sub-bands. Each sender-receiver pair is assigned its individual sub-set of 512 or more of these sub-bands. The sender sends on all of these sub-bands simultaneously by distributing the bits to be transmitted. The receiver listens on all sub-bands and re-assembles the bit sequence.

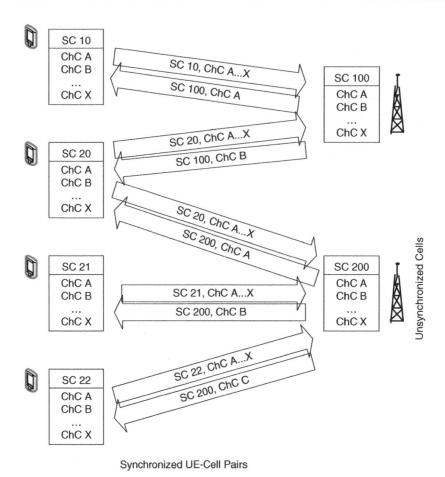

Figure 5.9 Usage of channelization codes (ChC) and scrambling codes (SC) by UEs and cells. Each UE and each cell are assigned one scrambling code and the whole set of channelization codes. For example, the topmost UE is assigned SC 10 and the topmost cell is assigned SC 100. Each downlink UE-cell communication and each uplink UE communication is identified by a unique combination of codes

OFDM is just as flexible as CDMA in its ability to accommodate short term and long-term load shifts between cells and to accommodate user's diverse bandwidth needs. Moreover, it is often more bandwidth efficient than CDMA, and it does not suffer from the near-far effect because senders are separated by the frequency band they use. WLAN IEEE 802.11a/g as well as WiMAX IEEE 802.16 are based on OFDMA, The next update of the 3GPP radio technology, E-UTRA/LTE, also adopts OFDMA on the downlink (network towards UE).

5.2.5.2 Single-Carrier Frequency Division Multiplex Access

The basic idea of **Single-Carrier Frequency Division Multiplex Access** (SC-FDMA) is similar to that of OFDMA: The available frequency band is divided, and each sender-receiver pair is assigned a set of sub-bands. The difference lies in how bits are transmitted over the sub-bands. Whereas in OFDMA, each bit is transmitted over exactly one band, in SC-FDMA, each bit is spread and transmitted over all bands! This is achieved by applying a Fast Fourier Transform on a sequence of n bits, effectively transforming the bit sequence from time space into one in frequency space.

While SC-FDMA is obviously more complex to implement than OFDMA, it has an important advantage: the sender consumes less power to achieve the same result because the peak-to-average power ratio can be reduced. Particularly in mobile senders, power consumption must be minimized. For this reason, 3GPP has decided to employ SC-FDMA on the E-UTRA/LTE uplink (UE towards network).

5.3 Summary

In this chapter we discussed the physical layer of the UMTS radio interface.

UMTS mostly uses QPSK as a modulation technique, i.e. for coding information on the electromagnetic waves. QPSK defines four distinct phase shifts, with each phase shift coding two chips. HSPA uses 16-QAM, which defines sixteen phase-amplitude shifts, with each shift coding four chips. HSPA + even employs 64-QAM.

UMTS originally used FDD or TDD for separating uplink and downlink. It uses CDMA for dividing radio resources between different users. CDMA works by the sender spreading each user signal by means of different codes. The receiver despreads the signal by multiplying it by the same code. Codes are orthogonal and quasi-orthogonal sequences of chips, the chips taking on values of plus and minus one.

Because of the near-far effect, power control is important in CDMA, making sure that more distant UEs send more strongly, so that the signals of all UEs arrive with the same power at the cell's antenna. Power control is performed by the Radio Access System. Straightforward power control, however, may cause the signal of distant UEs to interfere too strongly with the signal in other cells. Therefore macrodiversity is introduced. Here, the UE is communicating via several cells simultaneously. The combination of the signal received via different paths allows one to eliminate transmission errors and in turn to reduce the sending power.

Another technique for dividing radio resources, OFDM, is recently gaining increasing acceptance. As we will see in Part II, it is the basis for most 4G radio technologies. In particular, 3GPP will adopt OFDM and SC-FDMA for its next radio technology update, E-UTRA/LTE.

Another important technique is called Multiple-Input Multiple-Output (MIMO) where sender and/or receiver employ multiple antennae and this way increase the number of possible propagation paths. MIMO is employed by HSPA + and E-UTRA/LTE.

6

Packet-switched Domain— Architecture and Protocols

There now follows a series of chapters featuring a more detailed discussion of the architecture and protocols in UMTS. They provide a per-network-element overview of functionality. These chapters will follow a set of chapters on network control, which provide an "orthogonal view" of how a particular functionality is achieved through the collaboration of all the network elements.

In this chapter we commence with the Packet-Switched Domain and in subsequent chapters progress through all the functional groups introduced in Chapter 4 and Figure 4.5. The reader interested in greater technical detail may study the 3GPP documents [3GPP 23.002] and [3GPP 23.060]. The latter is the main specification for the PS Domain of both UMTS and GPRS.

Terminology discussed in Chapter 6:	
Gateway GPRS Support Node	GGSN
GPRS Mobility Management	GMM
GPRS Support Node	GSN
GPRS Tunneling Protocol - Control-Plane	GTP-C
GPRS Tunneling Protocol - User-Plane	GTP-U
Home Location Register	HLR
Logical architecture	
Mobile Application Part	MAP
Packet-switched Domain	PS Domain
Paging	
PDP Context	
Physical architecture	PEP
Policy Enforcement Point	
RAN Application Protocol	RANAP
Session Management	SM
Serving GPRS Support Node	SGSN
Signalling Connection and Control Part	SCCP
Signalling System Number 7	SS7
Transaction Capabilities Application Part	TCAP

6.1 Architecture

The PS Domain is the packet-switched part of the Core Network of UMTS. Its basic architecture is shown in Figure 6.1. This figure provides more detail than Figure 4.5 by depicting two functional subgroups located in the PS Domain: the **Serving GPRS Support Node** (SGSN) and the **Gateway GPRS Support Node** (GGSN) – the general term GSN refers to both SGSN and GGSN. Furthermore, it shows the **Home Location Register** (HLR) which is accessed from both the PS Domain and CS Domain. SGSN and GGSN are connected by an IP Network. From Release 5 onwards, the IP Network may additionally connect HLR and UTRAN. We will see in Section 6.2 that this IP Network employs a number of protocols that make it UMTS-specific. Figure 6.1 shows the basic architecture only. There are a small number of additional nodes with restricted functionality which will be introduced later as need be.

It is important to realize that the architecture in Figure 6.1 is a **logical architecture** in which each functional subgroup appears only once in each role that it may adopt in communication. For example, in Figure 6.1, the HLR is shown once, while both SGSN and GGSN are shown twice in order to clarify the communication relations: We see that the SGSN can communicate with other SGSNs in the same PS Domain, and with GGSNs in other PS Domains. However, we also see that GGSNs only ever communicate with SGSNs in the same PS Domain.

In contrast to a logical architecture, a **physical architecture** shows the actual, physical network elements of a specific real-life network and how they are connected, see Figure 6.2. In the physical architecture of a small network, functional subgroups may be co-located on one physical network element, as for example SGSN and GGSN in Figure 6.2a. In the physical architecture of a large network, there may be many SGSNs and GGSNs (Figure 6.2b), each a separate physical network element, distributed over the entire geographical area served by an operator. The number of SGSNs and GGSNs does not need to be identical. Furthermore, in most networks, each SGSN and GGSN has an identical back-up for reasons of reliability.

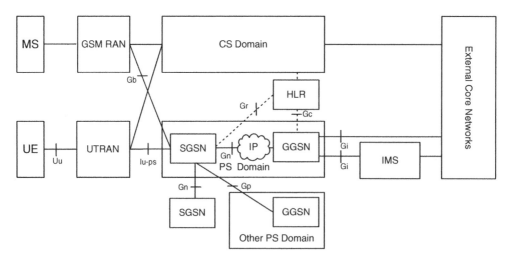

Figure 6.1 Logical architecture of the PS Domain. Solid lines depict user-plane and control-plane interfaces, dashed lines depict control-plane interfaces only

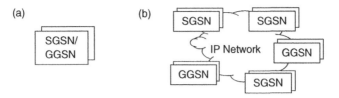

Figure 6.2 Example physical architectures of the PS Domain

Since, typically, each functional subgroup is a single physical network element, SGSN, GGSN, HLR, etc. are usually called "network elements" rather than "functional subgroup" which would be more precise. So this is what we will call them from now on.

The PS Domain is responsible for security, session control, QoS, mobility and charging for all UEs using packet-switched services. Furthermore, packets must be transported and routed. In the following we will see how these control- and user-plane functions are realized by the individual PS Domain network elements.

Figure 6.1 also shows that in addition to the UMTS-specific UTRAN, a GERAN can be attached to PS Domain and CS Domain. This feature was introduced in Release 5.

6.1.1 Serving GPRS Support Node (SGSN)

The SGSN is the main control node in the Core Network. When a UE powers up and connects to the network, it attaches itself to a specific SGSN which is determined by the physical location of the UE. User traffic is transported via the SGSN, and most important control functions are performed by the SGSN.

The SGSN participates in the following tasks:

- **Security**: The SGSN authenticates and authorizes UEs and their users based on subscription information which is stored in the HLR. In Figure 6.1 this is indicated by the control signalling interface Gr between SGSN and HLR.
- **Session control**: For establishing a session, a route must be found to the destination. While the SGSN is determined by the physical location of the UE, the GGSN is determined by the service which is the goal of a particular user session. The SGSN is responsible for finding the appropriate GGSN. When a UE requests a session, the SGSN establishes a tunnel, called **PDP Context**, through the IP Network from UTRAN via itself to the GGSN. Data and control traffic of the UE are routed through this tunnel.
- **QoS**: When the UE starts a new session it asks for the reservation of the corresponding resources. The SGSN determines whether the PS Domain and the UTRAN have these resources. If yes, the PDP Context is assigned these resources. If this is not the case it may negotiate with the UE and offer a lower QoS.
- **Mobility**: The SGSN together with the UTRAN stores the location of the UE. The choice of SGSN depends on where the UE is currently. When the UE moves too far, the SGSN may change. In this case SGSNs exchange user context via the Gn interface shown in Figure 6.1.

The SGSN is also involved in **paging**—when a new session for a particular UE comes in from an external network, it collaborates with the UTRAN to localize the UE.

- **Charging**: The SGSN collects charging-relevant information such as number of packets per session or Radio Access Network type and sends it to a charging gateway (not shown in Figure 6.1). When a pre-paid card runs out of funds, the SGSN terminates the session.
- **Packet transport and routing**: The SGSN transports IP-based user traffic just as an ordinary router.

6.1.2 Gateway GPRS Support Node (GGSN)

The GGSN is the gateway node from the PS Domain to other packet-based Core Networks, e.g. the Internet or the IMS. The choice of GGSN depends on the service the user intends to use. Put differently, when a UE is not actively using a service, there is no GGSN associated with this UE.

The GGSN is responsible for routing packets, and for converting UMTS specific protocols to the protocols of the external networks. It acts as a gatekeeper—the technical term is **Policy Enforcement Point** (PEP)—that, in collaboration with, e.g. the IMS, blocks undesired flows.

Finally, the GGSN, just as the SGSN, collects relevant charging information and sends it to a charging gateway.

6.1.3 Home Location Register (HLR)

The HLR is the central database in a UMTS Network. For each subscriber it has an entry with data such as identifier and telephone number as well as authentication and authorization information. When a UE is currently attached to the network, the HLR also knows which SGSN is responsible.

6.2 Protocols

In this section we introduce the protocols which the network elements introduced in the previous section use for communication. Their detailed functionality will be covered in later chapters. Protocols providing user-plane functions and control-plane functions are quite independent from each other, so we treat them in different subsections. In each section we compare the protocols of the PS Domain to corresponding protocols in a Computer Network. As discussed in Chapter 4, Section 4.4.1, the PS Domain is an IP Network, with however a telecommunication heritage and telecommunication demands regarding network control. In the last subsection we therefore discuss how in this particular case a telecommunication-style IP Network is realized.

6.2.1 User-Plane

We start by looking at the user-plane protocol stack for a generic network element in a Computer Network as depicted in Figure 6.3 (cf. also Figure 1.1b). The network element is generic in the sense that Computer Networks usually have no explicit architecture. We see a protocol stack with any Layer 1 and Layer 2 technology, e.g. WLAN, topped by IP, a transport layer, and finally the application. A second IP layer (in grey) may be present when mobility is supported by a protocol employing IP-in-IP tunnelling, e.g. Mobile IP [RFC, 3344 RFC 3775]. Greater detail will be provided in Chapter 11.

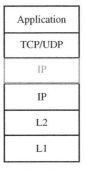

Application
TCP/UDP
IP
IP
L2
L1

Generic
Network Element

Figure 6.3 Typical user plane protocol stack of a Computer Network

The user-plane protocol stack for the PS Domain is shown in Figure 6.4. Since the SGSN communicates with both UTRAN and UE, the figure also depicts those parts of the network.

From top to bottom, Figure 6.3 shows an application on the UE running over an end-to-end IP layer. This part of the protocol stack is intercepted at the GGSN. The GGSN may shape the traffic or manipulate particular fields in the protocol headers (e.g. DiffServ Code Points, see Chapter 14). Before leaving the network, user traffic may be subject to a **Network Address Translator** (NAT) and a firewall which can also be integrated into the GGSN.

Below the end-to-end IP layer, there is a UMTS specific protocol, the **GPRS Tunneling Protocol—User-Plane** (GTP-U), which also runs over IP. Both GTP-U and the lower IP layer are employed only between RNC and GGSN. The GTP-U protocol provides the tunnel, i.e. the PDP Context, with a particular QoS from GGSN via SGSN to the UTRAN which was mentioned in Section 6.1.1. The lower IP layer hides the UE's mobility from the upper IP layer,

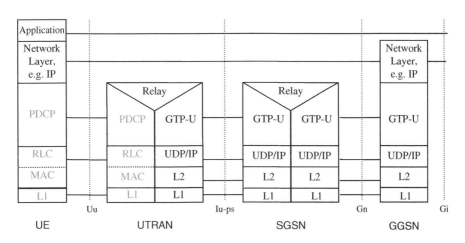

Figure 6.4 User-plane protocol stack of the PS Domain. Protocols in grey are discussed in other chapters

using the same principles as in a Computer Network (e.g. Mobile IP). More details will be provided in Chapter 12 on Mobility.

Layer 1 and Layer 2 in the PS Domain can be any suitable technology, e.g. **Asynchronous Transfer Mode** (ATM) (for more details see [Tannenbaum 2002] or [Kasera 2005]).

6.2.2 Control-Plane

By analogy with the previous section, we should start with a discussion of the control-plane of a Computer Network. It is however difficult to show a generic network element with a protocol stack composed of IETF Protocols, as the control-plane of Computer Networks is not defined coherently, cf. Chapter 4, Section 4.1.2. A large number of control protocols, however, has been defined by the IETF, some of which will be introduced in later chapters. In keeping with the spirit of the Internet, each control protocol provides a single function, e.g. Mobile IP [RFC 3344 RFC 3775] supports mobility, RSVP [RFC 2205] or NSIS [ID QoS NSLP] support QoS delivery, and Diameter [RFC 3588] or RADIUS [RFC 2865] support authentication and authorization. A Computer Network control node would implement the appropriate subset of control protocols.

The control-plane of the PS Domain is illustrated in Figures 6.5 and 6.6. The figures clearly show that the SGSN is the main control node in the PS Domain, communicating "to the left" with UE and UTRAN, and communicating "to the right" with other GSNs, i.e. SGSNs and GGSNs.

With the UE, the SGSN performs the control functions described in Section 6.1.1 using **GPRS Mobility Management** (GMM) and **Session Management** (SM). GMM is used for admission control and mobility control. SM is used for session management.

With the UTRAN, the SGSN communicates using the **RAN Application Protocol** (RANAP) running over **Signalling Connection and Control Part** (SCCP). RANAP/SCCP is used for all control functions between SGSN and UTRAN. Additionally, RANAP/SCCP encapsulates the GMM and SM signalling messages between SGSN and UE.

RANAP/SCCP are members of the **Signalling System Number 7** (SS7) protocol family. SS7 is a comprehensive control signalling suite for Telecommunication Networks. SS7 is vertically integrated in the sense that it defines an entire protocol stack from layer one

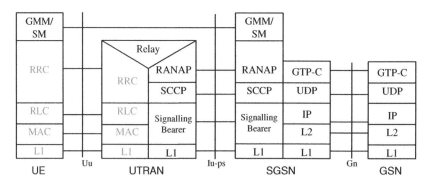

Figure 6.5 Control-plane protocol stack of the PS Domain between UE and SGSN resp. GGSN. Protocols in grey are not discussed in this chapter

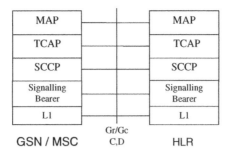

Figure 6.6   Control-plane protocol stack of the PS Domain between SGSN or GGSN and HLR

upwards. RANAP/SCCP can in principle run in normal "SS7 style" over the lower layers of the SS7 protocol stack. However, an adaptation layer ("Signalling Bearer" in Figure 6.5) was introduced in Release 99 of UMTS that allowed running RANAP/SCCP over ATM. In Release 5, the widespread usage of IP was acknowledged by designing an additional adaptation layer that allowed running RANAP/SCCP also over IP. The CS Domain takes this idea of "bearer independent" design even further. Chapter 7 on the CS Domain will provide greater detail.

With other GSNs, the SGSN communicates using the **GPRS Tunnelling Protocol— Control-Plane** (GTP-C), which complements GPT-U in the user-plane. With GTP-C, the user-plane tunnel, i.e. PDP Context, between SGSN and GGSN is controlled, as well as the change of SGSN due to UE mobility.

Figure 6.6 depicts the control-plane between a GSN and the HLR. The main control protocol is the **Mobile Application Part** (MAP) running over the **Transaction Capabilities Application Part** (TCAP). MAP transports all of the information which the HLR will ever need to exchange with either SGSN or GGSN. TCAP controls connections between pairs of nodes. As one can tell by the "AP" in their names, MAP/TCAP are also protocols from the SS7 family, which can run either natively over the rest of the SS7 protocol stack, over ATM, or over IP.

6.2.3 Discussion

The previous subsections illustrated a number of interesting points. It became obvious that the design of the PS Domain is architecture-centric in that the architecture defines network elements with well-defined tasks and communication relations. Each pair of network elements communicates based on a single control protocol over a generic lower protocol stack. This protocol serves for all functions. For example, GTP-C is the protocol between SGSN and GGSN and transports information on mobility, QoS, charging, etc. In contrast, in a Computer Network, there are multiple protocols between network elements, one for each function.

Additionally, we saw how the PS Domain employs a mixture of IETF protocols, UMTS specific protocols and SS7-derived protocols to achieve an IP Network that still allows telecommunication-style comprehensive control. We observe that the user-plane exhibits only one UMTS-specific protocol, GTP-U. The control-plane, however, employs a great number of protocols that are not understood outside UMTS.

6.3 Summary

This chapter presented the basic architecture of the PS Domain, and introduced the main network elements. User traffic entering the PS Domain from the UE, passes from SGSN to GGSN. The SGSN is the main control node of the PS Domain. It participates in authentication and authorization, session control, QoS control and provisioning, mobility support and charging. The GGSN is the gateway to external networks. It is a Policy Enforcement Point for the IMS, and converts between protocols in the PS Domain and protocols used outside. The HLR is the central data base of a UMTS Network.

This chapter also introduced the protocols employed in the PS Domain, illustrating the one-protocol-per-pair-of-network-elements paradigm of Telecommunication Network. We saw—particularly in the control-plane—a mixture of IETF Protocols, UMTS-specific protocols and SS7-derived protocols providing comprehensive control of the network.

7

Circuit-switched Domain— Architecture and Protocols

This chapter deals with the architecture and protocols of the Circuit-Switched Domain and proceeds by way of analogy to the previous chapter on the PS Domain. More information is available in the 3GPP documents [3GPP 23.002] and [3GPP 23.205]. The latter is the main specification for the CS Domain.

Terminology discussed in Chapter 7:	
Bearer	
Bearer-independent	
Bearer Independent Call Control	BICC
Call Control	CC
Circuit-switched Domain	CS Domain
Gateway Mobile Switching Center	GMSC
GMSC Server	
H.248	
Home Network	
ISDN User Part	ISUP
Media Gateway	MGW
Media Gateway Control Protocol	MEGACO
Message Transfer Part	MTP
Mobility Management	MM
Mobile Switching Center	MSC
MSC Server	
MTP3- User Adaptation Layer	M3UA
Signalling ATM Adaption Layer	SAAL
Signalling Gateway	SGW
Stream Control Transport Protocol	SCTP
Visited Location Register	VLR
Visited Network	

UMTS Networks and Beyond   Cornelia Kappler
© 2009 John Wiley & Sons, Ltd

7.1 Architecture

The CS Domain is the circuit-switched part of the Core Network of UMTS. Its architecture in Release 99 is depicted in Figure 7.1. It is instructive to compare the architecture of the CS Domain with that of the PS Domain (Figure 6.1): The design idea is the same and in fact was inherited from GSM. The **Mobile Switching Center** (MSC) corresponds to the SGSN and the **Gateway Mobile Switching Center** (GMSC) to the GGSN. Also, the distribution of functionalities between MSC and GMSC is almost the same as between SGSN and GGSN: the MSC is the main control node, and the GMSC is the gateway to external networks. There is a minor variation in the architecture of the CS Domain, in the form of the **Visited Location Register** (VLR) which is usually collocated with the MSC: when the UE is roaming in a **Visited Network**, it attaches to a local MSC. Authentication and authorization data, however are, of course, only available from the HLR of the **Home Network**. In order to reduce the inter-network traffic of sensitive data, the HLR usually sends the relevant set of authentication and authorization data to the VLR so it can be stored locally for the time being. A similar mechanism exists for the PS Domain, where the VLR functionality is integrated into the SGSN.

In Release 4 of UMTS, the architecture of the CS Domain was changed considerably, cf. Figure 7.2. In keeping with the spirit of modularity, both MSC and GMSC were split into a control node (**MSC Server** and **GMSC Server** respectively) and a user traffic transport node, the **Media Gateway** (MGW). The advantage of this modular design is that the control-plane and user-plane can be evolved independently. With the new design all control traffic is handled by the MSC Server and the GMSC Server, and all user traffic is handled by the MGW. In Figure 7.2 this is indicated by the MSC Server and GMSC Server featuring control interfaces only. Based on the control information exchanged "horizontally" with UE, UTRAN and external networks, the MSC Server and GMSC Server control and configure the MGWs via the Mc interface. A similar split into control and user traffic transport nodes was discussed with regard to the PS Domain. At the time, this discussion did not progress beyond a feasibility study. In the current debate on 4G and UMTS evolution (cf. Chapters 20 and 21), the topic is being taken up again.

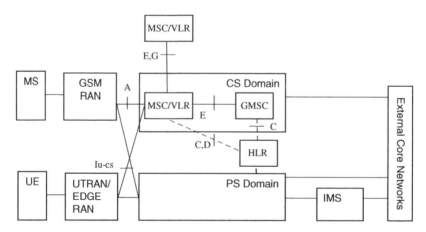

Figure 7.1 Logical architecture of the CS Domain in Release 99. Solid lines depict control and data interfaces, dashed lines depict control interfaces only

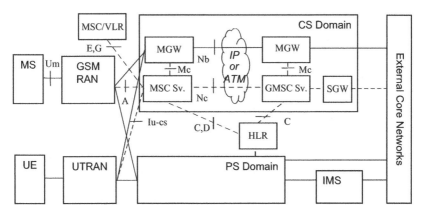

Figure 7.2 Logical architecture of the CS Domain with Release 4. Solid lines depict control and data interfaces, dashed lines depict control interfaces only

7.2 Protocols

As with the architecture, the protocols for the CS Domain have also evolved from GSM. With each release of UMTS we witness the slow but steady transformation of a true circuit-switched network into an IP Network transporting circuit-switched service. Whereas in the previous chapter we compared the PS Domain protocols to those used in Computer Networks, in this chapter we compare CS Domain protocols to those used in GSM.

GSM works by multiplexing both user data and control signalling onto a **Time Division Multiplex** (TDM). In UMTS Release 99, ATM was introduced in order to replace TDM. However, owing to time constraints (see Chapter 3, Section 3.2.1 on standardization in 3GPP: "Timely termination as a guiding principle"!), ATM was introduced in the UTRAN and on the Iu-cs interface between UTRAN and MSC only.

In Release 4, the entire CS Domain was enabled to run over ATM. And, at the same time, IP was also introduced. The result is three transport possibilities for CS Domain protocols: "native circuit-switched", ATM and IP. The aim of this exercise is better modularity, just as with the split of the MSC discussed above: the entire design of the CS Domain has become **bearer-independent**, the term **bearer** referring to the different transport options. The bearer independent design allows for an independent evolution of the signalling semantics and the transport technology: if some day another transport technology becomes *en vogue* it can easily be introduced. The bearer independence necessitates the introduction of **Signalling Gateways** (SGW), which translate between the different transport varieties, e.g. at the transition to other networks. SGWs are included as necessary; Figure 7.2 provides just one example of SGW location.

While, of course, circuit-switched aficionados remain sceptical as to whether the quality of circuit switched services can be maintained when running them over IP, it is also true that the simultaneous maintenance of two independent Core Networks, e.g. a CS Domain over TDM or ATM, and a PS Domain over IP, imposes a considerable burden on operators. The introduction of IP in the CS Domain now allows for the implementation of the two logically separated domains with a single physical IP Network.

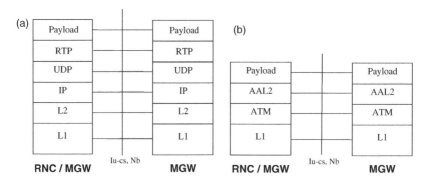

Figure 7.3   User-plane protocol stacks of the CS Domain, (a) over IP and (b) over ATM

7.2.1 User-Plane

The user-plane of GSM transports mostly voice which a codec transforms into frames multiplexed onto TDM. Service and transport networks are therefore closely coupled which is very efficient but also inflexible. The same technique was used in Release 99 for the UMTS CS Domain. With the introduction of ATM and IP-based transport in Release 4, alternative protocol stacks became possible. Figure 7.3a depicts the transport of a CS Domain payload between MGWs over IP. Since the CS Domain is used for real-time services, the payload is sent over the **Real Time Protocol** (RTP) [RFC 3550] and UDP. Figure 7.3b shows the transport of the real-time payload over ATM. In order to support the demands of different services, ATM defines a number of adaptation layers [Tanenbaum 2002] that are placed between the actual ATM layer and the higher layers. The **ATM Adaptation Layer 2** (AAL2) depicted in Figure 7.3b supports real-time services of variable bandwidth.

7.2.2 Control-Plane

The control-plane of UMTS is an evolution of GSM in the sense that the upper half of the protocol stack consists of a set of SS7 protocols (cf. Figures 7.4 and 7.5 for Release 99 and

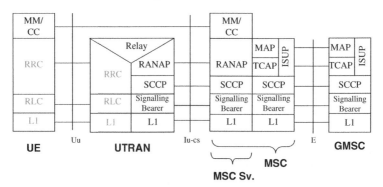

Figure 7.4   Control-plane protocol stacks of the CS Domain. Protocols in grey are not discussed in this chapter

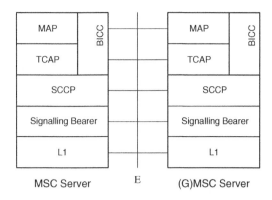

Figure 7.5 Control-plane protocol stack of the CS Domain between MSC Servers

Release 4 respectively), whereas the lower half provides the above mentioned three transport varieties, in this case native SS7, ATM or IP, see Figure 7.6.

The organization of the control-plane protocol stack of the CS Domain is quite similar to that of the PS Domain: The main control node of the CS Domain, the MSC, communicates with the UE using the **Mobility Management** (MM) and the **Call Control** (CC) protocols. These protocols correspond to GMM and SM in the PS Domain, respectively. Also, as in the PS Domain, RANAP over SCCP are employed for all control functions between Core Network and UTRAN, as well as for encapsulating MM and CC messages towards the UE.

A Release 99 MSC communicates with other MSCs using the **ISDN User Part** (ISUP) for call control—the protocol used in fixed-line ISDN, and MAP, which is also used with the HLR (cf. Figure 6.6) for other control aspects. With the introduction of the bearer-independent CS Domain and MSC Servers in Release 4, ISUP was replaced by **Bearer Independent Call Control** (BICC), see Figure 7.5.

While the upper half of the control protocol stack is bearer independent, the lower half obviously is not. Depending on which variety is chosen, a number of adaptation layers must be

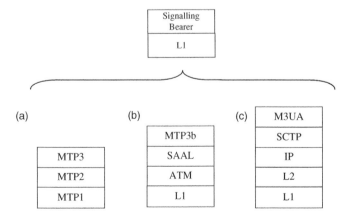

Figure 7.6 Signalling Bearer possibilities for control-plane signalling in the CS Domain: (a) native SS7, (b) ATM, (c) IP

included. The different options are given in Figure 7.6. Figure 7.6a shows the original SS7 protocol stack with the three layers of **Message Transfer Part** (MTP). Figure 7.6b depicts the corresponding protocol stack for ATM, which is slightly more complicated. An adaptation of MTP3 for broadband (MTP3b) runs over the **Signalling ATM Adaption Layer** (SAAL) which really consists of three sublayers, which runs over the actual ATM. Figure 7.6c shows the IP protocol stack with the adaptation layer **MTP3-User Adaptation Layer** (M3UA) and the recent Layer 4 transport layer protocol, the **Stream Control Transport Protocol** (SCTP). To extend the picture of possible combinations, IP in Figure 7.6c can, of course, also run over an ATM layer 2 which then introduces another group of adaptation layers.

The last protocol stack we need to mention is that between MSC Server and MGW. The IETF and the ITU together developed a protocol to this end, which is called **Media Gateway Control Protocol** (MEGACO) by the IETF and **H.248** by the ITU-T [ITU H.248]. Meanwhile the ITU-T is responsible for the further evolution of the protocol. By means of suitable adaptation layers MEGACO can run over IP and ATM.

7.3 Summary

This chapter presented the basic architecture of the CS Domain, and introduced the most important network elements. User traffic entering the CS Domain from the UE, passes from MSC to GMSC. The initial CS Domain architecture was closely modelled after the GSM architecture with the MSC and GMSC as the main control node. With Release 4, there was a split of the G(MSC) into a control-plane node, the G(MSC) Server, and a user-plane node, the MGW. The motivation for this split was improved modularity which allows for a flexible response to changes in technology.

This chapter also introduced the protocols employed in the CS Domain. We saw that the control-plane protocols, with the exception of MEGACO, are protocols from the SS7 family. Whereas in Release 99 the CS Domain essentially reused the CS technology from GSM, in Release 4 bearer independence was introduced: it is now possible to run both user-plane and control-plane over ATM and/or IP. Again, the aim was improved modularity that allowed for the flexible introduction of new bearers. Moreover, the unification of networking technologies across the CS Domain and PS Domain decreases administrative overheads for operators. On the other hand, as we saw in Figure 7.5, bearer independence implies a variety of adaptation layers and introduces considerable complexity in the protocol stacks.

The architecture and protocols in this chapter, while certainly not simple, are still simplified compared to the actual state of affairs. The reader interested in greater complication may consult [Kasera 2005] or 3GPP specifications such as [3GPP 23.205].

8

UMTS Terrestrial Radio Access Network—Architecture and Protocols

This chapter studies the architecture of the UMTS Terrestrial Radio Access Network (UTRAN). We also look at some of the lower layer protocols in the UTRAN and on the radio interface. These protocols are specific to UMTS and—as we will see in later chapters—their design approach to some extent influences how certain functions are provided. Further detail on UTRAN architecture is available in [3GPP 25.401] and on UTRAN protocols in [3GPP 25.301].

Terminology discussed in Chapter 8:	
Broadcast and Multicast Control	BMC
Iu-flex	
Logical Channel	
Node B	
Packet Data Control Protocol	PDCP
Physical Channel	
Radio bearer	
Radio Link Control	RLC
Radio Network Controller	RNC
Radio Resource Control	RRC
RRC Connection	
Stateful	
Transport Channel	
UMTS Terrestrial Radio Access Network	UTRAN

8.1 Architecture

The main task of the UTRAN is the provisioning and controlling of radio resources. To this end it collaborates with both the UE and Core Network. Furthermore, the small-scale mobility of the UE is handled in the UTRAN and hidden from the Core Network. Finally, the UTRAN is the

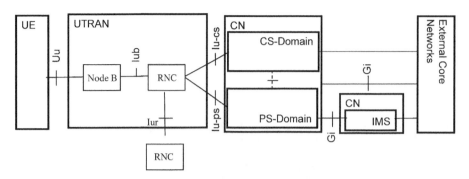

Figure 8.1   Logical architecture of the UTRAN

first contact point for a UE trying to attach to the network. As we saw in the previous chapter, ultimately the UE is anchored with a SGSN or MSC; however the UE and UTRAN first establish a signaling relation in order for UE and SGSN/MSC to find each other.

   The logical architecture of the UTRAN is shown in Figure 8.1. We see two network elements, **Node B** which is responsible for the actual radio transmission, and **the Radio Network Controller** (RNC), the main control node of the UTRAN. In Figure 8.2 we see a corresponding physical architecture. The most important feature to observe in Figure 8.2a is

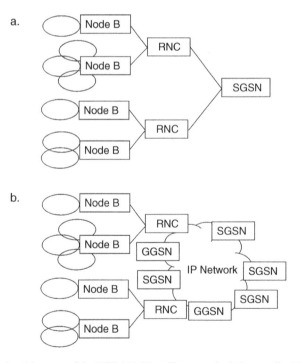

Figure 8.2   Physical architecture of the UTRAN. The ellipses on the left are cells. (a) pre-Release 5 and (b) Release 5

the hierarchical tree-like structure. One Node B is responsible for several antennae, i.e. cells, one RNC is responsible for several Node Bs, and one SGSN (or MSC) is responsible for several RNCs. This tree-like structure simplifies the control of the network—for example, once the cell of a UE is determined, Node B, the RNC and SGSN are determined as well. On the other hand, this hierarchical structure is inflexible: the geographical areas covered by SGSNs do not overlap, making load-balancing or redesign of the physical network difficult Therefore, in Release 5 the hierarchical relation between RNC and SGSN was abandoned. The so called **Iu-flex**—named after the interface affected, allows a many-to-many relationship between RNCs and SGSNs [3GPP 23.236]. In the same release, IP was introduced on the Iu interface in addition to ATM. As illustrated in Figure 8.2b, this allows for the integration of RNCs, SGSNs and GGSNs (as well as, possibly, MSC Servers, MGWs etc.) into a single IP Network.

8.1.1 Node B

Node B is active on the lower OSI layers. It is responsible for transmitting both control and user traffic destined to the UE over the radio interface, and for receiving traffic from the UE and putting it "on the wire". To this end it is responsible for (de)spreading and (de)modulation as well as for generation of CDMA codes (cf. Chapter 5).

Furthermore, Node B is involved in radio resource control: It measures the quality of the received signals and makes this information available to the RNC. Based on these measurements, the RNC may decide to attach the UE to another cell. Furthermore, Node B is involved in power control in order to counter the near-far effect, as the reader may remember from Chapter 5, Section 5.2.4.1: The further away from the antenna, the stronger the signal the UE must emit. Node B measures the power of the incoming signal and tells the UE whether it needs to adjust.

8.1.2 RNC

The RNC is the main control node in the UTRAN. It is the UE's first contact point in the network. When a UE powers up, it contacts the RNC responsible for its cell, establishes radio connectivity, and only then attaches to the SGSN or MSC. The UE maintains the relation to this RNC, its **Serving RNC**, as long as it is active, i.e. sending or receiving traffic. At the same time, the Serving RNC maintains information on the UE, e.g. to which cell(s) it is attached. One says, therefore, that the Serving RNC **keeps state**, or that it is **stateful**.

The RNC performs the following tasks:
- **Radio Resource Control and provisioning**: While Node B has a rather limited view of the world, and limited control over its own resources, the RNC has an overview of all radio resources attached to it. It is responsible for these resources and controls the set-up, maintenance and release of radio connections. These radio connections are also called **Radio Bearers**—another usage (cf. Chapter 7) of the concept of bearers which is rather popular in UMTS. Chapter 11 will shed more light on this.

  Radio bearer control also involves the planning of resources and the calculation of interference and utilization levels, as well as the control of CDMA codes.

Furthermore, the RNC is involved in power control. Power control actually has two stages: the "inner loop" performed by Node B as described in the previous section, and the "outer loop" performed by the RNC. As one might expect, the RNC outer loop controls the Node B inner loop. This means that the RNC determines the target power level the UE should achieve based on the overall radio resource picture, and Node B is responsible for enforcing this power level.

- **QoS**: The SGSN or MSC can only admit a new session once the RNC has confirmed that sufficient resources are a available in the UTRAN, i.e. the session can be established without compromising the quality of existing sessions.

- **Security**: Communication on the radio interface is protected, e.g. by encrypting all packets. The RNC performs the necessary operations.

- **Mobility**: The RNC, the Serving RNC to be precise, controls the small-scale mobility of the UE, i.e. mobility across a small number of cells, whereas the SGSN/MSC controls large scale mobility, including roaming. The Serving RNC decides, based on measurement reports received from both UE and UTRAN, whether a handover is necessary and then initiates this handover. The Serving RNC is also responsible for the control of macrodiversity (cf. Chapter 5, Section 5.2.4.2), i.e. for deciding whether the UE should attach to more than one cell, and if yes, which cells.

  When the UE moves a large distance since attaching to the network, it may have moved out of the coverage are of cells controlled directly by its Serving RNC. In this case the Serving RNC is relocated to another RNC. To this end, the old and the new Serving RNC exchange user context via the Iur interface shown in Figure 8.1. The reader by now should understand that Iur is a logical interface. Iur does not imply a physical wire connecting all RNCs directly.

- **User data transport and routing**: The RNC transports IP-based traffic in the same way as an ordinary router. Additionally, the RNC must protect the traffic against a variety of security threats on the radio interface by means of encryption and integrity protection.

In addition, the RNC is responsible for broadcasting system information on the radio interface. This system information ranges from the identifier of the network—so UEs can determine whether this is the network of their home operator—and location information, to information necessary for accessing and maintaining radio resources, e.g. synchronization signals and radio measurement criteria.

8.2 Protocols and Channels

This subsection concentrates on the protocols between the RNC and UE, as shown in Figure 6.4 for the user-plane and in Figures 6.5 and 7.4 for the control plane. In addition to protocols, a concept called **channels** plays a role. The attentive reader will notice that this section only covers protocols between the UE and UTRAN, as protocols between the UTRAN and Core Network were dealt with in previous chapters. Note that we skip the protocols involving Node B as they concern lower OSI layers, which are not the focus of this book.

Figure 8.3 provides a more detailed view of the protocols and channels between the RNC and UE. We see the control-plane protocol with the **Radio Resource Control** protocol (RRC) controlling the entire set of control-plane and user-plane protocols. Furthermore, we observe a

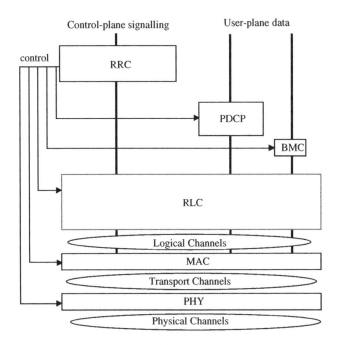

Figure 8.3 The protocols and channels between RNC and UE

variety of channels sandwiched between L1 and L2 sub-layers. The protocols specific to UMTS as well as the use of the different channels are now introduced in more detail.

8.2.1 User-Plane

Unicast user packets from the PS Domain usually arrive at the RNC in the form of IP packets. Since the link layer protocol serves both CS Domain and PS Domain data, PS Domain packets are encapsulated and adapted to the link layer via the **Packet Data Control Protocol** (PDCP), see also Figure 6.4. The PDCP ensures some independence of link layer and higher layers; it can, e.g. deal with both IPv4 and IPv6. Furthermore, it provides header compression. Multicast and broadcast data are encapsulated in the **Broadcast and Multicast Control** (BMC) protocol. The CS Domain payload (cf. Figure 7.3) is transported directly over the link layer as discussed in Chapter 7, Section 7.2.1.

8.2.2 Control-Plane

When a UE powers up and attaches to the network, it first aspires to its own **RRC Connection**. The RRC Connection is a logical concept that describes a static relation between the UE and Serving RNC that exists for as long as the UE is actively sending data. There is exactly one RRC Connection for each active UE. It controls the radio interface for all sessions of this UE. This is practical because, in the case of handover, all of the UE's sessions must be handed over and it is convenient to have a single protocol instance where all information is gathered. RRC carries all control information between the UE and RNC. Its functions include radio resource set-up and release, mobility management, outer loop power control and the transport—i.e. tunnelling—of Core Network control protocols. Additionally, the RRC is used by the RNC to broadcast system information such as network identifier and location information.

8.2.3 Lower Layers and Channels

The RRC, PDCP and BMC are carried over another protocol called **Radio Link Control** (RLC). The RLC carries out tasks related to the transmission of data, such as error protection, segmentation/reassembly and flow control. Layer 3 protocols of the CS Domain run directly on top of the RLC.

As illustrated in Figure 8.3, between RLC and the physical wire we find a sequence of channels offered by the individual protocol sub layers. The channels describe *how* the data from the layer above is transported. The channel concept allows for a better decoupling of the services offered by the individual layers and hence greater flexibility in how data is transported.

Logical Channels. The MAC layer offers a number of **Logical Channels**, each specific to the type of information being transported. The RRC protocol instance chooses which Logical Channel is appropriate and then hands over its data unit. There are two classes of logical channels, **dedicated channels** which are assigned to a single UE, and **common channels** or **shared channels**, which are shared by several UEs. Furthermore, there are channels for control information and channels for user data. Among the Logical Channels for control information, there are

- the **Broadcast Control Channel** (BCCH) for downlink broadcast data;
- the **Paging Control Channel** (PCCH) for paging;
- the **Common Control Channel** (CCCH) used by all UEs and RNCs for control signalling prior to RRC Connection establishment;
- once a RRC Connection is established, UE and RNC communicate via a **Dedicated Control Channel** (DCCH).

Logical channels for transporting user data include
- the **Common Traffic Channel** (CTCH) which is shared by UEs;
- a **Dedicated Traffic Channel** (DTCH) which is used by one UE only.
- **Transport Channels**. The MAC layer maps the data from the Logical Channels onto the Transport Channels offered by the physical layer. There is a large variety of Transport Channels. The choice of Transport Channel depends on the quality of transmission that should be achieved. The individual channels differ, e.g. in bit rate and error protection, and a given channel may offer different options.

There is one Transport Channel dedicated to a single UE,

- the **Dedicated Transport Channel** (DCH). It has a variable bit rate and can be used both uplink and downlink. The dedicated Logical Channels DCCH and DTCH can obviously be mapped onto DCH.
    Then there are a number of Transport Channels that are shared between UEs, among them
- the downlink **Broadcast Channel** (BCH);
- the **Paging Channel** (PCH);

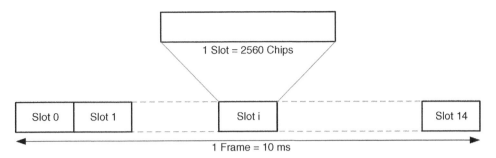

Figure 8.4 The structure of Physical Channels (Reproduced by permission of Dunod Editeur, from *UMTS, Les origines, l'architecture et la norme* (2nd ed) by Pierre Lescuyer.)

- the uplink **Random Access Channel** (RACH);
- the downlink **Forward Access Channel** (FACH).
- An interesting shared Transport Channel is the **Downlink Shared Channel** (DSCH), which was newly introduced in GPRS (cf. Chapter 4, Section 4.4): it allows the RNC to multiplex data destined for different UEs onto the same Transport Channel—in other words to multiplex several DTCHs and DCCHs onto one DSCH. This way the fixed bandwidth of a Transport Channel can be shared between applications that have volatile bandwidth needs, e.g. web surfing.
- **Physical Channels**. The Transport Channels are received by the physical layer and mapped onto Physical Channels. Physical Channels define how the Transport Channels use the physical medium (i.e. chips and codes). A Physical Channel is structured in **frames** and **slots**. A frame lasts 10 ms and consists of 15 slots. Since UMTS works with a chip rate of 3.84 Mega chips/s, this translates into 2560 chips per slot. The frame and slot structure allows organizing the information on the Physical Channels, e.g. into control information and payload. Figure 8.4 illustrates the structure of Physical Channels.

Examples of Physical Channels are the

- **Dedicated Physical Data Channel** (DPDCH) that receives input from DCH;
- **Physical Downlink Shared Channel** (PDSCH) that receives data from DSCH;
- **Primary Common Control Physical Channel** (P-CCPCH);
- **Secondary Common Control Physical Channel** (S-CCPCH); and
- **Physical Random Access Channel** (PRACH).

Also worth mentioning are a number of stand-alone Physical Channels which are not utilized by Transport Channels. The UE tunes into these stand-alone Physical Channels when it powers up and tries to find the network. The details will be covered in Chapter 11. Examples of such channels are

- **Primary Synchronization Channel** (P-SCH);
- **Secondary Synchronization Channel** (S-SCH); and
- **Common Pilot Channel** (CPICH).

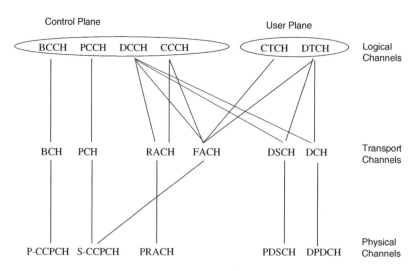

Figure 8.5 The mapping of Logical Channels onto Transport Channels onto Physical Channels (Reproduced by permission of Dunod Editeur, from *UMTS, Les origines, l'architecture et la norme* (2nd ed) by Pierre Lescuyer.)

Figure 8.5 illustrates how the different channels collaborate.

8.3 Summary

This chapter presented the basic architecture of the UTRAN, and introduced the main network elements. User traffic entering the UTRAN passes from Node B to the RNC. Node B is responsible for the lower-layer radio transmission. The RNC is the main control node in the UTRAN. It is involved in radio resource control, QoS, security and mobility. The control protocols utilized in the UTRAN are UMTS-specific. The lower protocol (sub)layers offer their services to the upper layer in the form of channels which come with a diverse set of layer-specific attributes. This way the treatment of data can be adjusted flexibly.

9

User Equipment—Architecture and Protocols

This chapter discusses the architecture of mobile User Equipment (UE). The protocols which the UE has to implement have been dealt with in previous chapters. Readers interested in greater technical detail may study the 3GPP documents [3GPP 23.002], [3GPP 27.001], [3GPP 31.102] and [3GPP 31.103].

Terminology discussed in Chapter 9:	
International Mobile Subscriber Identity	IMSI
IP Multimedia Services Identity Module	ISIM
Mobile Equipment	ME
Mobile Station International ISDN number	MSISDN
Mobile Termination	MT
Network Access Identifier	NAI
(Packet) Temporary Mobile Subscriber Identity	P-TMSI/TMSI
Private User Identity	
Public Identifier	
Public User Identity	
Terminal Equipment	TE
Uniform Resource Identifier	URI
Universal Integrated Circuit Card	UICC
Universal Subscriber Identity Module	USIM
User Equipment	UE

9.1 Architecture

The UE has a substructure, illustrated in Figure 9.1. It consists of the **Universal Integrated Circuit Card** (UICC), which communicates with the **Mobile Equipment** (ME). UICC and ME

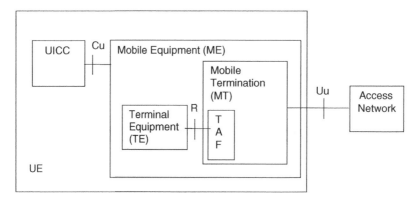

Figure 9.1   UE Architecture from [3GPP 27.001]

correspond to the UIM and Mobile Terminal, respectively, in the high-level architecture for 3G Networks discussed in Chapter 4, Section 4.1.2. The UICC is a smart card, like the SIM card in GSM, and contains subscriber-specific information. It is prepared and sold by the operator of a UMTS Network. The ME is an end system which is independent of the subscription. This clear division between subscriber-specific and subscriber-independent elements allows for the sale and purchase of end systems and subscriptions separately. Early mobile Telecommunication Networks did not make this distinction. It was introduced in GSM and is considered to be one of the factors contributing to its success.

Traditionally, the UE is a single piece of equipment, i.e. a single ME containing a UICC; when 3GPP started work on UMTS, this was a tacit assumption. Increasingly, however, UEs can consist of several networked MEs, e.g. all of the mobile devices of a subscriber, sharing the same UICC. When we deal with 4G, in Part II of this book, we will hear more about this.

The ME is composed of the **Terminal Equipment** (TE) and the **Mobile Termination** (MT) which we will explain below.

9.1.1 TE

The TE can be a specialized UMTS mobile phone. In principle, however, it can be any end system, e.g. a laptop. The TE manages the end system hardware such as display, camera, microphone, etc. It runs the applications, and performs session control by communicating with a peer TE at the other end of the communication session.

9.1.2 MT

The MT terminates the radio transmission, performs radio resource control, deals with security on the radio link and supports mobility and QoS for data received from the TE. It can be integrated with the TE in a mobile phone. However, it can also be a separate card that—together with the UICC—is included in a laptop to make it UMTS capable. In this case the MT contains a **Terminal Adaptation Function** (TAF) that allows it to interface with the TE. A MT can also support multiple TEs simultaneously, all on the basis of the same UICC, in other words on the basis of a single subscription. In this way **Personal Area Networks** (PANs) can be supported. This topic will be pursued further in Part II, when we discuss the evolution of UMTS towards 4G.

9.1.3 UICC

This section discusses the UICC, a core concept in 3G Networks, in more detail. The UICC is a smart card, typically of a size 25 mm × 15 mm. It is bought together with the subscription and inserted into the end system, e.g. mobile phone. In fact, the UICC normally resembles the SIM card for GSM that most readers are likely to have come across. The UICC, however, is more general than a SIM card: the SIM card holds subscriber-specific data for precisely one technology, namely GSM. The UICC, by contrast, may hold subscriber-specific data—so-called *applications*—for several technologies: it may hold a **SIM application** for GSM, a **Universal Subscriber Identity Module** (USIM) **application** for UMTS and an **IP Multi-media Services Identity Module** (ISIM) **application** for IMS usage, all on the same physical UICC. This way, the user may access GSM, and UMTS including IMS, all with one UE. This concept is illustrated in Figure 9.2.

A UICC application identifies a user, and, most importantly, identifies how to charge this user. Without it, only emergency calls are possible. A UICC application also contains the secret keys that allow for subscriber authentication. This information is also available on the network side. It is stored, of course, in the HLR.

One important feature is that the UICC is inaccessible to the user. The identifier and secret keys can only be manipulated by the operator who is selling the UICC. The UICC thus creates a secure environment (for the operator) which is essential to many commercial applications. One could speculate that the UICC will prove to be one of the key assets for UMTS operators when it comes to 4G and in competition with other mobile Communication Networks.

Technically, UICC applications contain the following data:
- USIM [3GPP 31.102]:
  - Identifier
    - **International Mobile Subscriber Identity** (IMSI), this is a unique identifier of the subscription. The IMSI is a number which is only used within the UMTS Network. It is not used as an identifier with regard to the outside.
    - **(Packet) Temporary Mobile Subscriber** Identity (P-TMSI/TMSI), these are the temporary identifiers of a subscription. For security reasons, temporary identifiers are

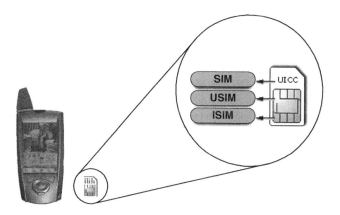

Figure 9.2 UICC and UICC applications: SIM, USIM and ISIM (Reproduced by permission of John Wiley & Sons, Ltd, from *The 3G IP Multimedia Subsystem* by G. Camarillo, M.A. Garcia-Martin (2004).)

used instead of the IMSI after the UE is attached to the network. The P-TMSI is used for
the PS Domain, the TMSI for the CS Domain.
- **Mobile Station International ISDN number** (MSISDN). This is the phone number
  under which this subscription can be reached. At the same time, the MSISDN is the
  publicly known identifier – called **public identifier** - of a subscription..

The identifier information is mirrored in the HLR.
- Secret keys. This information is mirrored in the HLR.
- Current location—the UE thus replicates the information about location that is also kept in
  the network (SGSN, MSC and RNC).
- Miscellaneous information such as preferred language, the list of preferred networks
  (e.g. the Home Network), etc.
- SMS, MMS, etc.

- SIM information is the GSM equivalent to the USIM data above.
- ISIM [3GPP 31.103]:
  - Identifier
    - A **Private User Identity**, this is a unique identifier of the subscription, which is only
      used for authentication. It is the IMS-equivalent to the UMTS IMSI and has the format
      user@realm, i.e. the format of a **Network Access Identifier** (NAI) [RFC 2486].
      "realm" thereby identifies the network to which the user has subscribed.
    - One or more **Public User Identities**, under which the user can be reached. This is the
      IMS equivalent to the UMTS MSISDN and has the format or a **Uniform Resource
      Identifier** (URI). It can resemble an email address or a phone number. Further detail
      will be provided in Chapter 10, Section 10.2.1.

The identification information is mirrored in the **Home Subscriber Server** (HSS), the IMS
extension of the HLR.
  - Home Network Domain URI.
  - Secret keys. This information is also mirrored in the HSS.

Additionally, the UICC contains an address book. The UICC includes a CPU that allows it to
access and process the information above, e.g. to perform authentication on the basis of the
secret keys.

9.2 Summary

The internal architecture of a UE consists of UICC, TE and MT. The UICC contains subscriber-
specific information, whereas TE and MT are subscriber independent. It creates a secure
environment which cannot be manipulated by the user. The TE can be a generic end-system. It
runs applications, participates in session control and controls the hardware. The MT performs
the UMTS specific tasks and may include a Terminal Adaptation Function for interfacing with
the TE. The MT is involved in radio resource control, mobility, security and QoS.

10

IP Multimedia Subsystem— Architecture and Protocols

The functional groups in a UMTS Network which have been discussed so far—UE, UTRAN, PS Domain and CS Domain—together provide the user with well-controlled connectivity, i.e. secure connectivity with QoS and mobility support. UMTS, however, is not just about connectivity. UMTS is about user services. This is where the IMS comes in. The IP Multimedia Subsystem (IMS) is a general platform that supports the delivery of IP-based multimedia services (cf. Chapter 4, Section 4.4.1).

It is important to note that the IMS does not necessarily deliver the multimedia services itself. It mainly *supports* their delivery. It is easy to see why service support is crucial. As discussed in Chapter 2, Section 2.3.2 on technical requirements, the previous generation of mobile Telecommunication Networks *integrated tightly* the services they delivered (e.g. telephony). UMTS thus abstracts from the service, and provides service support, e.g. finding the IP address of the communication partner, and negotiating the codec. This way UMTS is somewhat flexible with regard to the services it delivers, and it is accessible to services that may be defined in the future. The services may be offered by the UMTS operator or by third party operators.

This flexibility, however, goes further. From a 3GPP perspective, the IMS can be accessed via the PS Domain. It can, however, also be accessed from other IP Networks. This means that the IMS must not employ solutions that rely on PS Domain specifics.

To summarize, 3GPP aims to align the delivery of IMS-based multimedia services wherever possible with the delivery of general IP-based services. Therefore, the general approach for the IMS is to adopt non-3GPP specific solutions, in particular IETF Protocols.

In this chapter we will first look in more detail at the service support which the IMS provides. We will then cover the main architectural element and the relevant protocols. Reader interested in more technical detail may study [Camarillo 2005] or the 3GPP documents [3GPP 22.228] and [3GPP 23.228].

Terminology discussed in Chapter 10:	
Application Server	AS
Breakout Gateway Control Function	BGCF
CableLabs	
Call State Control Function	CSCF
Common Open Policy Service Protocol	COPS
Diameter	
Home Subscriber Server	HSS
Interrogating CSCF	I-CSCF
Instant Message Service	
IP Connectivity Access Network	IP-CAN
IP Multimedia Subsystem	IMS
Media Gateway	MGW
Media Gateway Control Function	MGCF
Media Resource Function Controller	MRFC
Media Resource Function Processor	MRFP
Open Mobile Alliance	OMA
Policy and Charging Rules Function	PCRF
Policy Decision Function	PDF
Proxy CSCF	P-CSCF
Presence Service	
Push Service	
Real Time Control Protocol	RTCP
Real Time Protocol	RTC
Serving CSCF	S-CSCF
Session Initiation Protocol	SIP
Signalling Gateway	SGW

10.1 IMS Service Support

What does it mean to support service delivery? Support starts with locating the service. Users should not have to worry about the IP address through which a service can be obtained. The IMS finds the IP address for them. Similarly, when a service (e.g. IP telephony or video conferencing) involves several users, they should not have to deal with the problem of finding each other's current IP addresses. Users should have a menu from which they choose service and communication partners, and from then on the set-up, maintenance and tear-down of the communication session is dealt with by the IMS. Beyond locating services and users, this involves the authorization of users, the negotiation of codecs, ports and QoS, as well as, of course, charging. This charging is coordinated with that performed by the PS Domain so that, in the end, a single bill is delivered.

Of course, for many services, an Application Server is necessary in addition to service support, e.g. a video streaming server. Such servers can be located inside or outside the IMS. In any event 3GPP is not responsible for specifying the services themselves.

10.1.1 Basic Service Support

The service support described above can be broken down into several sub-problems. The most basic problem is the following: a user would like to set up a communication session with an Application Server or another user. She knows the public identifier of the communication partner (e.g. a phone number), but does not know its current availability, location, IP address and so forth. The basic IMS architecture contains network elements called **Call State Control Functions** (CSCFs) that help resolve the public identifier into the currently preferred IP address. Furthermore, CSCFs support the user in determining other communication parameters that must be fixed before a session can start, e.g. ports, QoS and codecs. In this book only this basic service support is discussed in detail.

Another basic service support feature provided by the IMS is a network element transcoding between codecs and a server supporting multi-party conferencing.

10.1.2 Advanced Service Support

Basic service support, as described above, is covered by the first release of the IMS in Rel-5. The subsequent releases of UMTS add a number of additional features. However, service support for mobile networks in general, including UMTS, is the responsibility of another organization, the **Open Mobile Alliance** (OMA) [OMA]. For example, digital rights management or Push-to-talk services are specified by OMA rather than by 3GPP.

We provide only brief examples of advanced service support standardized by 3GPP. Further detail can be found in [Camarillo 2005]:

- **Presence Service**
  The Presence Service [3GPP 23.141] provides the infrastructure for informing other users about one's availability. For example, a user can set as default that she becomes available for communication sessions when she switches on her mobile device. However, the Presence Service also allows for the implementation of more complicated rules, e.g. "when I am in a meeting I am only reachable by email, except when it is an emergency call from my family", or "When travelling I cannot receive video conferences".
- **Instant Message Service**
  An Instant Message Service allows a user to send a message to another user in virtually real-time. Instant Message Services are popular today, and support of this feature was integrated into the IMS from the start [3GPP 23.228]. Of course, the Instant Message Service can use the Presence Service to determine whether the receiver of the message is available.
- **Push Service**
  The Push Service allows for the pushing of information from the network to the user without prior user action. Such information could include notification that mail has arrived, charging information, or—affording interesting business opportunities—advertisements. Work on the Push Service was abandoned for some time but has been taken up again lately.

10.2 Architecture

The basic architecture of the IMS is shown in Figure 10.1. This Figure also shows how IMS network elements interact with the network elements introduced in previous chapters. The most

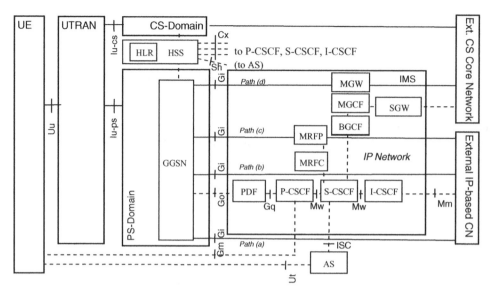

Figure 10.1 Basic logical architecture of the IMS (Rel-6). Solid lines depict control and user-plane interfaces, dashed lines depict control interfaces only. When lines cross this is due only to difficulties in two dimensional drawing. It does not mean that they interact

important network element in the IMS is the CSCF. A CSCF is responsible for controlling and policing multimedia services. When a UE wishes to use a multimedia service, it contacts a CSCF via the Gm interface—Gm is of course a logical interface; the physical connectivity is provided via UTRAN and PS Domain and this is the path which packets take. CSCFs come in three flavours, **Proxy CSCF** (P-CSCF), **Serving CSCF** (S-CSCF) and **Interrogating CSCF** (I-CSCF), whose different roles will be explained in Chapter 15 on Session Control. CSCFs need access to a central user-database in the UMTS Network, e.g. in order to authorize a user. To this end, the HLR is extended and henceforth becomes the **Home Subscriber Server** (HSS).

Figure 10.1 shows several paths for user traffic from GGSN to external networks via the Gi interface. From bottom to top, there is

(a) A direct path, not involving IMS services.
(b) A path utilizing CSCF services only.
(c) A path utilizing additional IMS services, in this case a media service for, e.g. multimedia conferencing or transcoding. Media services are supported by a Media Resource Function which is split into a control node (**Media Resource Function Controller** or MRFC) and a user traffic transport node (**Media Resource Function Processor** or MRFP). We therefore find that the division of a network element into a control plane-node and user-plane node which was first exercised in the CS Domain has been taken up in the IMS.
(d) A path involving a gateway that allows for accessing an external circuit switched network, e.g. a PSTN. By means of this path it is possible for a user to make or receive IP-telephony calls to or from a circuit-switched network. The gateway enabling the translation is also split

into a user-plane node (the **Media Gateway** or MGW) and a control-plane node (the **Media Gateway Control Function** or MGCF). The MGCF also translates the SS7 ISUP protocol into SIP. Additionally, a **Signalling Gateway** (SGW) is introduced for translating the signalling bearer, i.e. MTP or ATM, into IP (cf. Figure 7.6).The **Breakout Gateway Control Function** (BGCF) selects the appropriate network to interwork with.

An essential network element in Figure 10.1 is of course the **Application Server** (AS) which provides the actual multimedia services. The AS can be operated by a third party and is then located outside the IMS. It can also be operated by the operator of the UMTS Network, in which case it is located in the IMS and can have an interface to the HSS.

Figure 10.1 shows the basic architecture only. The IMS can be home to a rather large number of additional nodes with specialized functionality, e.g. to support Presence Services or Push Service. These nodes will be introduced later as needs be.

10.2.1 CSCF

CSCFs are the main control nodes in the IMS. When a user wishes to utilize IMS services, he registers with a CSCF using his Public User Identity. This identity is stored on the ISIM application of the UICC—see Chapter 9, Section 9.1.3—and takes the form of a **Uniform Resource Identifier** (e.g. sip:arthur.dent@earth.org [RFC 2396] or tel:+49-30-1234567 [RFC 2806]). Somebody trying to contact Arthur Dent (or the person with the phone number +49-30-1234567) can use the CSCF infrastructure to locate the corresponding UE and set up a communication session.

CSCFs thus perform the following tasks:

- **Security**: The CSCFs authorize UEs and their users based on subscription information which is stored in the HSS.
- **Session control**: The CSCFs support session establishment, maintenance and tear-down as described above.
- **QoS**: The CSCF, particularly the P-CSCF, collaborates with a **Policy Decision Function** (PDF) to authorize the QoS a user may reserve for a session. Via the Gq and Go interface the GGSN is instructed which QoS a user may reserve for a particular session. The GGSN only admits the corresponding PDP Context when its QoS matches this target. We will later see in Chapters 16 and 17 how in Rel-7 a more general policy infrastructure is introduced: the PDF is replaced by a network element called **Policy and Charging Rules Function** (PCRF), the Go interface is extended for generalized policing and is renamed Gx, and the Gq interface becomes Rx (cf. Figure 17.6).
- **Charging**: CSCFs collect charging-relevant information such as the start and stop time of a service and send it to a charging gateway.

Note that the CSCF is a pure control node. It does not transport user traffic.

10.2.2 IP Connectivity Access Network

Extensive access-independence is an important feature of the IMS; its services cannot only be used from the PS Domain but also from other IP Networks called **IP Connectivity Access**

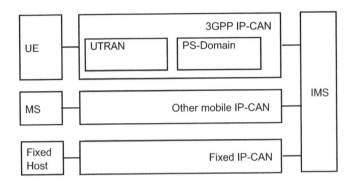

Figure 10.2   Access to the IMS from IP-CANs

Networks (IP-CANs), as illustrated in Figure 10.2. Of course, this means that the IMS is designed so that it makes few assumptions about the nature of the access.

This design allowed, e.g. ETSI to standardize access to the IMS from fixed Telecommunication Networks, and **CableLabs**, a consortium of cable operators—who were originally known as offering cable TV, to standardize access to the IMS via cable. Also, 3GPP2 took up the idea of IMS for cdma2000. They specify a "3GPP2 IMS" which is almost identical to the IMS specification—except where differences between UMTS and cdma2000 made deviations necessary [3GPP2 X.S0013], [Agrawal 2008]—the concept of access-independence seems to have limitations after all.

10.3 Protocols

The general approach for the IMS is to adopt non-3GPP specific IP-based solutions. In particular, the IMS employs IETF Protocols that are also employed outside UMTS. However, as the IMS had some requirements that went beyond what was available, 3GPP and IETF collaborated in order to provide extensions to existing protocols.

10.3.1 User-Plane

The protocol stack of the IMS user-plane is rather simple, namely IP together with a transport layer protocol (e.g. TCP, UDP) as appropriate. For real-time applications, the **Real Time Protocol** (RTP) [RFC 3550] is stacked on top of UDP. RTP adds an additional header containing a time stamp which allows the receiver to play the packets at the right point in time, in other words to remove the jitter which the IP Network may have introduced. RTP is always employed together with the **Real Time Control Protocol** (RTCP) which delivers monitoring data on QoS and reports on how data was received.

10.3.2 Control-Plane

The structure of the IMS control-plane is somewhat different from the structure of the PS Domain control-plane (and very different from the structure of the CS Domain control-plane). The IMS in fact illustrates a convergence of Telecommunication Networks and the Internet. The IMS control protocol stack consists of a limited number of IETF Protocols. These protocols

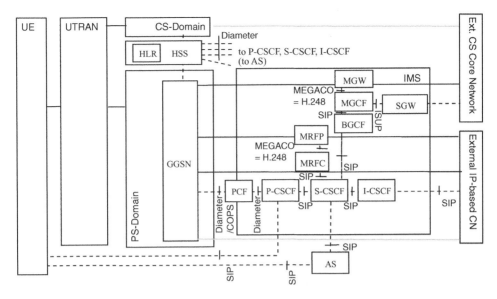

Figure 10.3   Control-plane protocols of the IMS

are specific to the functionality they carry out, as in a Computer Network; they are not specific to the communicating network elements, as in the PS Domain. Furthermore, the IMS control protocols run over a generic IP protocol stack, i.e. UDP/IP or TCP/IP, as the case may be. Therefore, the illustration of the IMS control-plane in Figure 10.3 depicts only the actual control protocol for each reference point, rather than entire protocol stacks.

The main protocol in the IMS is the **Session Initiation Protocol** (SIP) [RFC 3261] which is utilized for session control between UE, CSCFs, AS and other control nodes (e.g. MRCF and MGCF). **Diameter** is the protocol for authentication and authorization between HSS and CSCFs. QoS control between CSCF and GGSN is performed by the **Common Open Policy Service Protocol** (COPS) or Diameter. Finally, control-plane nodes control their corresponding user-plane nodes with the MEGACO protocol, which was introduced in Chapter 7.

10.4 Summary

The IMS supports the delivery of IP-based multimedia services. It can be accessed not only from the PS Domain, but also from other IP Networks. The services which the IMS provides are not standardized by 3GPP. The underlying idea is that services can be conceived and deployed dynamically, supported by the IMS infrastructure.

This chapter presented the basic architecture of the IMS, and introduced the most important network elements. The CSCF is the main control node of the IMS and supports session control, authorization, QoS control and charging. It comes in three varieties, P-CSCF, S-CSCF and I-CSCF. A CSCF collaborates with the central data-base of a UMTS Network, the HSS, which is an IMS-extended HLR.

The general approach for the IMS is to adopt IP-based solutions that combine easily with external non-UMTS Networks. This is reflected in particular in the choice of IETF Protocols. The IMS user-plane is a normal Computer Network user-plane. The control-plane also employs

Table 10.1 The distribution of control functions on network elements. "x° denotes actual control, whereas ªo° denotes collection/provisioning of information

		USIM	TE	MT	RNC	SGSN/ MSC	GGSN/ GMSC	CSCF
Ch. 12	Mobility control			x	x	x		
Ch. 13	Session or call control		x			x		x
Ch. 14	Admission control	x				x		
Ch. 15	QoS control			x	x	x	x	x
Ch. 16	Charging control					o	o	o
	Service control		x					
	Radio resource control			x	x			

exclusively IETF Protocols such as SIP, DIAMETER and COPS. To satisfy the requirements of the IMS, 3GPP-specific extensions were defined for these protocols.

10.4.1 Introduction to Chapters 11–17

Chapters 6 to 10 introduced the basic architecture, functionality and the protocols of UMTS, i.e. of PS Domain, CS Domain, UTRAN, UE and IMS. In Chapters 11 to 17, we will adopt an orthogonal view and discuss in detail how particular functionality is provided by the collaboration of all network elements. Emphasis will be on control functions and on packet-switched services.

Table 10.1 summarizes the distribution of control-plane functions in the main network elements, and indicates the chapter in which each function is described. We thereby distinguish network elements that participate only in control by collecting or providing information, and network elements which actually make decisions. It is always difficult to categorize anything outside mathematics; in this sense the table below should be regarded as a simplification, as a number of grey areas exist.

One more word on the overall organization of this book: up till now we have treated the circuit-switched and packet-switched aspects of UMTS with equal emphasis. We will, however, henceforth concentrate on packet-switched functionality since it is expected to have a more important role in the evolution towards 4G Networks.

11

Basic UMTS Functionality

This chapter explains how the basic UMTS functionality is realized, namely the user switches on the UE and starts a communication session; the communication terminates and the user switches off the UE. As it turns out a large number of steps is involved in between, which can be structured roughly as follows:

- **UE preparation**. After the UE is switched on, it first works on a stand-alone basis, listening to system broadcasts, and finds a suitable access point in a suitable network.
- **Establishing radio connectivity (RRC Connection set-up)**. The UE can now contact the network, the RNC to be precise, and establish a dedicated radio connection—the RRC Connection described in Chapter 8, Section 8.2.2—for control signalling.
- **Attaching to the network (GPRS Attach/IMSI Attach)**. Subsequently, the UE determines its location on the basis of system broadcasts. It uses the RRC Connection to make its presence and location known to the SGSN and/or MSC. It is authenticated, and may initiate a communication session. This procedure is known as GRPS Attach for the PS Domain, and as IMSI Attach for the CS Domain.
- **Establishing IP connectivity (PDP Context establishment/Call set-up)**. When the UE wants to start a communication, it must establish a route through the network to the appropriate gateway (GGSN or GMSC). In the PS Domain, it invokes the establishment of a tunnel, called **PDP Context**, through which the user- and control traffic of this UE is routed. In the CS Domain, it triggers the establishment of a circuit-switched call.

The steps listed above lead to the establishment of IP connectivity. They do not cover how a service is started over this connectivity, e.g. video telephony! In technical terms, this chapter illustrates PS Domain functionality (and—to a lesser extent—CS Domain functionality) whereas basic IMS functionality will be covered in Chapter 15 on Session Control.

This chapter starts with the introduction of two important UMTS concepts. Firstly, we introduce the concept of UMTS network identifiers—this is necessary as a UE must identify a suitable network to which to attach. Secondly, we explain in more detail the concept of **bearer** that we already encountered in some of the previous chapters. We then go through the individual steps identified above which enable a UE to communicate over a UMTS Network.

UMTS Networks and Beyond   Cornelia Kappler
© 2009 John Wiley & Sons, Ltd

Subsequently, we describe how the corresponding functionality is realized in a mobile Computer Network, in particular how to attach to a 802.11 WLAN Extended Service Set (ESS, cf. Chapter 4, Section 4.7). Finally, we compare the UMTS approach and the WLAN approach and discuss their differences.

As usual, the descriptions in this chapter are simplified. Further material is available in [3GPP 25.304] (Chapter 11, Section 11.3), [3GPP 25.331] (Chapter 11, Section 11.4), [3GPP 23.060] (Chapter 11, Sections 11.5 to 11.7) and [Gast 2005] (Chapter 11, Section 11.9).

Terminology discussed in Chapter 11:	
Access Point Name	APN
Association	
Beacon	
Bearer	
Call set-up	
Core Network Bearer	
End-to-End Bearer	
GPRS Attach	
GPRS Detach	
GPRS Roaming Network	GRX
Home PLMN	HPLMN
Home Routed	
IMSI	
IMSI Attach	
IP Roaming Exchange	IPX
Local Breakout	
Location Area	LA
Mobile Country Code	MCC
Mobile Network Code	MNC
Mobile Subscription Identification Number	MSIN
Mobility Management State	
Opportunistic authentication	
PDP Context	
PDP Context activation	
PDP Context deactivation	
PDP Context establishment	
PDP State	
PLMN identifier	
PMM-CONNECTED	
PMM-DETACHED	
PMM-IDLE	
Public Land Mobile Network	PLMN
Radio Access Bearer	
Radio Bearer	
Roaming Agreement	
Routing Area	RA

Routing Area Identifier	RAI
RRC Connection set-up	
Service Set Identity	SSID
Soft-state design	
UMTS Bearer	
UTRAN Registration Area	URA
Visited PLMN	VPLMN

11.1 Public Land Mobile Network (PLMN)

In the context of establishing basic UMTS functionality, we need to introduce the term **Public Land Mobile Network** (PLMN). A PLMN is a mobile Telecommunication Network operated by a single operator. It includes a Radio Access Network and a Core Network (cf. Chapter 4). A PLMN can be based on various technologies; in this book, we are concerned with UMTS PLMNs.

The UICC in the UE of the user is always associated with a particular **Home PLMN** (HPLMN), namely the PLMN operated by the operator that sold the UICC. This means that the identifiers and secret keys on the UICC are known to a HLR in the HPLMN. When a UE is not in the coverage area of its HPLMN, e.g. because the user is roaming abroad, the UE can still access a UMTS Network, a **Visited PLMN** (VPLMN), if this VPLMN has a **Roaming Agreement** with the HPLMN. As illustrated in Figure 11.1, a Roaming Agreement is a contract that allows the VPLMN to obtain information from the HPLMN, enabling it to authenticate and authorize the UE. A Roaming Agreement also settles how the HPLMN compensates the VPLMN for its services: The user is always charged by its HPLMN operator. The VPLMN is reimbursed by the HPLMN.

PLMNs have an identifier which is composed of **Mobile Country Code** (MCC)—unique to each country and **Mobile Network Code** (MNC)—unique to each PLMN in a country.

$$PLMN\ identifier = MCC + MNC$$

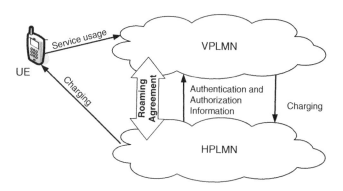

Figure 11.1   Relationship between UE, HPLMN and VPLMN

As we learned in Chapter 9, each UMTS subscription is associated with a unique identifier, the IMSI, which is stored on the UICC. The IMSI contains the PLMN identifier together with a **Mobile Subscription Identification Number** (MSIN). When a UE attaches to a UMTS Network, the network can thus analyse the IMSI in order to determine the Home PLMN of the subscriber.

$$IMSI = PLMN\ identifier + MSIN$$

11.2 The Bearer Concept

The second concept which we need to introduce is that of a **bearer**. In 3GPP, one says that the UE has a bearer when one would like to express that the UE is provided with well-controlled connectivity, i.e. a bidirectional, secured transport service for packets with a certain QoS. Often, a bearer provides mobility control. To put it briefly, when a UE needs to send or receive information it needs to set up a bearer.

More generally, a bearer is a service offered by one layer to another layer between two end-points, as illustrated in Figure 11.2. For example, for the UE to send a packet to a receiver, it needs to set up an end-to-end service. The **End-to-End Bearer** uses the service of a **UMTS Bearer** between UE and GGSN. The UMTS Bearer in turn builds on a **Radio Access Bearer** between UE and SGSN and a **Core Network Bearer** between SGSN and GGSN—and so forth down to the physical bearers. We will encounter further examples of bearers in the course of this chapter.

11.3 UE Preparation

Upon being switched on, the UE must find a suitable cell in a suitable network. A suitable cell is a cell offering the right technology and acceptable radio reception. A suitable network is the HPLMN of the UE, or a UMTS PLMN that has a Roaming Agreement with its HPLMN. How does the UE perform this task?

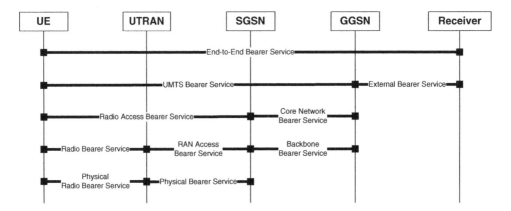

Figure 11.2 The bearer hierarchy in UMTS

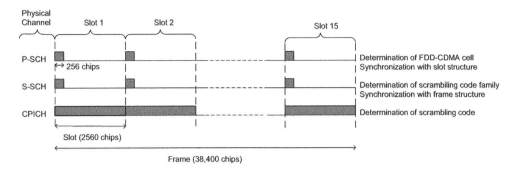

Figure 11.3   Identifying a suitable cell and synchronizing with this cell

11.3.1 Searching for a Suitable Cell

The first step is the identification of a suitable cell, i.e. a cell with the right technology, say FDD-CDMA, and acceptable radio reception. The procedure is illustrated in Figure 11.3.

All UMTS cells based on FDD-CDMA broadcast a standardized characteristic sequence of 256 chips on the Primary Synchronization Channel (P-SCH, one of the stand-alone Physical Channels, see Chapter 8, Section 8.2.3). This broadcast is repeated at the beginning of each slot, i.e. every 0.67 ms. The UE thus scans all possible frequencies and searches for strong signals with the characteristic chip sequence. The detection of the characteristic chip sequence assures the respective cell is a UMTS/FDD-CDMA cell.

After the UE has identified one or more suitable cells it must listen to system broadcasts before making the final decision. To this end, the UE must synchronize with the slot and frame structure of the cell (cf. Chapter 8, Section 8.2.3) and it must learn the cell's scrambling code (cf. Chapter 5, Section 5.2.4.3). We will cover the procedure for FDD-CDMA; for TDD-CDMA, an analogous procedure applies.

The detection of the characteristic chip sequence has already enabled synchronization with the slot structure. Next the UE synchronizes with the frame structure and makes some headway in determining the scrambling code. Scrambling codes are grouped into 64 families. Each family is assigned a distinct characteristic chip sequence, also of a length of 256 chips. Cells emit this second characteristic chip sequence at the beginning of each frame on the Secondary Synchronization Channel (S-SCH). After evaluating three successive chip sequences the UE is able to synchronize with the cell's frame structure and learn the chip sequence characterizing the scrambling code.

Finally, the UE determines the full scrambling code. To this end, it listens on the Common Pilot Channel (CPICH), where a standardized bit sequence is scrambled with the cell's scrambling code. The UE tries descrambling the signal by applying the scrambling codes of the family determined in the previous step, and in this way it can finally determine the scrambling code which is used in this cell.

11.3.2 Searching for a Suitable Network

In order to find a suitable network, the UE now listens to system information broadcast in regular intervals on the BCCH Logical Channel. This information includes the identifier of the

PLMN to which the cell belongs. It also includes information on neighbouring cells (of the same PLMN) such as frequency and scrambling codes—when the UE needs to change cells because of mobility within the same PLMN, the procedure described in the previous section is much simplified.

As described in Chapter 9, Section 9.1.3, the UICC contains a list of preferred PLMNs, with the HPLMN typically topping the list. The UE thus ranks the suitable cells determined in the first step according to this list. This list does not list exhaustively all VPLMNs with which the HPLMN has a Roaming Agreement. Thus, the ranking may produce no result. In this case cells are ranked according to reception quality. Finally the UE picks the best cell. When the chosen cell is not in the PLMN with highest priority, the UE periodically re-performs the search for a suitable cell.

11.4 RRC Connection Set-up Procedure

The UE now contacts the network, the Serving RNC to be precise, and establishes a **Radio Bearer** (cf. Figure 11.2) for signalling—the RRC Connection described in Chapter 8, Section 8.2.2. It is used for the subsequent control signalling. The RRC Connection is equivalent to the assignment of a dedicated Logical Channel (the DCCH) for the UE.

11.4.1 Message Flow for RRC Connection Set-up

In order to perform a RRC Connection set-up, the UE initiates a signalling exchange with the RNC using the RRC protocol. Readers with an "IP-mind" may wonder how the UE knows the address of the RNC. As a matter of fact, the UE does not know the RNC's address. Rather, the UE sends its messages down the tree-like structure of the UTRAN (cf. Figure 8.2a) and the RNC receives it by filtering out all messages of the RRC protocol.

The detailed (yet simplified) procedure is illustrated in Figure 11.4.

(i) The UE sends a **RRC Connection Request** message to the RNC, using the RRC protocol over the common Logical Channel (the CCCH). This message includes the subscription identifier, i.e. the IMSI, and a measurement report on the quality of radio reception. The measurement report is included for obvious reasons; the IMSI serves the RNC in order to perform a weak form of so-called **opportunistic authentication** of the UE. This means that the RNC is able to recognize, based on the IMSI, when it talks again to the same UE. It will, however, not check subscription information with the HLR, it does not care whether the user has a subscription.

(ii) RNC communicates to the Node B that radio resources are to be allocated for the UE, in other words that a Radio Bearer should be established. The RNC includes a number of radio parameters in its message such as the CDMA code for the UE and the power level. Node B establishes the Radio Bearer, and then sends to the RNC a message with another set of radio parameters.

(iii) RNC informs the UE that the RRC Connection set-up was successful and includes in its message the radio parameters which are of relevance for the UE, such as CDMA code and power level. It also includes a temporary identifier for the UE which is used henceforth for communication between UE and RNC, instead of the IMSI. This is in order to reduce the

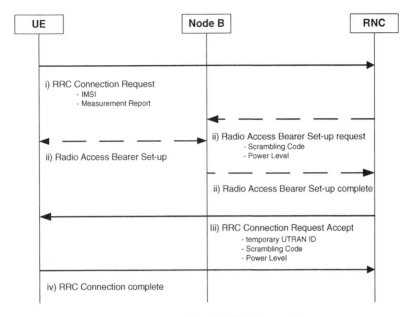

Figure 11.4   Message flow for RRC Connection set-up

risk that a malicious person may eavesdrop upon the communication and "steal" the IMSI. In Chapter 13 on Security we will hear more about this.
(iv) The UE acknowledges the completion of the RRC Connection set-up.

11.5 GPRS Attach Procedure

The UE is now in a position to make its presence and location known to the Core Network. Depending on whether it intends to use PS or CS services, it contacts SGSN or MSC, and attaches to PS Domain and/or CS Domain. As a result of the procedure, the UE is able to use the services of the respective domain. For example, it can start surfing the Internet when attached to the PS Domain, or receive telephone calls, when attached to the CS Domain.

The UE can, however, also idle after attaching to the network. In this case, however, the RRC Connection would be torn down. Radio resources are valuable, the RRC Connection is therefore only maintained when the UE utilizes it. It is re-established when necessary.

11.5.1 Mobility Management States

Technically speaking, the attach procedure alters the **Mobility Management State** of the UE. In both CS-Domain and PS Domain, three Mobility Management States are defined. For the PS Domain, they are shown in Figure 11.5. The UE is at all times in one of the three states. Before the GPRS Attach procedure, the UE is in PMM-DETACHED state. The GPRS Attach procedure moves it into PMM-CONNECTED state. When no PS Domain services are used, the RRC Connection is torn down and the UE is moved to PMM-IDLE state. The Mobility Management State of a UE is stored on both UE and SGSN. The details in Figure 11.5 will be

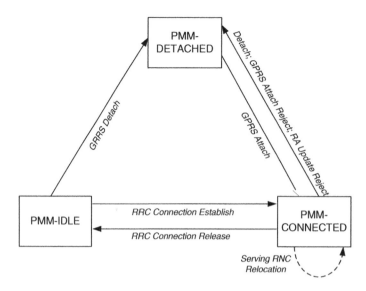

Figure 11.5   Mobility Management States of UE and SGSN. The dashed state change is only valid for the SGSN

covered in Chapter 12 on Mobility. The CS Domain has its own set of Mobility Management States.

11.5.2 Determining the Location of the UE

In UMTS—and GSM for that matter—the UE is responsible for keeping track of its location. For efficiency reasons, the accuracy of the location information depends on whether the UE is currently sending or just idling, i.e. whether it is in PMM-CONNECTED state or in PMM-IDLE state. Thus, a hierarchy of location information is defined, illustrated in Figure 11.6. In the PS Domain, one or several cells are subsumed in an **UTRAN Registration Area** (URA), and one or several URAs are subsumed into **Routing Areas** (RAs). The most precise location information is thus an identifier of the cell to which a UE is attached. More imprecise information is provided by the identifier of the UE's current Routing Area. The equivalent to RAs in the CS Domain is called **Location Area** (LA)—there is no equivalent to URAs in the CS Domain. Routing Areas are identical to or a subset of Location Areas. While a single RNC is

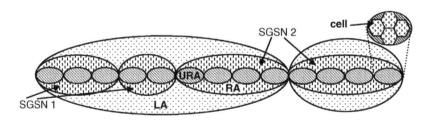

Figure 11.6   Hierarchical organization of the location information in UMTS

responsible for one or several cells, a single SGSN is responsible for one or several Routing Areas; a single MSC is responsible for one or several Location Areas.

Cells, Routing Areas and Location Areas have globally unique identifiers. The composition of the identifiers reflects the hierarchical nature of the location information.

$$Location\ Area\ Identifier = PLMN\ identifier + LA\ Code$$

$$Routing\ Area\ Identifier = Location\ Area\ Identifier + RA\ Code$$

$$Cell\ Global\ Identifier = Routing\ Area\ Identifier + cell\ identifier$$

The **Routing Area Identifier** (RAI) is broadcast in each cell on the BCCH along with other system information. The UE thus notes the RAI and is now finally ready to contact the network.

11.5.3 Message Flow for GPRS Attach

The detailed procedure for attaching to the PS Domain, the so-called **GPRS Attach** procedure, is explained below and illustrated in Figure 11.7. Between UE and SGSN, the protocol GMM is used. Between SGSN and HLR, the protocol MAP is used. As above, the question arises as to how the UE addresses the SGSN. And, as above, the UE simply sends its message to the UMTS Network, where it travels down the tree-like structure: the receiving Node B is governed by exactly one RNC, which is governed by one SGSN which terminates all GMM protocol messages.

Strictly speaking, this tree-like structure only applies up to Release 5. In Release 5, Iu-flex was introduced, i.e. the many-to-many relation of RNCs and SGSNs, cf. Chapter 8, Section 8.1. In the case of Iu-flex, however, a RNC has its internal algorithm which determines to which SGSN it sends the incoming GPRS Attach messages.

(i) The UE sends a **GPRS Attach Request** message to the SGSN. This message includes the the IMSI. When the message passes the RNC, the Routing Area Identifier (RAI) is included. The IMSI serves the SGSN so as to inform the HPLMN of the subscriber. Subsequently, the SGSN asks the HLR in the HPLMN for authentication information. Based on this information, SGSN and UE authenticate each other and other security functions such as encryption and integrity protection are activated. Further detail on this exchange will be provided in Chapter 13, Section 13.3.

(ii) If the authentication was successful, the SGSN informs the HLR that it is now handling the UE with this IMSI. The HLR stores the fact that the UE is reachable via this SGSN.

(iii) The HLR informs the SGSN about the subscription details related to this IMSI, the so-called **GPRS Subscription Data**. This data includes the phone number (the MSISDN) and **Charging Characteristics** (e.g. whether the IMSI is associated with a pre-paid card). Furthermore, it contains information as to what the subscriber is allowed to do in the network, e.g. access restrictions, etc. This information is known as **authorization** information.

(iv) The SGSN acknowledges reception of the subscription data.

(v) The HLR acknowledges that the SGSN is responsible for the UE.

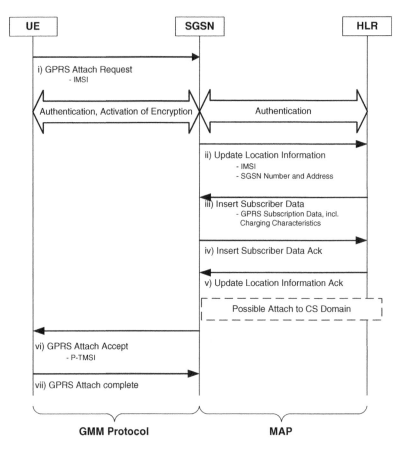

Figure 11.7 Message flow for GPRS Attach

(vi) The SGSN informs the UE if its access request has been granted. Access could, e.g. be denied because there is no Roaming Agreement between the subscriber's HPLMN and the PLMN of the SGSN. If access is granted, the SGSN assigns a temporary ID, the P-TMSI. The UE stores the P-TMSI on its UICC. For security reasons the P-TMSI is now used for communication between UE and SGSN instead of the IMSI.

(vii) The UE acknowledges that the GPRS Attach procedure has been completed.

11.5.4 Combined GPRS/IMSI Attach

After the GPRS Attach procedure is completed, the UE is attached to the PS Domain; as is obvious from the message flow, the CS Domain has as yet no notion of its presence. If the UE also wishes to use CS Domain services it must attach to the CS Domain–it must perform a so-called **IMSI Attach**.

To this end, the UE could start an independent IMSI Attach procedure, going through another authentication. However, the UE can also short-cut the procedure by indicating in the GPRS Attach Request message that it would also like to perform an IMSI Attach. In this case the SGSN, after completing its own conversation with the HLR, triggers the responsible MSC/VLR to perform IMSI Attach. The MSC/VLR then go through the equivalent of steps (ii) to (v) above. It is only after the MSC/VLR has acknowledged completion of these steps that the SGSN informs the UE that its access request was granted. In Figure 11.7 this is depicted by the dashed box.

We still must consider how the SGSN knows which MSC/VLR is responsible. It is the MSC/VLR handling the current Location Area of the UE. As illustrated in Figure 11.6, knowledge of the Routing Area uniquely determines the LA. Thus, the SGSN can determine the VLR/MSC based on the RAI.

11.6 PDP Context Establishment Procedure

Even after GPRS Attach, the UE cannot yet send or receive packets. At this point it usually does not even have an IP address.[1] The user first needs to choose a service, e.g. web surfing or IP telephony. This service determines the destination network and a gateway to this network, i.e. the GGSN. The UE then needs to initiate the set-up of the UMTS Bearer (cf. Figure 11.2). An important building block of the UMTS Bearer is a tunnel extending from the RNC, via the SGSN, through the PS Domain to the GGSN. Both uplink and downlink user-plane traffic is routed through this tunnel. The tunnel is called **PDP Context** (cf. Chapter 6, Section 6.1). Establishing this tunnel is called **PDP Context establishment** or **PDP Context activation**. In the course of the PDP Context establishment procedure, the UE also receives a dynamic IP address—a prerequisite for sending and receiving IP-based packets. This IP address, called **PDP Address**, remains unchanged for the duration of the PDP Context, regardless of UE movement. Figure 11.10a illustrates a PDP Context for a user attached to her HPLMN.

The PDP Context, just as the RRC Connection, has a time-out value: when not used, the network tears it down. However, the time-out value of the PDP Context usually is much longer than the time-out value of the RRC Connection: network resources are more abundant than radio resources. It is therefore possible to maintain a PDP Context while the corresponding RRC Connection has already timed out.

11.6.1 The PDP Context

Prior to establishing the PDP Context, the UE must settle its characteristics.

End Point. The end point of a PDP Context is a particular GGSN. It is identified by the **Access Point Name** (APN). The APN, and thus the GGSN, is selected based on the destination network. Each GGSN provides access to particular IP-based Core Networks. For example, one GGSN may provide access to the Internet and the IMS, while another provides access to other PS Domains. In fact, there is usually also a GGSN providing access to "itself": communication between UEs attached to the same PS Domain also are routed via PDP Contexts, and thus would be routed via this GGSN (note that in this example there are two PDP Contexts, one for each UE.). Thus, when a UE wants to entertain several simultaneous communications, it must establish one PDP Context for each GGSN it is targeting—a feature which though standardized is usually not implemented today. It can, however, use the same PDP Context for all communications that share the same end point and QoS.

---

[1] The standard also allows UEs to have static IP addresses. However, in practice this is rarely done. We therefore disregard this case.

Figure 11.8 The PDP States in the PS Domain

QoS. Packets routed via a PDP Context are provided with a specific QoS. For example, if the UE plans to establish an IP-based video session it would have to reserve the appropriate bandwidth. If, however, the UE wants to surf the Internet, it would not bother about reserving resources.

11.6.2 PDP States

Just as there are Mobility Management States (cf. Section 11.5.1), there are PDP States. Quite simply, when a UE has a PDP Context, it is in PDP State ACTIVE. If it does not have a PDP Context, it is in PDP State INACTIVE.

Both UE and SGSN store the PDP State of a UE. A UE can only establish a PDP Context when it is attached to the PS Domain. In other words, it cannot go into PDP State ACTIVE when it is in Mobility Management State PMM-DETACHED. This is illustrated in Figure 11.8.

11.6.3 Message Flow for PDP Context Establishment

The message flow for PDP Context establishment is explained below and illustrated in Figure 11.9. Between UE and SGSN, the protocol GMM is used. Between SGSN and GGSN, GTP-C is used.

(i) The UE sends an **Activate PDP Context Request** message to the SGSN. This message includes an identifier for the desired destination network (i.e. the APN) and the desired QoS. It may also contain a filter for packets that should travel in this PDP Context (**Traffic Flow Template** (TFT)). When the UE has more than one PDP Context, the TFT is used, e.g. by the GGSN to match packets to the appropriate tunnel. The TFT can include fields such as IP sender address, IP destination Address, destination/source port range, etc.

An alert reader will notice that the UE does not include an identifier for itself, e.g. P-TMSI, in the message. This is because on the lower layers of the control protocol stack – the SS7-inherited ones – the GPRS Attach Request message triggered the establishment of a dedicated connection for the UE. The arrival of the Activate PDP Context Request on this connection uniquely identifies the sending UE.

The SGSN now determines the GGSN corresponding to the APN by a DNS look-up. It also determines whether the subscriber is entitled to both the request (remember the APN also codes the service requested) and the required QoS—this is called **authorization**, cf. Chapter 13. It also may check whether sufficient resources are available in the network to accommodate the request; performing such admission control and reserving the necessary resources is, however, not prescribed by the standard. Depending on the outcome of the various investigations, the SGSN can downscale the desired QoS. In Chapter 14 on QoS we will hear more about this.

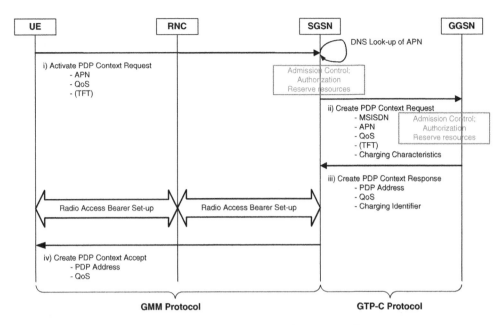

Figure 11.9   Message flow for PDP Context establishment

(ii) The SGSN sends a Create PDP Context Request message to the GGSN. The message includes
      the MSISDN of the subscriber, the APN, the (possibly downscaled) QoS as well as a TFT
      if available, and also the **Charging Characteristics**. We will see in Chapter 16 how the GGSN
      needs the Charging Characteristics in order to perform the appropriate charging procedures.
      The GGSN may now perform its own admission control and resource reservation.

(iii) The GGSN acknowledges the successful establishment of the PDP Context and sends
      the—again possibly downscaled—QoS it is able to accommodate. It also generates a
      dynamic IP address, called **PDP Address**, and a **Charging Identifier** and includes both
      in the message to the SGSN. The usage of the Charging Identifier will be explained in
      Chapter 16.

      Upon reception of this acknowledgement, the SGSN communicates with the RNC to
      establish a Radio Access Bearer (cf. Figure 11.2) with the currently valid QoS for the
      request. If the requested QoS is not available, it may, of course, be decreased again. When
      this happens, the SGSN and GGSN exchange a further two messages in order to adapt the
      QoS of the already established PDP Context, when sufficient QoS is not available (not
      shown in Figure 11.9).

(iv) Finally, the SGSN informs the UE that the PDP Context was established. It conveys the
      PDP Address as well as the final value for the QoS assigned to the PDP Context.

Now that the PDP Context is established the UE is in the position to send or receive packets.

11.7 Detaching from the Network

At some point, invariably, the user tires of communicating, or the UE's battery runs out. Or the
UE moves into an area without sufficient radio coverage. It is then time for the UE to detach

from the network. If possible, the UE triggers this detachment. We have already seen, however, that the network is in a position to time-out the UE's records and resources if it does not hear a keep-alive signal on a regular basis. Since one cannot rule out that UEs run into problems and disappear, this so-called **soft-state design** is an important feature. It protects the network from state table overflow and saves resources. Note that fixed phones do not have a record of disappearing and therefore classical fixed telephony networks can live with stateful design.

Detaching the UE from the network involves stepping backwards through the process described above. First, all active PDP Contexts are deactivated. As a result the PDP State of the UE becomes inactive (cf. Figure 11.9) and the corresponding PDP Context information at GSNs and RNCs is deleted. Next, a GPRS Detach is performed. As a result the Mobility Management State of the UE becomes PMM-DETACHED (cf. Figure 11.5) and UE-related information is deleted at the SGSN. The SGSN also sends a message to the HLR that it is no longer handling the UE—alternatively, the SGSN may decide to keep the UE data for some time and notify the HLR later, in the expectation that the UE will come back and the data can be reused. Finally, the RRC Connection is torn down, unless it has timed out earlier.

11.8 Basic UMTS Functionality in Roaming Scenarios

When the UE is not attaching to the HPLMN but roaming to a VPLMN it is interesting to ponder the location of network elements involved in a communication session. As we saw in Section 11.1, the HLR is always in the HPLMN, because this is where the subscriber data is stored. On the other hand, Node B(s), RNC(s) and SGSN are always located in the VPLMN. However, the end point of the PDP Context, the GGSN, may be located in HPLMN (this is called **Home Routed**) or in the VPLMN (called **Local Breakout**), depending upon the choice of the operators. A GGSN in the HPLMN gives the home operator better control and simplifies authorization and charging, as the corresponding data stays in the HPLMN. A GGSN in the VPLMN, however, results in more optimal routing. Today, typical deployments configure the GGSN in the HPLMN.

Figure 11.10 depicts the location of network elements in different scenarios. It also shows the dedicated backbone network connecting PLMNs called **GPRS Roaming Network** (GRX) or its successor **IP Roaming Exchange** (IPX). GRX and IPX are independent networks that can be operated by an entity independent of the PLMN operators.

GRX and IPX are specified by an organization called **GSM Association** (GSMA) [GSMA]. General UMTS roaming guidelines are described in [GSMA IR.33]. The GRX and IPX are described in [GSMA IR.34].

11.9 Basic WLAN Functionality

The four steps from switching on the UE to having IP connectivity—UE preparation, establishing radio connectivity, attaching to the network and establishing IP connectivity— have approximate equivalents in a WLAN network which are described in the following subsections.

11.9.1 Mobile Station Preparation

When a WLAN Mobile Station (MS), e.g. a laptop, is switched on, it does not automatically try to connect to a WLAN. Usually, the user explicitly switches on this functionality. Once this is

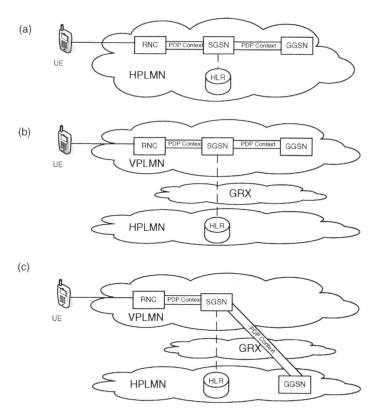

Figure 11.10   Location of network elements in different roaming scenarios. (a) The UE is attached to the HPLMN. (b) The UE is roaming and attached to a VPLMN, Local Breakout. (c) The UE is roaming and attached to a VPLMN, Home Routed. PLMNs are connected via the GRX

done, the Mobile Station must perform the same task as a UMTS UE: find a suitable Access Point belonging to a suitable network. A suitable Access Point is an Access Point offering the right technology and acceptable radio reception. A suitable network is a WLAN Extended Service Set (ESS, see Chapter 4, Section 4.7), which is identified by a suitable **Service Set Identity** (SSID).

The Mobile Station performs this task by scanning a WLAN frequency-band (e.g. 2.4 GHz). The scanning can be passive, i.e. by searching for good-quality **beacon** frames which are emitted by the Access Points, or it can be active, i.e. by transmitting probe requests to which nearby Access Points must reply. Beacon frames are preceded by a technology-specific preamble (i.e. bit pattern or chip pattern) and contain the SSID. The Mobile Station compiles a list of eligible SSIDs and prompts the user to pick one, or it consults a user-configured table of "preferred SSIDs". Subsequently, the Mobile Station uses timer information which is also contained in the beacon in order to synchronize with the Access Point.

As an aside, the beacon interval is not standardized in 802.11 but configured by management operation. A typical value is 100 ms, i.e. over 100 times longer than the corresponding UMTS "beacon interval". This illustrates once more how UMTS is designed for real-time applications.

11.9.2 Establishing Radio Connectivity

The Mobile Station can now make a request to **associate** with the Access Point (cf. the architecture shown in Figure 4.7). The Access Point returns an **Association ID** and registers the Mobile Station's MAC address with the Access Router. The Access Router thereupon knows that frames destined to the Mobile Station should be delivered to the Access Point.

11.9.3 Establishing IP Connectivity

At this point, the Mobile Station has Layer 2 connectivity. It now needs to authenticate. In Chapter 13, Section 13.4 we will see how this is done. Subsequently, the Mobile Station establishes IP connectivity. This means that it acquires an IP address and informs the other hosts in the subnet, and particularly the Access Router, about its presence. This procedure is not specific to the WLAN, but is based on standard IETF protocols. Several possibilities exist, see [Tanenbaum 2002, Gast 2005]. For example, the Mobile Station could ask the **Dynamic Host Configuration Protocol** (DHCP) server [RFC 3315] in the ESS to assign it an IP address. Finally, it informs the other hosts in the subnet about the IP address—to—MAC address mapping by sending a gratuitous **Address Resolution Protocol** (ARP) [RFC 826] request.

11.10 Discussion

Comparing how equivalent functionality—enabling a Mobile Station to communicate—is achieved in UMTS and WLAN we observe a number of differences.

An obvious difference is that the WLAN procedures are much simpler. We also see that the user has more configuration options in WLAN and that she is more involved in establishing the connectivity. Finally, we recognize that the architecture and procedures in WLAN are much more flexible.

Why do these differences exist? To some extent they are based on their different heritage: WLAN designers—and WLAN users—start from the idea of a mobile—or rather a nomadic—Computer Network. The WLAN standard therefore only defines the "add-on" required on Layer 2 to support the radio access. In contrast, UMTS designers and UMTS users start from the idea of a circuit-switched 2G Network becoming packet-switched. A UMTS Network is therefore designed as a complete system. From this also comes the idea of a PDP Context, which from many perspectives is equivalent to a "circuit" in the PS Domain.

Many differences, however, are intrinsically based on the different business models and on the different approaches to network design, which were presented in Chapter 1, Section 1.2: the goal of the UMTS operator is selling high-quality services, whereas the goal of the WLAN operator is simply selling connectivity to the Internet. This manifests itself, for example, in the following:

- The UMTS user does not just attach to a single Access Point, but to a potentially globally reachable network, which implies organizational overheads. A WLAN ESS, by contrast, is designed as an isolated island, without cooperation with other WLAN ESSs.
- User involvement in UMTS is minimized and procedures are automated as much as possible. For example, the UE connects to the network upon start-up, the network is selected without user intervention, and the authentication is performed based on the UICC. The user of a WLAN Mobile Station is, however, prompted several times during the process.

- A UMTS UE is given a dedicated radio channel for signalling which will not be disturbed by other UEs. The WLAN radio medium is shared.
- Prior to communicating, the UMTS UE must set-up a tunnel with a particular QoS, the PDP Context whereas QoS is not an issue in WLAN.
- The UE triggers procedures such as RRC Connection set-up and PDP Context establishment which are then carried out and controlled by the network. Its active involvement is minimized. WLAN 802.11, however, specifies only the establishment of link-layer connectivity. The remainder is left to standard procedures defined by the IETF, which usually leave control to the Mobile Station.

11.11 Summary

In this chapter we described the basic UMTS procedures from switching on the UE until the set-up of IP connectivity. A number of steps are necessary in order to achieve this. First, the UE prepares itself, listening to system broadcasts and identifying a cell with acceptable radio reception of a preferred PLMN. It synchronizes with this cell and then contacts the RNC to perform RRC Connection set-up. The UE now has a dedicated radio connection for signalling. It uses this connection to perform GPRS Attach. As a result, the UE is in Mobility Management State PMM-CONNECTED. It is registered with the SGSN, and the HLR knows how to find the UE. However, in order to actually send packets, the UE, as a final step, has to activate a PDP Context. The PDP Context is a UE-specific tunnel between RNC and GGSN with a particular QoS. The GGSN is chosen based on the destination network. Note that several sessions of a UE can use the same PDP Context if they need the same GGSN and QoS. Once the UE has activated a PDP Context, it has an IP address and is in PDP State ACTIVE.

When the UE is switched off, the steps above are performed in reverse order: PDP Context deactivation and GPRS Detach. The RRC Connection is torn down independently, and, usually, earlier.

Regarding information handling, we see that the UE maintains Mobility Management State and PDP State, the SGSN maintains Mobility Management State, PDP State and location information, the GGSN maintains PDP State and the HLR knows static user subscription data as well as current SGSN.

This chapter also introduced two important concepts:

- A PLMN is a mobile Telecommunication Network operated by a single operator. A UICC is always associated with a particular PLMN, its Home PLMN. When the subscriber is accessing a Telecommunication Network that is not his Home PLMN, he is said to be roaming to a Visited PLMN.
- A bearer is a bidirectional transport service for packets that offers a certain QoS. When a UE would like to send information, it needs to set up a bearer to the receiver of the information.

We also described how IP connectivity is achieved in a 802.11 WLAN network. It turned out that the procedure is simpler and requires greater involvement from the user. We explained how these differences are down to the different business models and to the different approaches to network design.

12

Mobility

This chapter concentrates on mobility, a crucial UMTS function. Mobility is handled by UTRAN and PS/CS Domain. The IMS is not aware of the fact that the UE is mobile, one says that mobility is transparent to the IMS. PS Domain mobility is described in detail in [3GPP 23.060], UTRAN mobility in [3GPP 25.303, 3GPP 25.304].

The chapter starts with a discussion of the general problems. After that mobility support in both UTRAN and PS Domain are covered in detail. Then we discuss how mobility is supported in a WLAN, and, generally, how mobility is supported by protocols standardized in the IETF, and finally we compare the different approaches.

Terminology discussed in Chapter 12:	
Active Set	
Binding Update	
Care of Address	
Context transfer	
Correspondent Node	CN
Drift RNC	DRNC
Fast Handoff for Mobile IP	FMIP
Foreign Agent	
Global mobility	
Hard handover	
Home Address	
Home Agent	
Host Identity Protocol	HIP
Localized mobility	
Make-before-break	
Mobile IPv4, Mobile IPv6	MIPv4, MIPv6
Mobility Anchor Point	MAP
Mobility Management Context	MM Context

Multi-homed	
Paging	
Power Saving Mode	
Reassociation	
Routing Area Update	
Seamless Handover	
Serving RNC	SNRC
Soft handover	

12.1 Description of the Problems

What does it mean to support mobility? The general problem is illustrated in Figure 12.1. On the one hand, somebody, either the network or MS, must detect when the MS moves out of the coverage of one Access Point into the coverage of a neighbouring Access Point. Then the decision to handover must be taken, and the handover must be performed. To this end, the MS contacts the new Access Point. Connectivity along the new section of the path is established, and connectivity on the old section of the path is removed or times out. This change of connectivity implies adaptations on the radio link (e.g. new CDMA codes), updating of routing tables and possibly the updating of QoS reservations and the IP address.

Mobility support can be offered with varying degrees of convenience. A basic scenario is as follows: the MS detects a new Access Point with better signal quality and decides to handover. In this basic scenario the MS can be associated only with one Access Point at a time, and therefore abandons the old Access Point and registers with a new Access Point—without reference to the previous Access Point. Connectivity along the new section of the path is now established. During this process no user-plane packets can be delivered to the UE. Obviously this may lead to packet delay or packet loss which can be a problem for some applications.

More advanced mobility support can offer the following additional features:

- Exchange of MS-specific **context information** between old access and new access. This context information could include control information such as authentication and

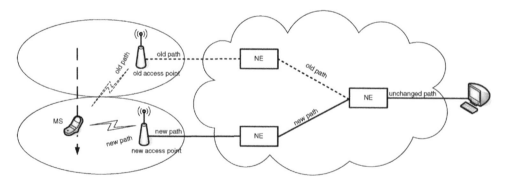

Figure 12.1 Generic mobility scenario of a MS moving into the coverage of a new Access Point. Connectivity along the new path and the old path through network elements (NEs) must be updated

authorization credentials and QoS parameters. Clearly, this re-use of context information is efficient, and in particular relieves the air interface. On the other hand, exchanging such information produces some coordination overhead on the network side.

- **Seamless handover**, in which both handover time and packet delay/loss are minimized. Handover time is reduced by a technique called "**make-before-break**", i.e. by establishing connectivity and QoS along the new section of the path before the MS actually hands over and says good-bye to the old Access Point. Packet loss is reduced by enabling the old access to buffer and forward user-plane packets destined for the UE to the new access. GSM, for example, supports a seamless handover: the user simply perceives a background "click" noise when the antennae are changed.

- **Power-saving mode**, where a mobile network may support a power-saving **sleep mode** for the MS, in which the MS does not actively expect packet delivery. In some radio technologies, e.g. WLAN, the MS wakes up regularly in order to check whether packets have arrived and whether it needs to register with a new Access Point. In other technologies, e.g. UMTS, a MS in power-saving mode is not registered with a particular Access Point. Rather, it merely listens to the broadcast channels and keeps the network updated about its current paging area, which usually includes several Access Points. When the network receives a packet destined for the MS, it issues a paging request. Upon reception of a paging message, the MS wakes up.

  The advantage of the supporting sleep mode is that resources can be saved: MS power, air interface bandwidth and processing resources—these savings, however, carry again the costs of organizational overhead.

12.2 Mobility in UMTS

Mobility support in UMTS is—of course—of the advanced type, offering network-internal context transfer, seamless handover and a power-saving mode with paging. However, UMTS mobility is always intra-PLMN mobility. Regarding mobility, PLMNs do not collaborate. When a UE moves out of the coverage area of one PLMN, it automatically performs GRPS Attach to the PS Domain (or IMSI Attach to the CS Domain) of the new PLMN, however all calls and PDP Contexts are dropped.

In Chapter 11, Section 11.5.1 we encountered the Mobility Management States, PMM-DETACHED, PMM-IDLE and PMM-CONNECTED. Depending on a UE's Mobility Management State, different mobility features are supported:

- When the UE is switched off, it is in PMM-DETACHED state. In this state, the location of the UE is unknown. Obviously, no mobility support is offered.
- When the UE is switched on but idling, it is in PMM-IDLE state. It has performed GPRS Attach and is registered with an SGSN. It may also have an active PDP Context and, associated with a PDP Context, an IP address. It is, however, in a power-saving mode and does not have a RRC Connection (unless it is actively using the CS Domain), and no RNC has knowledge about the UE. The UE tracks its own location by recording the current Routing Area identifier —it does not record the cell ID (cf. Figure 11.6). When the Routing Area changes, the UE updates the SGSN. In this state the UE can be paged by the SGSN. The Routing Area thus corresponds to the generic paging area introduced in the previous section.

- When the UE is sending or receiving packets, it is in PMM-CONNECTED state. It has an active PDP Context, an IP address and an RRC Connection. The UE keeps tracking its Routing Area and updates the SGSN when the Routing Area changes. It additionally tracks its location with the accuracy of a cell or URA, and synchronizes the Serving RNC with this information. Note that the UE's IP address is not affected by the movement. In this state, seamless mobility is supported.

We now look into mobility handling in PMM-IDLE state and PMM-CONNECTED state in more detail.

12.2.1 Mobility in PMM-IDLE State

The rationale for introducing the PMM-IDLE state is that for an idling UE mobility support can be more relaxed. There are no ongoing communication sessions, thus "handover" is reduced to the UE and network knowing how to contact each other if need be. The UE thus updates the network about its location with the accuracy of a Routing Area instead of with cell or URA accuracy, thereby saving air interface bandwidth and processing resources. Of course, when the network receives a packet for the UE it must broadcast a paging request into the entire Routing Area. We therefore see that there is a trade-off when designing the Routing Areas: The bigger a Routing Area, the less frequently must the UE update its Routing Area with the network, but also the bigger the area that must be covered by paging broadcasts.

Next, we cover the two mobility procedures that are important in PMM-IDLE state: updating the Routing Area and paging.

12.2.1.1 Routing Area Updates

In PMM-IDLE state only the UE's Routing Area is known to the SGSN. The UE does not have a RRC Connection, it may, however, have a PDP Context and IP address. It sleeps and listens only to broadcast channels. One of the broadcast channels is the BCCH on which the local Routing Area Identifier (RAI) is announced. From the perspective of the UE we can distinguish two cases. Either the RAI is the same as in the last broadcast, or the RAI changes because the UE moved out of the Routing Area.

If the RAI does not change, the UE must still send a keep-alive signal to the SGSN regularly in the form of a **Routing Area Update** message. With this message it informs the SGSN that it still resides in the same Routing Area. If the SGSN does not receive a Routing Area Update, it moves the UE into PMM-DETACHED state unilaterally. This soft-state mechanism protects the network from "lost UEs" and state overflow (cf. Chapter 11, Section 11.7).

If the RAI changes, the UE also performs a Routing Area Update, albeit a more interesting one. As discussed in Chapter 11, Section 11.5.2, a SGSN controls one or more Routing Areas. A change of Routing Area may thus entail a change of SGSN, as illustrated in Figure 12.2. In this case, the new SGSN pulls information from the old SGSN, which it uses to authenticate the UE. This is an example of the context transfer between old access and new access mentioned earlier. Subsequently, the new SGSN informs the HLR that it is now responsible for the UE.

If the UE has active PDP Contexts the procedure becomes slightly extended (not shown in Figure 12.2). The information pulled from the old SGSN then also includes the active PDP Contexts. Based on this information the new SGSN prompts the affected GGSNs to update

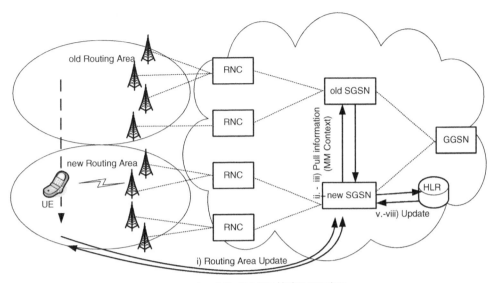

Figure 12.2   Change of Routing Area entailing change of SGSN. The roman numerals refer to the
message flow explained in Section 12.2.1.1 and Figure 12.3

their corresponding PDP Contexts. The PDP Context "tunnels" through the PS Domain are
rerouted.

12.2.1.1.1 Message Flow for Inter-SGSN Routing Area Update

Figure 12.3 shows in more detail the message flow for an **Inter-SGSN Routing Area Update**,
i.e. a Routing Area Update in which the SGSN changes. Note that we assume that the UE does
not have an active PDP Context. Messages between UE and SGSN are sent with the GMM
Protocol, messages between SGSN and HLR are sent with the MAP Protocol (cf. Chapter 6,
Section 6.2.2)

(i)   The UE sends a Routing Area Update message containing P-TMSI and old RAI to the
new SGSN.

(ii)   The new SGSN derives the identity of the old SGSN from P-TMSI and RAI (each
SGSN has its own pool of P-TMSIs). It sends a message to the old SGSN and requests
the **Mobility Management Context** (MM Context) for the UE with this P-TMSI.

(iii)   The old SGSN sends the MM Context to the new SGSN. The MM Context contains
the IMSI and security information such as keying material. Note how this procedure
avoids sending the sensitive IMSI over the more insecure air interface.

(iv)   The new SGSN authenticates the UE based on the security information received from
the old SGSN and acknowledges receipt of the MM Context.

(v–viii)   The new SGSN updates the HLR and receives subscription information.

(ix–x)   The new SGSN informs the UE that the Routing Area Update was successful, and the
UE acknowledges receipt of this message.

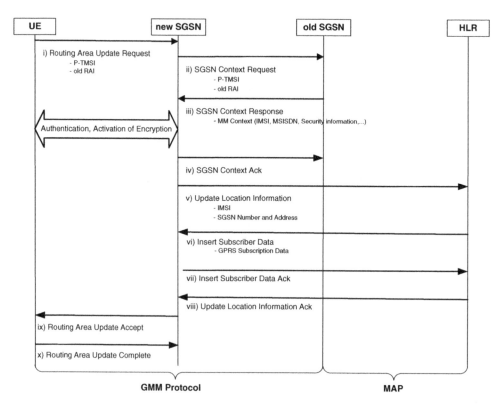

Figure 12.3   Message flow for Inter-SGSN Routing Area Update in PMM-IDLE mode

It is interesting to note that the message flow just described is almost identical to the message flow for GPRS Attach in Figure 11.7! The main difference is that the (new) SGSN, after being contacted by the UE, pulls MM Context from the old SGSN, rather than communicating directly with the HLR. Another important difference is that the UE identifies itself to the SGSN with the P-TMSI rather than the IMSI, in order to decrease the risk of identity theft.

12.2.1.2 Paging

When a packet arrives for a UE in PMM-IDLE state, the UE must be found and woken up. Let us see how this works in detail. An IP packet for the UE arrives at the GGSN. This can in fact only happen if the UE has an active PDP Context, because otherwise it usually does not have an IP address. The GGSN thus forwards the packet to the SGSN at the end of the PDP Context. The SGSN detects that the UE is in PMM-IDLE mode and issues a paging request that is broadcast on the Paging Channel by all Node Bs in the current Routing Area of the UE. The UE, upon reception of the paging request, contacts the RNC in order to set-up a RRC Connection. The UE moves into PMM-CONNECTED state and the SGSN finally delivers the packet to the UE.

12.2.2 Mobility in PMM-CONNECTED State

A UE in PMM-CONNECTED state has an active PDP Context and a **Serving RNC** (SRNC). The SRNC knows the UE's location with the accuracy of cell or URA. Additionally, and independently, the SGSN is up-to-date regarding the Routing Area of the UE.

We now cover the most important mobility procedures in PMM-CONNECTED state: the actual handover and relocation of the SRNC.

12.2.2.1 Handover

It is interesting to ponder what "handover" means in a network that supports macrodiversity. Macrodiversity, introduced in Chapter 5, Section 5.2.4.2, allows a UE to be attached to more than one cell simultaneously. In Chapter 5, macrodiversity was introduced as a means of countering the near-far effect. It turns out, however, that macrodiversity can do more: it enables **soft handover**, a particular way of performing seamless handover.

A simple scenario for soft handover works as follows (see Figure 12.4): picture a UE that is being connected to a single cell X. The UE moves towards the border of cell X's coverage area, and the reception quality deteriorates. The SRNC, upon reception of the regular report on reception quality from Node B, decides that it needs to act. It therefore asks the UE to connect to an *additional* cell Y. This is also called "adding the radio link to Y to the **Active Set** of the UE". Cell Y can be governed either by the SRNC itself, or by a different RNC—the so-called **Drift RNC** (DRNC). Once the connection to Y is added to the Active Set, UE communicates in a "local multicast" fashion via two radio paths. The path via cell Y leads over the DRNC to the SRNC. In the SRNC, the paths from X and Y are joined. Once the reception quality with cell X becomes sub-threshold, the SRNC decrees that the connection with cell X be torn down. This way the communication of the UE is never interrupted.

Figure 12.5 illustrates the message flow for adding a radio link to the Active Set of a UE. Upon reception of a measurement report from the UE, the SRNC contacts the DRNC in order to establish a radio connection (of course, if the SRNC itself governs cell Y this step is skipped). When the DRNC successfully sets up the new radio connection, the SRNC instructs the UE to add this radio connection to its Active Set. The UE configures the new connection and then sends an acknowledgement message to the SRNC.

UMTS is in fact also capable of performing "**hard handovers**", i.e. handovers where the UE is communicating with only one cell at a time. Hard handover is the fall back method when macrodiversity is not possible, e.g. when handing over to GSM, or when the cells in question are governed by two RNCs that do not have an Iur interface between them.

12.2.2.2 SRNC Relocation

As we saw in the previous sections, the UE moves and keeps updating its Active Set of radio links. At some point, all active connections will be via one or more DRNCs. However, all communication is routed via the SRNC (see Figure 12.4c), which is clearly inefficient. Therefore, the RNC acting as SRNC may contact one of the DRNCs and ask it to take on the role of the SRNC. If the new SRNC is connected to a different SGSN than the old SRNC this may also entail a change of SGSN. Once the SRNC relocation is performed, the old SRNC drops out of the signalling path. Note that while the SRNC relocation procedure is standardized, the algorithm for deciding when to perform SRNC relocation is not—it is a decision local to the

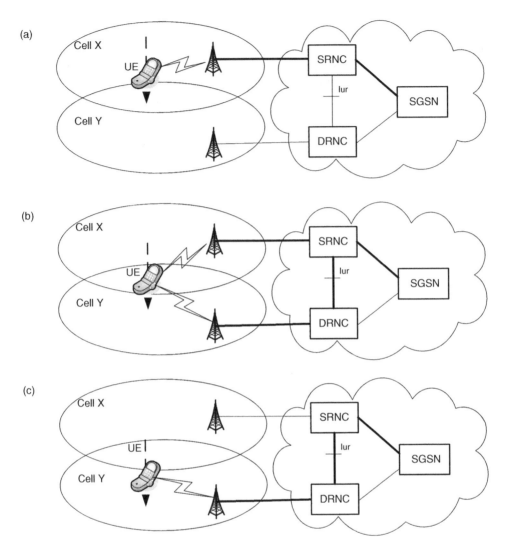

Figure 12.4   Soft handover. (a) the UE is still in cell X. (b) the UE moves from cell X to cell Y. (c) the UE arrives in cell Y

SRNC which does not require interworking. Note also that SRNC relocation timing is independent of the timing of other mobility procedures such as soft handover.

12.3 Link-Layer Mobility in a WLAN

The 802.11 standard for WLAN offers basic mobility support on the link layer in the form of *(re)associations*. We learned in Chapter 11, Section 11.9.2 that a MS in a WLAN is associated with exactly one Access Point. However, even in associated state it continues to monitor the quality of the signals from other Access Points. If the MS finds a new Access Point offering

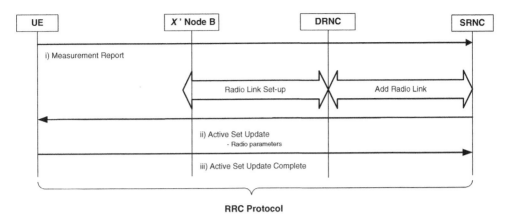

Figure 12.5 Message flow for Active Set Update

better quality than its current Access Point, it decides to switch. This switch is performed in a very simple fashion: the MS associates with the new Access Point starting from zero, and the association state in the old Access Point eventually times out.

Obviously, this is very basic mobility. More advanced mobility support is also possible: the MS sends the new Access Point a re-association message containing the address of the old Access Point. If the old and new Access Points belong to the same ESS, the new Access Point can inform the old Access Point of the handover, and verify that the MS is authenticated — the reader should recognize this as context transfer (cf. Section 12.1).

Furthermore, the new Access Point registers the new association of the MS with the Access Router. The old Access Point de-associates with the MS and forwards all frames destined for the MS which it may still be buffering to the new Access Point. Obviously, the old Access Point and new Access Point need a protocol for this communication, the **Inter Access Point Protocol** (IAPP). While IAPP was standardized in 802.11f [802.11f], in practice inter-Access Point communication is based on proprietary, vendor-specific protocols, and in 2006 the standard was withdrawn.

802.11 also supports a power-saving mode: unless the MS is sending or receiving packets, it falls asleep. At regular intervals it wakes up in order to find out what is going on. These intervals are multiples of the Access Point's beacon period. The Access Point knows when the MS is asleep and buffers all incoming packets destined for the MS. When the MS wakes up, the Access Point sends a message that packets are waiting. When the MS wakes up it also determines whether it is still in the coverage area of the Access Point, or whether it needs to re-associate.

12.4 Mobility in Computer Networks

Mobility in a Computer Network follows the toolbox idea and can range from basic to advanced, depending on which protocols and link layers are applied and combined. The link layer must support mobility, and additional support can be provided from the IP-layer upwards. In this section we present a number of mobility protocols developed by the IETF. When IP was designed originally, mobility was not an issue. Additional protocols for supporting mobility were designed later on for both IPv4 and IPv6, they are called **Mobile IPv4** (MIPv4) [RFC 3344], and **Mobile IPv6** (MIPv6) [RFC 3775], respectively. In this section, we cover MIPv6 in detail and a number of protocols for optimization. We also introduce MIPv4 briefly because it plays a role in 4G Networks. Finally, we discuss how advanced mobility may be achieved by the collaboration of these protocols.

12.4.1 Basic Mobility Support by the IETF

From the perspective of IP, it is only the aspects of mobility from Layer 3 and upwards that are visible. Mobility is an issue because — in a coarse-granular fashion — the IP address codes the location of a network element. The fact that the MS re-associates with a new Access Point on Layer 2 is in itself not of interest: when a MS re-associates to an Access Point within the same ESS it usually keeps its IP address. When the MS moves to a new ESS, however, it changes its IP address, upsetting routing and addressing.

A Computer Network has an additional problem: the IP address actually codes two things, the location and the identity of a network element. This was fine in the early days of the Internet when network elements were stationary and IP addresses were permanent. However, a network element's location can now change, and hence also its IP address. As a side-effect, its identifier changes, also. Therefore, when a **Correspondent Node** desires to send a packet to a mobile MS, it does not know how to identify it. This basic problem with Computer Networks has become a topic of much research and discussion in recent years, with the **Host Identity Protocol** (HIP) [RFC 4423, RFC 5205] as a possible solution.

12.4.1.1 Mobile IPv6

MIPv6 solves these problems as follows. A MS has a static IP address, its **Home Address**. The MS can always be identified and contacted by this address. When the MS moves, it additionally acquires a temporary IP address, the **Care of Address**, from the Visited Network using, e.g. IPv6 address autoconfiguration [RFC2462] or by asking a DHCP server [RFC3315]. The Mobile IP protocol subsequently changes the global end-to-end routing of packets destined for the MS to the new Care of Address. Mobile IP is therefore also called a **global mobility** protocol.

MIPv6 achieves the re-routing of packets by defining a new network element which is in charge of mobility control in the Home Network, the **Home Agent** (HA). The HA can be integrated into a router. When the MS acquires its Care of Address, it sends a message, the **Binding Update**, to the Home Agent, informing it of its new address. The Home Agent now enters the Care of Address-to-Home Address mapping in a table. One says that it installs state. This state is soft, i.e. it times out in order to prevent state overflow when a problem occurs. Therefore the Binding Update is resent on a regular basis.

Now let us look at how IP packets are addressed between MS and Correspondent Nodes. In a basic mode, illustrated in Figure 12.6, Correspondent Nodes keep using the Home Address of the MS, independent of its current location. The Home Agent intercepts all packets destined for the Home Address of the MS and tunnels them to the current location of the MS. Therefore "tunnelling" effectively means that the Home Agent takes the entire IP packet as received from the Correspondent Node and adds an additional IP header with the MS's Care of Address as the destination address. In the opposite direction, the MS composes packets destined for the Correspondent Node, with the MS's Home Address as the source address. It tunnels this packet to the Home Agent. The Home Agent removes the outer IP header, discovers the encapsulated packet and forwards it to the Correspondent Node. The reader may note that the mobility of the

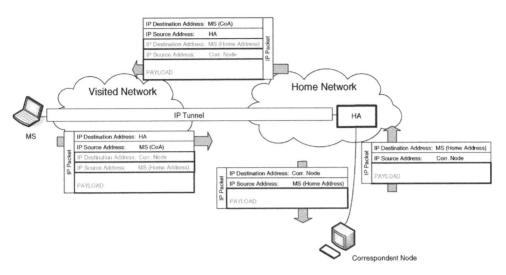

Figure 12.6   Mobile IPv6 with triangular routing

MS is transparent to the Correspondent Node. This also implies the Correspondent Node does not need to implement MIPv6.

What happens when the MS performs a handover? After establishing Layer 2 connectivity, the MS obtains a new Care of Address and sends Binding Updates to its Home Agent. The Home Agent henceforth addresses all packets to the new Care of Address. In a simple environment, in which the MS can only maintain one Care of Address at the one time, packets will be lost that were sent to the MS between acquiring the new Care of Address and arrival of the Binding Update at the Home Agent. This packet loss would typically happen for a time period of the order of seconds.

The basic mode of IPv6 shown in Figure 12.6 is not very efficient since all packets are always routed via the Home Agent. This is also called triangular routing. Therefore an alternative **route optimization** mode can be employed if the Correspondent Node is also capable of MIPv6. In this case, the MS also sends a Binding Update with the Care of Address to the Correspondent Node, and the two of them now communicate directly, without tunnelling via the Home Agent. This is illustrated in Figure 12.7. A problem with this approach is that location privacy is compromised, because the Correspondent Node can deduce the current location of the MS from the Care of Address.

Recently, the IETF has been working on a further generalization of MIPv6 that allows the MS to acquire dynamically both its own Home Address and the address of its Home Agent. This process is called **MIPv6 bootstrapping**[RFC 4640].

12.4.1.2 Mobile IPv4

The basic idea of Mobile IPv4 is the same as in Mobile IPv6: The MS has a Home Address and a Care of Address. The Home Agent intercepts packets destined for the MS and tunnels them to the Care of Address.

MIPv4 is different from MIPv6 because, apart from running over IPv4 rather than over IPv6, it can use an auxiliary network element, the **Foreign Agent**, in the Visited Network. When the

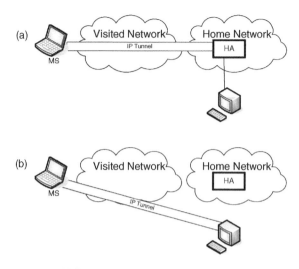

Figure 12.7   Mobile IPv6 with routing optimization

MS arrives in the Visited Network, it registers with the Foreign Agent and receives the Care of Address. Interestingly, the Care of Address is in fact the IP address of the Foreign Agent, and is thus identical for all visiting, mobile MS in this network. The Home Agent thus tunnels packets destined for the MS to the Foreign Agent. The Foreign Agent removes the outer IP header from the packet, looks up the link-layer address of the MS in its registration table, and forwards the packet to the MS.

12.4.2 Advanced Mobility Support by the IETF

At the beginning of this chapter we defined advanced mobility support so as to include context transfer, seamless mobility and support of power-saving mode. MIPv6, however, offers just basic mobility support: each time the MS performs a handover, the communication session is interrupted until movement detection, new Care of Address configuration and the Binding Update procedure are performed. MIPv6 therefore does not support seamless mobility or any of the other advanced mobility mechanisms. It should be added that the situation is similar for MIPv4. A number of optimizations both on the IP layer and on the radio link are, however, available. In the following subsections we will see that advanced mobility support with IP-based mechanisms is feasible to a considerable degree. However, the operators of a mobile network must design a concrete solution.

12.4.2.1 Context Transfer Between Old Access Router and New Access Router

The IETF has standardized a protocol, the **Candidate Access Router Discovery** (CARD) protocol that allows the MS to contact its current Access Router and inquire as to the properties of potential new Access Routers [RFC 4066]. Subsequently, the **Context Transfer Protocol** [RFC 4067] may be used to transfer control state information between the old and new Access Router, e.g. authentication credentials and QoS information. Both protocols have, however, not yet seen serious deployment; one of the problems encountered during their standardization was that Computer Networks—as opposed to Telecommunication Networks—are not designed for cooperation between different functionalities, which is exactly what could be achieved with context transfer. Thus, while protocols for context transfer exist, it is still unclear as to which context information to transfer.

12.4.2.2 Seamless Mobility

The IETF has addressed the problem of seamless mobility with several enhancements to **Mobile IP**, which we will now study. These enhancements result in **localized mobility** protocols that operate locally and only change the routing locally. This is in contrast to a global mobility protocol involving global routing changes, as performed by Mobile IP.

12.4.2.2.1 Hierarchical Mobile IPv6
In MIPv6, any movement of the MS that translates into a change of IP address must be reported to the Home Agent in the Home Network and possibly to all Correspondent Nodes. In **Hierarchical Mobile IPv6** (HMIPv6) [RFC 4140], "local" movement is handled locally. To this end a "local mobility agent" called **Mobile Anchor Point** (MAP) is introduced. The MAP is located in the Visited Network. The MS now has two Care of Addresses: the **Regional Care of Address**, whose mapping to the Home Address is registered with the Home Agent (or Correspondent Node), and the **Local Care of Address** whose mapping to the Regional Care of Address is registered with the MAP. The Home Agent tunnels packets destined for the MS with the Regional Care of Address. The MAP intercepts these packets and wraps them in another IP packet, this time addressed to the Local Care of Address. The result is a tunnel-in-a-tunnel. Both tunnels are terminated at the MS which then unwraps the packets.

The advantage of this procedure is that mobility within an area controlled by the same MAP results only in a change of the Local Care of Address, whereas the Regional Care of Address remains the same. Upon change of the Local Care of Address, the MS sends a Binding Update only to the MAP, and thus local mobility is handled locally. Only when the MS moves out of the coverage area of the MAP does it need to acquire a new Regional Care of Address and must inform its Home Agent. The advantages are that signalling overhead is reduced, and that handover can be performed more quickly so that packet loss is reduced. Figure 12.8 illustrates the topology in a network employing HMIPv6. It is worth noting the similarities with the UMTS topology shown in Figure 12.2.

12.4.2.2.2 Fast Handoff for Mobile IPv6
Even when Hierarchical Mobile IP is used some packets are bound to be lost. To counter this effect, **Fast Handoff for Mobile IPv6** (FMIPv6) [RFC 4068] was specified. FMIPv6 specifies how a tunnel is established between the old Access Router and new Access Router so that packets addressed to the old Care-of-Address are buffered and forwarded to the new Care-of-Address. Remember from Section 23.3 that WLAN offers a similar feature on the link layer, albeit based on proprietary protocols.

Compared to Mobile IP, mobility with HMIP and FMIP offers a decrease in handover time and packet loss. These protocols, however, do not support make-before-break explicitly. Make-before-break refers to the ability to maintain connectivity via the old Access Point while already having connectivity with the new Access Point. This can be achieved if the link-layer technology supports association and frame delivery via multiple Access Points—we have already seen that WLAN does not support this. Alternatively, the MS may feature two

LCoA ← RCoA   RCoA ← Home Address

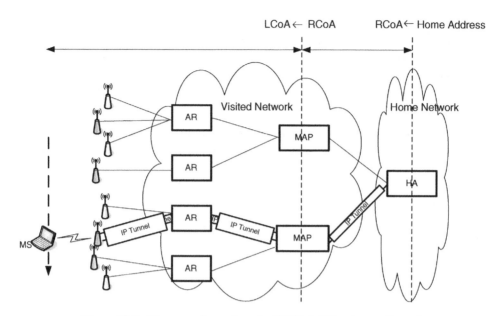

Figure 12.8   The network topology for HMIPv6 (AR—Access Router)

independent radio interfaces, e.g. two WLAN cards, each with its own IP address. Such a MS is called **multi-homed**.

Make-before-break, however, not only refers to the ability to maintain IP connectivity via two Access Points. It also refers to the ability to move a QoS reservation from the old path to the new path. We will cover this aspect in Chapter 14 on QoS.

12.4.2.3 Power-Saving Mode and Paging

Whether or not the power-saving mode is supported is first of all a matter for the radio technology. From an IP perspective, a MS is reachable under its Care-of-Address. It is invisible to the IP layer as to whether a MS is currently associated with an Access Point, unless a change of Access Point entails a change of Care-of-Address.

One might now think of scenarios where the Care-of-Address changes frequently, e.g. because the MS moves at high speed or because it moves between subnets (e.g. WLAN ESSs). In this case it may be advantageous to update the Care-of-Address and/or routes in the network only when the MS is indeed sending or receiving packets. In particular, the IETF investigated whether a technology-independent paging protocol over IP should be standardized [RFC 3132]. It is, however, unclear whether such a mechanism, which would be employed in addition to a link-layer power-saving mode, would offer considerable benefits [Kempf 2003]. In any event, the IETF has so far not pursued the issue further.

12.5 Discussion

When comparing the mobility support in UMTS with the mobility support in a Computer Network we observe the typical differences that have already been described in Chapter 1:

mobility support in UMTS is designed as an entire system ("cathedral"), based on architectural considerations, whereas mobility support in a Computer Network is designed as a toolbox ("bazaar") based on protocols.

Furthermore, mobility in UMTS is controlled by the network—the UE only provides location information and reception quality measurements, the actual decision to handover is made in the network. In contrast, mobility and handover decisions in a Computer Network are traditionally controlled by the MS. We should add that recently the IETF has developed a mobility protocol, **Proxy Mobile IP**, which places mobility control in the network. We will hear more about this protocol in Part II.

We may also observe that many differences derive from the different business models. UMTS operators sell services. UMTS mobility support therefore intrinsically links other control features such as authentication and authorization or QoS. UMTS is expected to deliver real-time services, and thus handover is seamless, and the reaction time of UEs in power-saving mode (PMM-IDLE) is minimized by paging. The operator of a mobile Access Network to the Internet such as a WLAN, by contrast, sells connectivity. Mobility support does not automatically include support for re-authentication at new Access Points or QoS, nor is seamless handover supported. MSs go into power-saving mode and wake up on their own time schedule in order to check whether new packets have arrived.

It is also interesting to look at the collaboration between the different layers. Since UMTS is designed as an entire system encompassing almost all layers, it was possible to optimize this cooperation. For example, seamless mobility is achieved by make-before-break on the link layer, and additionally is supported on the network layer because the IP address of a UE never changes unless it deactivates all of its PDP Contexts.

12.6 Summary

In this chapter support for mobility was discussed. Mobility requires extra functionality, both in the MS and in the network: it must be detected when a MS moves out of the coverage area of one Access Point into the coverage of a new Access Point. Then the decision to handover must be taken, and the handover must be performed. This implies adaptation both on the link layer and on the higher layers. For example, radio resources may need to be reassigned and routing tables may need to be updated.

We distinguish basic mobility support, where the relation between MS and new Access Point must be built from zero after the handover, and advanced mobility support which can additionally offer seamless handover and context transfer between Access Points. Another useful feature is support of the power-saving mode, which may be combined with the paging capability.

UMTS offers advanced mobility support, designed as usual as an entire system with optimized collaboration between layers. UMTS mobility is handled in UTRAN, PS Domain. and CS Domain (the latter was not described in detail). The IMS assumes that mobility is handled elsewhere.

Mobility support in UMTS depends on the Mobility Management State of the UE. In PMM-idle state, when the UE is not sending or receiving packets, it is in power-saving mode and does not have a RRC Connection. The UTRAN does not have a record of the UE. The UE listens to system broadcasts in order to track its Routing Area and keeps the SGSN updated with this information. When a packet arrives for the UE, the SGSN pages the UE in the Routing Area and the RRC Connection is re-established.

In PMM-Connected state, the UE has an RRC Connection and a SRNC. The UE keeps updating the SGSN about its Routing Area just as it does in PMM-idle state. Additionally, the SRNC monitors the situation of the UE and may order it to add or remove RRC Connections to particular Node Bs, thereby performing soft-handover.

Mobility support in a Computer Network follows the toolbox idea and can range from basic to advanced, depending on the protocols employed, and also on the support from the underlying radio technology.

The WLAN standard supports a basic handover on the link layer and power-saving mode. Proprietary protocols are normally used for context transfer.

Mobile IP offers basic handover support on the network layer. More seamless handover could be achieved by additionally employing HMIP and FMIP. The IETF also standardized protocols supporting context transfer. An additional paging protocol on the IP layer has not, so far, been considered necessary.

13

Security

This chapter covers another important control function of UMTS, security. Security is a multi-facetted, complex topic, and we will only be able to inspect the tip of the iceberg. While security is pervasive in UMTS, we will concentrate mostly on network aspects, i.e. Layer 3. The basic security mechanisms are explained in [Tanenbaum 2002]. A basic specification of UMTS security, particularly PS Domain security, features and mechanisms is [3GPP 33.102]. IMS security is described in [3GPP 33.203]. A summary is provided in [Boman 2002].

The chapter starts with a description of the general problems and solutions. We then address the security mechanisms in UMTS. After that, we discuss security in a WLAN, and generally in the Computer Networks. As usual, we close with a comparison of the different approaches.

Terminology discussed in Chapter 13:	
AAA Server	
Authentication	
Authentication and key agreement	AKA
Authentication, Authorization, Accounting	AAA
Authentication Client	
Authentication Server	
Authentication Vector	
Authenticator	
Authorization	
Back end	
Challenge-response	
Cryptographic hash function	
Cryptographic Key	CK
Denial of service	
Diameter	
Diameter application	

Diameter Client	
Diameter Server	
Digital signature	
EAP Authentication Method	
EAP over LAN	EAPOL
Eavesdropping	
Extensible authentication protocol	EAP
Encryption	
Front end	
Identity spoofing	
Information disclosure	
Information forgery	
Integrity Key	IK
Integrity protection	
Internet Key Exchange	IKE
IPsec	
IPsec Security Association	SA
Man-in-the-middle attack	
Master Key	
Message Authentication Code	MAC
Mutual Authentication	
Network Access Server	NAS
Public/private key	
Remote Authentication Dial-In User Service	RADIUS
Replay attack	
Repudiation	
Secret key	
Security Gateway	SEG
Security threat	
Shared secret	
Supplicant	
Symmetric integrity protection	
Temporary identity	
Temporary session keys	
Theft of Service	
Threat analysis	
Transport Layer Security	TLS

13.1 Description of the Problems

Communication Networks, particularly mobile Communication Networks, are imperilled by a multitude of **security threats**. In Table 13.1 we list the typical threats and look at the range of attacks that an adversary could carry out in order to realize those threats. In order to restore our peace of mind, the final column of the table provides measures to counter these attacks. The following subsections describe the individual items in Table 13.1 in more detail.

Table 13.1   Selected security threats, attacks and countermeasures

Threat	Example Attacks (how to realize a threat)	Countermeasures
Information disclosure - User identity - Location - Data packet content	Eavesdropping (passive attack)	{ Encryption Temporary identies
Information forgery	"Man-in-the-middle attack" (active attack)	{ Mutual authentication Authorization Integrity protection
Theft of service	Identity spoofing Session hijacking Repudiation Repley	{ Authentication and authorization Encryption Integrity Protection Temporary identities Time-variant parameterization of messages
Denial of service (DoS)	Flooding with messages that need to be processed, or cause set-up of state	Authentication Soft state Low computational overhead Stateless protocols

13.1.1 Information Disclosure

One threat is **information disclosure**. Our adversary, let us call her Alice, may be interested in learning any kind of information about a user, let us call him Bob. This information could be his location, who he is calling, which web-sites he is visiting, or generally the content of his conversations. The attack which Alice carries out consists of **eavesdropping**. She sits somewhere on the communication path and listens to all of the packets Bob is sending and receiving. The radio interface is especially vulnerable to this kind of attack.

The most important defence against eavesdropping is **encryption** of data. Alice can still eavesdrop, but all she sees is alphabet soup. An additional measure is for Bob to make it more difficult for Alice to identify who is sending packets. He could therefore use **temporary identities**. We have already encountered a temporary identity used in UMTS, the P-TMSI.

13.1.2 Information Forgery

Going beyond just passively eavesdropping upon information, Alice may actively **forge information**. This threat may be carried out by a **man-in-the-middle attack**, which is also most easily carried out on the radio interface, as illustrated in Figure 13.1. This attack consists of Alice introducing herself on the communication path between Bob and network and forging packets.

How would Alice do this? She would veil her identity and would convince Bob that she is a rightful Access Point of his network, and convince the network she is Bob. Uplink, Bob sends her all his packets; she manipulates them, and sends them on to the network. Downlink, she receives all of the packets destined for Bob, manipulates them and sends them to Bob.

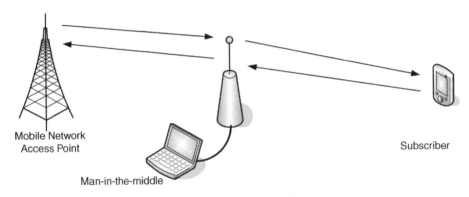

Mobile Network
Access Point

Subscriber

Man-in-the-middle

Figure 13.1   Man-in-the-middle attack

A man-in-the-middle attack can be countered by a three-step process:

- Firstly, the network must ascertain that Bob is indeed Bob. Vice versa, Bob must ascertain that the Access Point belongs to his HPLMN. This step is called **mutual authentication**.
- Secondly, both Bob and the network must each clarify what the other is allowed to do. The network is allowed to receive and relay Bob's packets, and Bob is allowed to send packets. Bob is, however, not allowed to relay the packets of another subscriber (although Bob, as we know him, would never forge packets). This step is called **authorization**.
- Thirdly, the network must be able to verify that the packets which it receives indeed originate from the Bob which it has just authenticated, and that nobody has tampered with the packets in between. Conversely, Bob must verify that the packets which he receives were sent in this exact form from the network he which he has just authenticated as his Home Network. This last step is achieved by carrying out **integrity protection** for all packets. This means that both Bob and the network can recognize who originated the protection and whether the packet was manipulated by a third party.

13.1.3 Theft of Service

Another possible threat is **Theft of Service**. It can, for example, be achieved by building on eavesdropping and man-in-the-middle attacks. Theft of Service consists, for example, of Alice pretending to be the virtuous Bob: Alice eavesdrops upon Bob's identifier, attaches to the network declaring that she is Bob, uses system services and Bob receives the bill. This kind of attack is called **identity spoofing**. This attack can be countered by performing encryption and authentication, and by using temporary identifiers.

Alice might, however, deceive the authentication mechanism by performing a **replay attack**. To this end, Alice records an authentication exchange between Bob and the network. She then authenticates with the network by replaying Bob's original messages. Note that this attack works even when the authentication exchange is encrypted and/or integrity protected.

The counter measure against replay attacks is the usage of time-variant parameterization of important message exchanges. For example, Bob could include a sequence number; let us say

"100", in his authentication message to the network. The next time that Bob authenticates he would have to include a higher number, e.g. "101". When Alice replays Bob's authentication message featuring the "100", the network would become suspicious and reject it.

Instead of impersonating Bob, Alice could also attach as herself and **hijack** Bob's **session**. She could, for example, ask the network boldly to send her all packets destined for Bob. This attempt may be impeded by proper authorization.

The resourceful Alice still does not give up. An alternative to session hijacking is intercepting the messages with which Bob initiates the session and replacing his source address with her own address. This attack can be barred by integrity protection for the packets.

Alice's final attempt might consist of attaching to the network as herself and using her own services, but later, when she is billed, **repudiate** that she ever used them. This attack may also be countered by integrity protection: integrity protection at the same time offers authentication of the protecting party such that repudiation is made difficult.

13.1.4 Denial of Service

A purely destructive threat is **Denial of Service**. In this case, Alice is only interested in paralyzing the network. The attack consists, for example, of flooding the network with meaningless messages, e.g. fake authentication requests. The network is kept busy with processing these messages and cannot serve anybody else. Alternatively, Alice could send messages that cause the set-up of vast amounts of state in network nodes.

An important counter-measure against Denial of Service attacks is, as usual, the authentication of Alice, so that it is at least possible to trace who is causing the attack and then to stop her. But what if Alice performs her attack without being authenticated, e.g. by flooding the network with authentication requests? This kind of attack can only be counteracted by following special design guidelines for those network processes that involve unauthenticated users: they must have low computational overhead, they must ideally not cause any state to be installed, and if there is state, it must be soft-state.

13.2 General Approach to Solutions

The measures against the various security attacks encountered in the previous section can be classed broadly in two groups: authentication, authorization, encryption and integrity protection might be called basic security mechanisms. The other measures—temporary identities, stateless protocols, time-variant parameterization of messages, etc.—are of a type called "design approach". We will look at the basic mechanisms in more detail.

Secure communication between Bob and his network always originates from a secure cornerstone such as a password, a **shared secret**, a certificate or a smart card. Authentication, encryption and integrity protection all build on this cornerstone. In this book we will concentrate on the shared secret, also called a **secret key**.

13.2.1 Secret Keys

The shared secret may be established by manual configuration, which is however not always practical and can be error-prone. The alternative is the automatic establishment of the secret key with the help of a trusted third party. For example, Bob could maintain a certificate associated with a pair of **public/private keys**. Only the private key is secret, whereas the public key is available to anybody, and furthermore it is certified as belonging to Bob by the trusted third party. By exercising some cryptographic magic on this public/private key it is possible for Bob and the network to arrive at a shared secret key.

A shared secret between Bob and the network is of course exceedingly valuable and must be protected. Therefore, the shared secret is usually used only for authentication and for generating one or more **temporary session keys**. The temporary keys are then used for encryption and integrity protection.

As an aside, the public/private key pair could serve as cornerstone for secure communication in its own right; in this book we concentrate, however, on the shared secret.

13.2.2 Integrity Protection

Integrity protection can be applied to any message, e.g. entire emails or single packets. It can also be applied on any OSI layer. For example, integrity protection applied as part of the application layer protects the entire Layer 7 data. One must consider, however, that in this case headers added on lower layers remain unprotected. Thus, the correct choice of layer on which integrity protection is applied is an important issue.

Integrity protection on the basis of a shared secret is called **symmetric integrity protection**. In addition to the secret, the sender and receiver of the data—in our case user and network—share a a mathematical function, a **cryptographic hash function H**. The cryptographic hash function **H** takes a message **M** and the secret key **s** as arguments and produces a fixed-length output, a **Message Authentication Code** (MAC). It is an important property of a cryptographic hash function that a message **N** which differs from message **M** only in a single bit produces a very different MAC.

As illustrated in Figure. 13.2, the sender calculates the MAC, appends it to the message and sends the result to the receiver. The receiver takes the message and secret key and applies the same function **H**. If the result of this calculation equals the MAC received the receiver can safely assume the message was unaltered. An attacker would have to know the secret key in order to meaningfully amend the MAC.

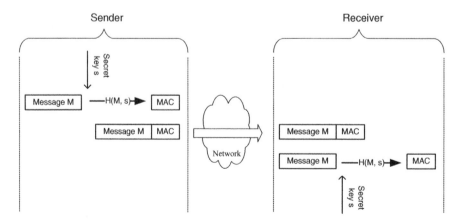

Figure 13.2   Integrity protection with a Message Authentication Code

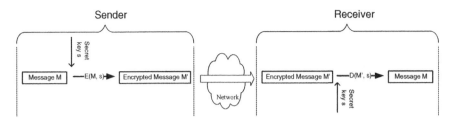

Figure 13.3   Encryption and decryption of a message

It is noteworthy that the security of this method lies in the secrecy of the secret key rather than in the secrecy of the cryptographic hash function. It is indeed simpler to keep keys secret than to keep functions and algorithms secret, especially when these functions are standardized. The underlying principle was recognized in the late 19th century by Kerckhoff. It states that a cryptographic system should stay secure even if the adversary knows how it works. A different design principle would be "security by obscurity", i.e. securing the system by keeping secret how it works.

Integrity protection at the same time offers authentication because the sender of the message must be the owner of the secret key. Integrity protection is therefore a form of **digital signature**.

13.2.3 Encryption

Just as with integrity protection, encryption can be applied on any OSI layer. We concentrate on **symmetric encryption**, on the basis of a shared secret.

Applying the same idea as for symmetric integrity protection, for encryption a calculation is performed on the data and the secret key. This time the sender knows the encryption function **E** and the receiver knows its inverse, the decryption function **D**. As illustrated in Figure. 13.3, the sender encrypts his message by running the function **E** over message **M** and secret key **s**. The result is an unintelligible message **M'**. The sender sends **M'** to the receiver. The receiver applies the decryption function **D** to **M'** and **s** and regains the original message **M**. Also, in this case, the sender of the message is authenticated.

13.2.4 Authentication

We have already seen how integrity protection authenticates individual messages or packets as originating from a particular user. However, we still need to solve the problem of authenticating the user. This can be performed by the network, or the computer, by asking for a password from the user. Obviously, when the password is sent over the network, it is vulnerable to eavesdropping and replay attacks. Another drawback is that, with passwords, it is only the network which authenticates the user. The user has no idea whether the entity asking for the password belongs to the rightful network.

User authentication based on a shared secret and a shared cryptographic function has the advantage that the secret does not need to be transmitted. Instead, the network sends a **challenge** to the user. The user encrypts the challenge and sends back the result as the **response**. The network performs the same calculation, and since it knows the secret key, it arrives at the same response. It compares its own response with the response of the user, and if both are identical, the user is authenticated, i.e. the network can be sure that the user knows the shared secret. Of course, the network sends a different challenge each time the user tries to authenticate, in order to obviate replay attacks.

But how can the user authenticate the network? This is quite simple. The network digitally signs the challenge, e.g. by adding an integrity protection, and the user verifies the signature.

13.2.5 Authorization

Network users usually only have limited access to network resources. For example, they are not allowed to shut down network nodes or to modify the traffic of other users. Furthermore, the access rights may vary from user to user, depending on their subscription. For example, only some users may be allowed to reserve the bandwidth for a video streaming application. Once a user has been authenticated, the network therefore performs authorization: it determines the user profile in a data base and restricts access rights accordingly.

13.2.6 Discussion

In the previous subsections we became acquainted with the typical measures taken against security attacks. However, it is important to realize that absolute security does not exist, for several reasons.

One reason is that security is not something you have or you do not have. Rather, one performs a **threat analysis** and designs the security to protect against those threats considered likely. For example, when you travel to the North Pole, you would pack clothing to protect yourself against the threats of "cold" and "wind", but you would not bring a sombrero and a fan to protect yourself against the threat of excessive heat. Thus, security mechanisms protect against only those threats for which they were designed.

Furthermore, given enough time and resources, any security system can be broken. As an example, consider a bicycle lock. It may take time and special tools, but eventually any lock will surrender. The security question is thus not "Is it possible to break the system?" but rather "How much effort is necessary to break the system?" In a Communication Network, a short secret key is simple and a long secret key is difficult to break—the key length is chosen so that it is considered unlikely that the attacker will have sufficient computing power or time to break it.

Another point to consider is that security measures, invariably, add overhead. A system may be secured to the point of self-paralysis; just as a knight's armour may offer fabulous protection against blows, but at the same time obstruct his movements considerably. We therefore often have to balance security and usability. For example, secret keys must be long enough to offer adequate protection and at the same time be short enough to allow a quick response time.

13.3 Security in UMTS

UMTS security protects against eavesdropping on the air interface, information forgery, Theft of Service and Denial of Service by employing the means illustrated in Table 13.1.

We apply the reasoning from the previous section and recognize that the threat analysis for UMTS identified eavesdropping in the fixed part of the network is unlikely. It is also instructive to compare UMTS security with the GSM security from which it was evolved. Compared to GSM, UMTS employs longer secret keys because computing power has increased. Furthermore, GSM does not protect against man-in-the-middle attacks, while UMTS does:

at the time when GSM was conceived, man-in-the-middle attacks where considered unlikely because the technology to build fake Access Point was too hard to obtain.

13.3.1 Secret Keys

UMTS security builds on a secret key with a length of 128 bit which we call the **Master Key**. It is shared between the operator and subscriber. The Master Key is stored on the USIM (or ISIM) and in the HLR (or HSS), respectively (cf. Chapter 9, Section 9.1.3). In a UMTS environment, key exchange is thus rather simple: the operator configures the Master Key before the USIM/ISIM is sold.

The secrecy of the Master Key is obviously crucial to the security of UMTS. The risk of disclosure must therefore be minimized. Only information derived from the Master Key is passed through the network. The Master Key itself never leaves USIM/ISIM and HLR/HSS, and even the subscriber does not know it. The Master Key is used only for authentication and for agreeing on the temporary session key. Afterwards, for integrity protection and encryption, the temporary keys are employed.

13.3.2 Authentication and Authorization in the PS Domain

The subscriber and the network mutually authenticate each other with a challenge-response mechanism. As part of the process they derive the temporary keys. The procedure is called **Authentication and Key Agreement** (AKA). AKA is performed as part of the GPRS Attach procedure, immediately after the UE sends the request to attach to the SGSN, cf. Chapter 11, Section 11.5 and Figure 11.7. After receiving the attach request, the SGSN requests authentication information from the HLR. This authentication information is a 5-tuple called the **Authentication Vector**. The HLR calculates the Authentication Vector based on the Master Key. The Authentication Vector contains the following items, as illustrated in Figure 13.4:

- RAND, a random number used as an authentication challenge.
- AUTN, an **Authentication Token**. The token has a substructure containing a sequence number SEQ for the time-variant parameterization of the Authentication Token (cf. Section 13.2.2), a field called AMF which can be defined as being operator dependent, e.g. to indicate which set of key generation functions to use, and finally a Message Authentication Code XMAC over the sequence number, AMF and RAND (cf. Section 13.1.3).
- XRES, the expected response to the challenge.
- CK, the encryption key to be used in this session with the UE.
- IK, the integrity key to be used for integrity protection in this session.

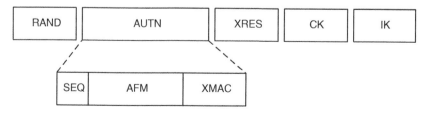

Figure 13.4   The structure of the Authentication Vector

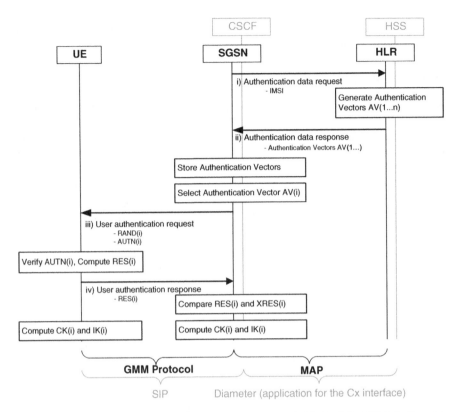

Figure 13.5   UMTS Authentication and Key Agreement (AKA) in black. IMS AKA is indicated in grey

We note that the SGSN is indeed simply relaying the authentication information produced by the HLR. It is not involved in any calculations and it does not need to have knowledge of the Master Key. One reason for this design is that the SGSN is located in the VPLMN in roaming scenarios (see Figure 11.10), and therefore should be involved as little as possible in sensitive processes.

Figure 13.5 illustrates the message flow for authentication in the PS Domain.

(i) The SGSN requests authentication information for a particular IMSI from the HLR using the MAP protocol.

(ii) The HLR generates one or several Authentication Vectors AV(1),..., AV(n) and sends them to the SGSN. The reason for sending more than one vector is that it is most likely that the same SGSN will need to authenticate a particular subscriber more than once.

(iii) The SGSN picks one vector and sends the Authentication Token AUTN and the random number RAND as a challenge to the UE. The communication between SGSN and UE is performed using the GMM protocol, as usual.

(iv) By inserting XMAC as part of AUTN, the HLR integrity protected and signed the AUTN. The UE verifies the authenticity of the token by recalculating XMAC with its Master Key. If successful, the UE is convinced that the HLR has actually computed the token. The UE then verifies that the SQN is in the right range and is thereby assured that the HLR has recently generated the token and the token is not reused. The UE then calculates a response RES to the challenge and sends it to the SGSN. The SGSN compares RES and XRES. If they are equal, the UE is authenticated.

Finally, the UE performs another calculation in order to derive the temporary keys CK and IK from RAND and the Master Key. They are, if all goes well, the same as those sent by the HLR to the SGSN as part of the Authentication Vector. As a result the UE and SGSN share keys for integrity protection and encryption.

The cryptographic functions involved in performing mutual authentication and for calculating the temporary keys are not standardized. The UMTS standard only suggests default algorithms as guidance. It suffices for HLR and USIM to know the algorithms. Since this is a PLMN-internal issue there is no need for standardization.

After completion of AKA, the SGSN proceeds to authorize the subscriber. To this end it pulls subscriber data from the HLR, as shown in Figure 11.7.

13.3.3 Authentication and Authorization in the IMS

As we have seen in Chapter 10, the IMS can be accessed not only from the PS Domain, but also from other access networks. Furthermore, an IMS subscription is conceptually independent from an UMTS subscription. The subscriber has different Master Keys for each subscription and different identifiers—the Private User Identity for the IMS and the IMSI for UMTS, cf. Chapter 9. Therefore, the IMS maintains its own security, and a subscriber accessing the IMS is authenticated, a previous authentication in the PS Domain notwithstanding.

The standardizers, however, avoided re-inventing the wheel. They therefore again prescribed the AKA mechanism. In this case, a CSCF pulls the Authentication Vector from the HSS using the Diameter protocol, discussed in more detail below in Section 13.5.1.4.1. The CSCF then performs the challenge-response exchange with the UE based on the SIP protocol. This is illustrated in Figure 13.5. Subsequently, the CSCF pulls the user's Service Profile from the HSS for authorization. In Chapter 15 on Session Control we will see how this procedure is embedded in other IMS procedures.

13.3.4 Integrity Protection

The integrity protection of messages adds overhead. Therefore, in UMTS it is only applied on the air interface and when messages cross operator boundaries. When they travel in the fixed part of a single PLMN they are considered secure. Furthermore, it is only signalling messages that are integrity protected.

Unlike the cryptographic functions for AKA, those for integrity protection and encryption are standardized. This is because a roaming UE must collaborate with network elements in the Visited PLMN.

13.3.4.1 Integrity Protection on the Air Interface

For protecting messages on the air interface in UMTS, the SGSN passes the integrity key IK to the RNC and integrity protection on the RRC layer is performed, i.e. below the IP layer. In fact, it is not only the air interface which is protected; the protection is applied up to the RNC, because RNC and Node B might be connected by a vulnerable microwave link.

The IMS, which may be accessed from different access networks, does not believe that sufficient protection is provided on the air interface. Therefore, integrity protection is also applied on the IP layer between UE and P-CSCF, the first contact point of the UE in the IMS. This integrity protection utilizes IPsec, a mechanism standardized by the IETF, which we will introduce in Section 13.5.

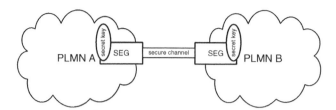

Figure 13.6   Two PLMNs have established a secure channel between Security Gateways SEG

13.3.4.2 Integrity Protection in Inter-PLMN Scenarios

Messages must cross the boundaries between different PLMNs whenever a subscriber is roaming to a VPLMN, because UTRAN, SGSN and, possibly, P-CSCF are located in the VPLMN, whereas HLR, HSS, S-CSCF as well as, possibly, GGSN, are located in the HPLMN, cf. Figure 11.10. Signalling messages crossing operator boundaries travel across the GRX (cf. Chapter 11, Section 11.8) and must be integrity protected. This protection is, of course, independent of the protection applied on the air interface.

Integrity protection for messages travelling between PLMNs is best applied on the IP layer. What key should be used to this end? The alert reader will notice that the secret key shared between UE and HLR/HSS does not help here, that PLMNs also need to share a secret key. This is achieved by introducing **Security Gateways** (SEGs) located at the PLMN boundaries, see Figure 13.6 and [3GPP 33.210]. The SEGs have a relationship of trust. This means that if two operators A and B have a Roaming Agreement, they have also established a secret Master Key. All SEGs in PLMN A and PLMN B have a copy of the key. They run a protocol to establish temporary keys for integrity protection and encryption. An example of such a protocol is the **Internet Key Exchange** (IKE) standardized by the IETF (cf. Section 13.5) [RFC 2409, RFC 4306].

13.3.5 Encryption

Just as with integrity protection, encryption is also applied on the air interface (up to RNC and P-CSCF, respectively) and when travelling across the GRX. However, user-plane traffic is also encrypted, not just signalling messages. Between UE and RNC, encryption is performed on the RLC or MAC layer. The IMS adds its own encryption, based on IPsec, between UE and P-CSCF. Inter-PLMN encryption also uses IPsec.

13.4 Security in a WLAN

The original security designed into 802.11 was **Wired Equivalent Privacy** (WEP). WEP is considered to be somewhat flawed so we will discuss its successor, defined in 802.11i [IEEE 802.11i]. This standard offers unilateral or mutual authentication and authorization of Mobile Station and WLAN, as well as encryption and integrity protection on the radio link layer.

13.4.1 Secret Keys

Unlike UMTS PLMNs, WLANs do not necessarily have a long-term relationship with their users, and there is not just one way of arriving at a shared secret key. Today, a typical procedure is to communicate the secret key to the user orally, or in printed form via a voucher. The user is requested to type in the key during the network attachment procedure. The 802.11i standard, however, also allows for the application of techniques for automatic key establishment with the involvement of a trusted third party, e.g. based on private/public key pairs.

As usual, the application of the original secret key is somewhat restricted. Instead, temporary keys are derived for encryption and integrity protection.

13.4.2 Authentication and Authorization

For authentication, 802.11i builds on a standard valid for IEEE 802 networks in general, 802.1x [IEEE 802.1x]. Figure 13.7 illustrates the underlying security architecture.

A Mobile Station, here called **Supplicant**, has communicated with an Access Point. It now requests authentication from this Access Point, which for this exchange assumes the role of the **Authenticator**. This component of the architecture is called **front end**. The Authenticator in turn communicates with an **Authentication Server** where authentication and authorization information is stored. The Authentication Server component is called the **back end**. We may recognize the similarity of this architecture to UMTS, where the equivalent to the Authenticator —SGSN or P-CSCF —are located in the Core Network rather than in the RAN, and where the Authentication Server corresponds to the HLR/HSS.

In small networks, the Authenticator has available locally all information necessary for authentication and authorization, including the secret key, and thus the Authenticator and Authentication Server coincide. In larger networks, the Authentication Server is a separate entity where sensitive information is stored centrally.

The cornerstone for security in a WLAN can be a password, a certified public key, a secret key stored on a (U)SIM, etc. —the WLAN standard is open in this respect. Based on this cornerstone, Supplicant and network mutually authenticate and derive temporary session keys. This flexibility in the Authentication Method is realized by employing the **Extensible Authentication Protocol** (EAP) [RFC 3748], which also allows for negotiation as to which so-called **EAP Authentication Method** should be used. An example of an EAP Authentication Method is EAP-TLS, a method based on a public key. EAP-AKA also exists, which allows for authentication based on secret keys and the AKA method used in UMTS. We will see how EAP-AKA is used when studying how a UE can access UMTS via a WLAN in Chapter 18.

EAP cannot run directly on top of the link or network layer. 802.1x thus decrees that on the front end, EAP be adapted to the link layer by encapsulating it into "EAP over LANs" (EAPOL).

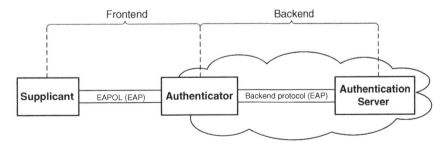

Figure 13.7   WLAN Security Architecture

Regarding the back end protocol for transporting EAP to the Authentication Server, 802.1x does not prescribe anything. A typical back end protocol would be the **Remote Authentication Dial-In User Service** (RADIUS) [RFC 2865] protocol developed by the IETF.

13.4.3 Integrity Protection and Encryption

As part of the EAPOL exchange, the temporary keys for encryption and integrity protection are derived. The encryption and integrity protection protocols work on the link layer. They are 802.11 specific, and build on standard mechanisms. Normally, all of the packets are encrypted and integrity protected.

13.5 Security Computer Networks

As usual, the Internet community standardizes building blocks for security in the form of more or less independent protocols and a small number of architectural elements, instead of a ready-made solution.

13.5.1 Authentication and Authorization

13.5.1.1 General Authentication Scenario

The general Internet authentication scenario (see Figure 13.8a) consists of one network Node A authenticating another network Node B; for example the **Network Access Server** (NAS) of an Internet Service Provider authenticates a residential host connecting by fixed line access, or a

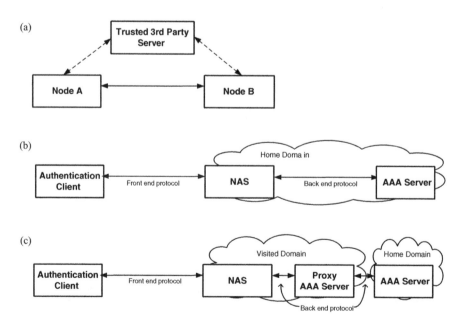

Figure 13.8   Authentication scenarios in the Internet. (a) General scenario, (b) Network access scenario, (c) Network access scenario with roaming

web client authenticates a web server. In some scenarios, the two parties also mutually authenticate each other. As part of the exchange, they may derive temporary keys for the encryption and integrity protection of future messages. The general scenario does not have as a prerequisite a shared secret key between the two nodes, and can build on certification by a trusted third party. An example of a protocol between Node A and Node B is the **Internet Key Exchange** (IKE) [RFC 4306] which authenticates and prepares for encryption or integrity protection on the IP layer using the IPsec protocol suite [RFC 4301]. Another example is **Transport Layer Security** (TLS) [RFC 4346], which provides authentication and key agreement as well as encryption or integrity protection above the transport layer.

13.5.1.2 Network Access Authentication Scenario

As we have already seen for UMTS and WLAN, a very typical authentication scenario is that of a host desiring access to a network based on a pre-shared secret. We can generate this from the general authentication scenario by dropping the trusted third party, which is no longer needed. Additionally, we introduce an Authentication Server for central maintenance of subscriber data—see Figure 13.8b, which apart from terminology has a striking similarity with Figure 13.7. In the Internet world, the Authentication Server additionally maintains accounting information and therefore is called **AAA Server**, AAA standing for authentication, authorization and accounting. The equivalent to the Supplicant may be called the **Authentication Client**, and the equivalent to the Authenticator may be called NAS. In fact, the correct terminology depends on the protocols involved.

In the general authentication scenario, the authentication was performed between the Authentication Client and NAS. Now it is between the Authentication Client and Authentication Server. Since we would like to continue using the authentication protocols, e.g. IKE and TLS, as above, EAP is introduced: As we saw in the previous section, EAP allows one to encapsulate these protocols (henceforth called EAP Authentication Methods) and to tunnel them from Supplicant to Authentication Server. And lo and behold, from the IKE or TLS perspective, the situation is still the same as in the general authentication scenario.

The NAS, nevertheless, is not a silent box. It needs to be involved in the communication. For example, it needs the temporary keys in order to perform encryption later. It therefore uses a front end protocol to communicate with the Authentication Client, and a back end protocol to communicate with the Authentication Server. Front end and back end protocol messages, furthermore, encapsulate the EAP messages. These multiple encapsulations may be confusing and are therefore illustrated in Figure 13.9.

13.5.1.3 Network Access Authentication Scenario with Roaming

We will now go one step further and consider what happens to the network access authentication scenario when the user is roaming. In this case, the NAS is located in the visited domain and the AAA Server is located in the home domain [RFC 2607]. We realize that the NAS and AAA Server now require a relationship of trust: the NAS and AAA Server must have authenticated each other and presumably would also wish to encrypt and integrity protect the data which they send each other. Of course, they also need such a relationship of trust in the scenario without roaming, but when the NAS and AAA Server are located in the same administrative domain (in UMTS we would say PLMN) this may often be taken as a given.

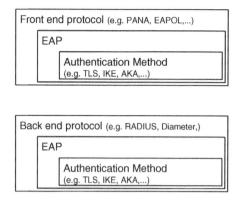

Figure 13.9 Front end and back end protocol encapsulating EAP encapsulating the Authentication Method

Since the number of independent networks with NASs and AAA Servers can be very large one would like to avoid acquainting all AAA Servers with all NASs of all the potential visited domains. Rather, only the AAA Servers of home domains and visited domains establish relationships of trust. The resulting authentication scenario is illustrated in Figure 13.8c: The NAS communicates with the AAA Server in its own domain, who proxies the request to the AAA Server in the home domain of the Authentication Client. The additional advantage of this set-up is that the NASs do not need to understand the concept of roaming. Rather, they always send authentication requests to the same AAA Server.

Note that this roaming architecture is different from UMTS (cf. Figure 11.10), where the SGSN (the equivalent of the NAS) in the VPLMN pulls authentication information directly from the HLR (the equivalent of the AAA Server) in the HPLMN. Since the number of PLMNs is traditionally not too large it is still feasible to establish the full set of SGSN-HLR relations. While this solution saves the additional proxy on the data path it is also less modular because an understanding of the roaming concept is also required from the SGSN (i.e. the NAS equivalent). 3GPP thus chose, as usual, efficiency over modularity.

13.5.1.4 Front end Protocols and back end Protocols

The classic front end protocol is the **Point-to-Point Protocol** (PPP) which provides authentication as a by-product of connection establishment, e.g. between a dial-up host and an Internet service provider. Before EAP was introduced, PPP in fact encoded the Authentication Methods directly and had to be updated each time a new method was invented. EAPOL as used in WLANs is, of course, also front end protocol, albeit a link-layer specific one. The IETF is currently working on a more generic front end protocol that runs over IP, the **Protocol for Carrying Authentication for Network Access** (PANA) [RFC 5193].

The classic back end protocol is RADIUS [RFC 2865]. It performs authentication, authorization and—additionally—accounting (cf. Chapter 16), and is therefore an AAA protocol. RADIUS is quite widespread and is used by Internet Service Providers for AAA-ing users connecting via dial-up modem or broadband access. It is also a common back end protocol in WLANs.

13.5.1.4.1 *Diameter*

The original use case of the RADIUS protocol was far more restricted than what is needed today. For example, roaming and mobility were not foreseen, and security requirements were not yet as high. RADIUS has therefore been updated and extended continuously, taking into account these developments. In parallel, the IETF developed **Diameter** [RFC 3588] as a more general and extensible protocol to replace RADIUS.

Diameter consists of a base protocol and a set of so-called **Diameter Applications**. The Diameter base protocol must be implemented by all Diameter nodes. It provides basic functionality such as reporting of accounting events from a **Diameter Client**, e.g. the NAS, to the **Diameter Server**, e.g. the AAA Server. A protocol exchange between two Diameter nodes almost always consists of a request and an answer. For example, a NAS would send the message Accounting Request (ACR) to the AAA Server in order to report the time when the user accessed the network. The AAA Server sends back an Accounting Answer (ACA) for acknowledgement. When the user disconnects from the network, the NAS sends another ACR message, which is likewise answered by the message ACA from the AAA Server. This exchange is illustrated in Figure 13.10.

Diameter Applications define new, additional request-answer message pairs. They are extensions to the base protocol that can be developed and implemented as needs be. An example of a Diameter Application is the "Network Access Server (NAS) application" [RFC 4005]. It specifies how NASs can perform authentication and authorization using, e.g. the Authentication and Authorization Request (AAR) and Authentication and Authorization Answer (AAA) messages. Another example is the "Diameter Application for the Cx interface" defined by 3GPP—not the IETF. It is used between CSCFs and HSS for transporting the Authentication Vector (cf. Section   as well as Figures 10.2 and 13.5). 3GPP thus adopted Diameter in the IMS. Outside 3GPP, however, Diameter seems to have experienced a similar fate to IPv6: the large deployment base of RADIUS, and IPv4, respectively, make a switch to

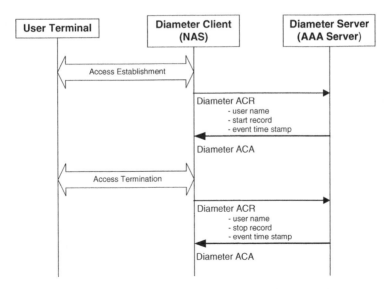

Figure 13.10   Exemplary Diameter exchange for accounting

the new protocols very difficult. Instead, effort is being invested into the further evolution of the existing, deployed protocols.

13.5.2 Integrity Protection and Encryption

As we have seen, encryption and integrity protection can be performed on different OSI layers, and the IETF therefore offers solutions for different layers, e.g. **IPsec** [RFC 4301] for protecting packets up to the network layer, or TLS [RFC 4346] with protection which includes the transport layer.

IPsec assumes that authentication has been performed and temporary keys have been established, e.g. by IKE, between the two nodes desiring to perform encryption and/or integrity protection between them. Such two nodes are said to have an **IPsec Security Association** (SA). IPsec works by encrypting the payload (if encryption is desired) and by appending an additional header to the packet that contains information such as a pointer to the SA on which the exchange is based, the MAC, etc. IPsec can work with any cryptographic algorithm, the nodes having the SA simply have to agree on which one to use.

TLS does not delegate the negotiation of algorithms and temporary keys to a separate protocol; this is instead part of TLS. TLS can also work with any cryptographic algorithm.

13.6 Discussion

Security in UMTS, as compared to security in a WLAN or in Computer Networks, exhibits the same differences as other control functions:

- Security is an inherent part of the UMTS cathedral; it is not conceivable to set up a UMTS network without the security mechanisms described above. In contrast, WLANs without access control, integrity protection and encryption are possible—in fact they are quite commonplace—but at the same time, a WLAN with fully fledged security is also possible. In the Internet, anything is possible, anyhow.
- Security in UMTS collaborates with the other UMTS functions such as mobility and roaming in order to achieve seamless service to the user. The UMTS standard describes in detail how to transfer Authentication Vectors between HPLMN and VPLMN for roaming, and between SGSNs for re-authentication of the UE after a change of Routing Area (see Chapter 12, Section 12.2). WLANs, of course, do not intend to sell seamless services, nor do WLANs have Roaming Agreements, and therefore they do not bother with such support. As concerns the Internet, the back end protocols, Diameter and RADIUS indeed support roaming. Support for collaboration of security and mobility is, however, only just being developed.
- The security design in UMTS starts from a fixed architecture and the main concern is the arrangement of functionality in network elements. The information exchange between network elements, e.g. for authentication, derivation of temporary keys, encryption, etc. is integrated into the existing protocols. There are no dedicated security protocols. For example, the UE authenticates with the SGSN using GMM, which is also used for all other communication between UE and SGSN. A WLAN allows for flexibility in its architecture, for example the Authentication Server may or may not exist, and furthermore employs dedicated security protocols. As far as possible the WLAN re-uses protocols already designed by the

IETF. In fact, the same can be said about many of the more recently devised parts of UMTS: in the IMS, Diameter is used as a back end protocol and IKE and IPsec are employed for securing the inter-PLMN traffic. For Computer Networks, as usual, protocols are designed that will fit almost any architecture.

Despite these differences, it is interesting to detect a number of commonalities between UMTS and Computer Networks. For example, the basic concept, i.e. authentication based on shared secrets and the derivation of temporary keys for encryption and integrity protection, is the same. Also, the underlying architecture is fairly similar so that a convergence will presumably be simpler here than in other areas.

13.7 Summary

This chapter began with an analysis of security threats, the attacks made in order to realize these threats and the counter measures. We identified authentication and authorization of the user, and integrity protection as well as encryption of data packets as the most important measures for securing a Communication Network. The cornerstone on which these measures build is a shared secret between user and network, or, alternatively, public/private key pairs certified by a trusted third party.

In UMTS, the shared secret is stored on USIM (ISIM) and in the HLR (HSS). Mutual authentication is performed between UE and SGSN as part of the GPRS Attach procedure to the PS Domain, or is performed between UE and C-PSCF as part of the equivalent attach procedure to the IMS. During the authentication, called the AKA method, temporary keys are derived for encryption and integrity protection. The UMTS requires the integrity protection of signalling messages and encryption of all messages between UE and RNC as well as between UE and P-CSCF.

In a WLAN, the shared secret usually is of a more short-term nature, and is passed, e.g. on a voucher. Authentication based on certified public/private keys is also allowed by the standard. The MS—called Supplicant—authenticates with the Access Point—called Authenticator—after the association procedure, using a front end protocol encapsulating EAP, encapsulating the Authentication Method. The Authentication Method itself is not standardized. The Authenticator may pull authentication and authorization information from an Authentication Server rather than storing it locally using a back end protocol. The WLAN security standard also describes integrity protection and encryption methods for the radio link layer. Generally, the WLAN standard relies on IETF-defined security protocols as much as possible, i.e. as soon as something can be done above the link layer.

For security in Computer Networks, a number of protocols and architectural elements have been defined. The basic authentication scenario is similar to the WLAN scenario. RADIUS and Diameter are standardized as back end protocols. The IETF Protocols for encryption and integrity protection are TLS and IPsec, the latter relying on IKE for authentication and establishment of temporary keys.

14

Quality of Service

An important requirement for UMTS is the support of multimedia applications, especially real-time applications. Such applications often consume high bandwidth and rely on being delivered in a timely, uniform fashion. For example, it is preferred that packets belonging to an IP telephony stream are rendered in the order in which they were generated. A single delayed packet is not particularly useful and must be discarded. The ability of a network to guarantee bandwidth, upper bounds of delay, etc. is called QoS support. We therefore conclude that UMTS must support QoS.

Traditional Telecommunication Networks have always supported QoS: it is inherent to circuit-switched technology since each call is given an end-to-end connection with a guaranteed bandwidth all to itself.

The traditional Internet, by contrast, offers a **best-effort** service. All packets are treated equally, independent of the necessities of the application to which they belong. When the network is serving only a few packets, every packet will be delivered quickly. When the network is crowded, long queues result and all of the packets are delivered late. For traditional Internet applications such as email or web-browsing this unpredictability can be a nuisance. For multimedia applications, it is a problem. The IETF has therefore since the 1990s developed protocols and mechanisms for the support of QoS. However, QoS support has not seen large-scale deployment in the Internet. UMTS, however, resorts ultimately to many IETF mechanisms, as we will see in this chapter.

We begin, as usual, with a description of the problems. We continue with a discussion of QoS support as standardized by the IETF, proceed with QoS in the packet-switched part of UMTS and close with QoS in WLANs. The reader will notice that we handle the topics in this chapter in a slightly different order than in the previous "control" chapters, treating Computer Network mechanisms first because UMTS QoS support, except in the CS Domain, relies on them.

Further reading on QoS in Computer Networks is available in [Tanenbaum 2002] and several RFCs that will be referred to below. QoS in UMTS is treated in [3GPP 23.107] and [3GPP 23.207].

Terminology discussed in Chapter 14:	
Admission control	
Allocation priority	
Bandwidth Broker	BB
Best effort	
Delay	
Differentiated Service	DiffServ
DiffServ Code Point	DSCP
Double reservation problem	
Flow	
Flow identifier	
Integrated Service	IntServ
Jitter	
Label Switched Path	LSP
Multiprotocol Label Switching	MPLS
Next Steps in Signalling	NSIS
Off-path signalling	
On-path signalling	
Overprovisioning	
Per-Hop Behaviour	PHB
Queue starving	
QoS NSLP	
QoS provisioning	
QoS signalling	
Resource Reservation Protocol	RSVP
RSVP for Traffic Engineering	RSVP-TE
Scalability	
Service Level Agreement	SLA
Session Initiation Protocol	SIP
Strict priority queueing	
Traffic Class	

14.1 Description of the Problems

If we want to configure a network so as to deliver QoS, we ought to have a precise idea of what QoS is. How does one parameterize or measure QoS, and how much of it is necessary for a given application? The answer to these questions is much debated. The "Communication Networks community" has not agreed on a single answer and, in the end, there is more than one answer. We look at QoS parameterization in the first subsection. We then look at how to make the network deliver the QoS which we are now in a position to describe.

14.1.1 QoS and Scalability

When solving a QoS problem, the **scalability** of the solutions is a major concern. Generally, a solution must be able to handle a large increase in input without "undue" additional resource usage. For example, a scalable QoS signalling solution should react to duplication in reservation requests (the input) with approximately a duplication of the number of signalling messages (the resource usage). A solution where the number of signalling messages is proportional to, e.g. the square of the number of reservations would be regarded as non-scalable. In the following sections we will find numerous examples of solutions that raise scalability concerns.

14.1.2 QoS Parameterization

Ultimately, QoS is something experienced by the user. Many QoS parameterizations therefore quantify QoS on this basis. On the other hand, what is technically more tangible is a parameterization on the network level, and this is what we resort to in this book.

QoS is normally delivered to something known as a **flow**. A flow is a stream of packets having some header characteristics in common, e.g. the same source and destination IP address and the same source and destination transport level port. In other words, packets belonging to the same application often belong to the same flow, e.g. the packets of a Voice over IP application. All packets belonging to a flow will receive the same QoS. A single application can, however, also generate several flows, e.g. a video streaming application could generate one flow for the actual video stream and one for the accompanying audio stream. Each may need a different QoS.

Almost all QoS parameterizations contain the parameter bandwidth. Given enough bandwidth, any application will play with good quality. Looking at it closely, however, the bandwidth parameter reveals greater complexity. Because data travels in packets, bandwidth is used in chunks rather than continuously. Bandwidth is therefore always an averaged quantity. Many parameterizations use the overall **average bandwidth** to describe the desired QoS for a flow. For applications with oscillating bandwidth demands the maximum or **peak bandwidth** is also of interest. Of course, average bandwidth and peak bandwidth together do not provide a complete description of the bandwidth needs. For example, it is not yet specified how frequently the peak bandwidth is needed. Figure 14.1 illustrates the "momentaneous" bandwidth of two rather different flows (obviously, even the "momentaneous" bandwidth is calculated by averaging over some small time interval), although both flows have the same average and peak bandwidth. An alternative to the description by bandwidth parameters is a description with the **token bucket**, which allows one to describe also the burstiness of a flow [Tanenbaum 2002].

For real-time applications, furthermore, the **delay** a packet accumulates as it travels through the network is an important parameter. For example, in a phone conversation a round trip delay greater than 400 ms becomes noticeable. Another important quantity is the variance of the delay, which is also called **jitter**. The constant delay of all packets might just be tolerable for a phone conversation. A great variance of delay, however, would have to be compensated by a buffer in the receiving phone, effectively adding more delay.

So far we have been looking at QoS as something provided to the individual packets of a flow while en route. Some QoS parameters, however, relate to the decision process of whether to admit a new flow into the network. For example, the flow for an emergency call should always be admitted, even if other flows must be pre-empted. The corresponding parameter is called **allocation priority**.

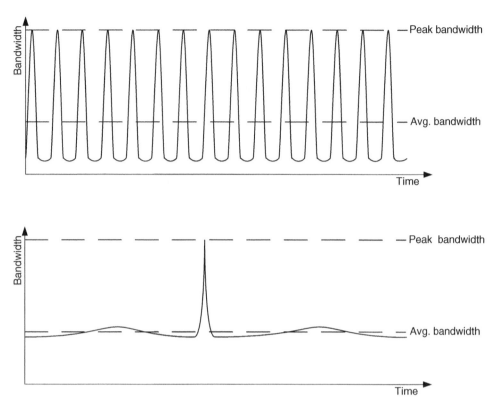

Figure 14.1   The bandwidth of two flows with identical average and peak bandwidths

14.1.3 QoS Signalling and QoS Provisioning

Let us assume that we know how much QoS is desired for a particular flow, and express it in terms of peak bandwidth, average bandwidth, maximum delay, etc. We now must make the network elements treat packets belonging to this flow in a particular fashion. The most complete procedure would appear as follows.

First we must inform all relevant network elements about the new flow—we do this by **QoS signalling**. A QoS signalling protocol tells the network elements how to recognize packets belonging to a flow by means of a **flow identifier**, e.g. their source and destination IP address and source and destination port. Furthermore, it signals what QoS is needed for this flow.

When a network element receives the signalling request, it must decide whether to admit the flow. There are two aspects to this admission: On the one hand, a flow can only be admitted when sufficient resources are available. If the network is already rather crowded, the flow is denied access. On the other hand, the user originating the flow must have administrative permission to make a reservation. If the user is not authorized, the flow is also denied access.

When the flow is admitted to access, packet treatment in network elements must be configured properly—this is called **QoS provisioning**. Many mechanisms are available for QoS provisioning. Ultimately, however, they come down to the bookkeeping of the reserved resources and to configuring router queues. In an extreme case, one could configure one queue

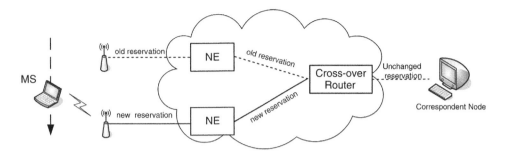

Figure 14.2   The change of QoS reservations for a MS after a handover

for each flow and serve each queue so that the flow's potential peak bandwidth can always be accommodated. Of course, this is both complex and wasteful. Therefore, in practice, flows with similar QoS needs are normally put into the same queue and the queue is served so that the resulting bandwidth is larger than the average bandwidth, but smaller than the peak bandwidth. The downside of this aggregation is of course that flows in the same queue compete for the available bandwidth.

14.1.4 QoS and Seamless Mobility

Because for UMTS we are interested in seamless handovers (cf. Chapter 12), we must look into the effect mobility has on QoS; see also [Chalmers 1999] for a view on this issue. Figure 14.2 depicts a MS having just experienced a handover.

Previously, the MS had a reservation along the old route. Owing to the handover, the reservation was updated and is now along the new route. The old reservation must either be torn down or times out.

Note that the old and the new path are not entirely different. They meet at the crossover router. It is therefore a good idea to perform the new reservation only up to the crossover router and to re-use the existing reservation between crossover router and Correspondent Node. Otherwise, if the old reservation is maintained, a new reservation end-to-end may be blocked due to lack of resources which are tied to the old reservation. Or, alternatively, if the old reservation is removed before the new reservation is issued, the resources occupied previously may meanwhile be assigned to somebody else. This is called the **double reservation problem**. The QoS signalling prototcol should thus be able to deal with partial signalling up to the crossover router, and moreover identify the old and new reservations as belonging together.

Generally, the collaboration between QoS and mobility in order to achieve seamless mobility is a signalling problem rather than a provisioning problem: the mobility signalling must trigger the QoS signalling, the QoS signalling must join the new and old reservations at the crossover router and, finally, in order to achieve make-before-break (cf. Chapter 12), the switch to the new Access Point should only be performed when the new reservation is in place.

14.2 QoS in Computer Networks

In today's Computer Networks, there is normally no explicit support given to QoS. The reasons are manifold. One reason is that the traditional popular Internet applications, e.g. web-surfing and emailing, work well with just a best-effort service. Additionally, the support of QoS signalling and resource provisioning only makes sense when QoS is available end-to-end. The explicit support of QoS end-to-end would require an update of many Internet routers, which is not feasible. On the other hand, the Internet appears to have sufficient resources for even the most extravagant application, and the bottleneck is usually the Access Networks or even just a radio interface. Therefore, explicit support of QoS is often only necessary in the Access Network. Technically speaking, the backbone of the Internet in this case employs the simplest QoS provisioning method, **overprovisioning**.

Another reason for the absence of QoS deployment in Computer Networks is that it is possible to work around the QoS problem more or less elegantly by employing mechanisms only in the end devices. For example, many applications switch to another codec with a lower bandwidth when packets are lost or delayed. Interestingly, the IP telephony application Skype pursues the opposite strategy: it responds to network congestion with an increase of its sending rate, which effectively pushes away other flows [Hoßfeld 2006]. Obviously, this strategy is only successful if not applied by all flows. For the time being, however, it works for Skype, and we thus have an example of a QoS-needy application that runs successfully over the Internet.

QoS in Computer Networks is therefore still an open issue. The IETF has been developing mechanisms for QoS provisioning and QoS signalling which we will study in this section. Additional information, e.g. on router-level enforcement of QoS is available from [Guerin 1999]. In conclusion, we will look at how the different provisioning and signalling methods collaborate in order to achieve end-to-end QoS.

14.2.1 QoS Provisioning

14.2.1.1 Overprovisioning

The simplest mechanism for QoS provisioning is **overprovisioning**. It consists of dimensioning the network such that it can handle all possible traffic without congestion. In an over provisioned network, QoS signalling is not necessary.

Let us do a very rough calculation of "how much" overprovisioning is needed, using the example of two different network topologies. Figure 14.3a illustrates a star-topology network. Grey edge nodes connect the network to the outside world. Each of the edge nodes has a capacity of $C_{ingress} = N$ bits/s for inbound traffic. From the edge nodes, all network traffic travels to the white "hub" node in the centre of the network. The maximum traffic which this hub has theoretically to handle is the sum of the capacities of the ingoing interface of the edge routers—multiplied by, let us say, 1.5, to account for fluctuations, i.e. $C_{hub} = 9\,N$. Moreover, the capacity C_{egress} of the outgoing interface of each edge router must also be $C_{egress} = 9\,N$— just in case the entire network traffic conspires to leave the network via the same egress. Most of the time, of course, the capacity $C_{egress} = 9\,N$ is just wasted.

Figure 14.3b shows a less simple topology. The single hub has been replaced by three interior routers, and the grey edge nodes have been connected among themselves. From a load-balancing perspective, this topology is certainly preferable. How must the routers be dimensioned in this case? Each router must be able to handle the traffic generated by all of its neighbours when they operate at full capacity. For example, the egress interface of all edge routers should be able to handle $9\,N$, as before.

(a)

(b)

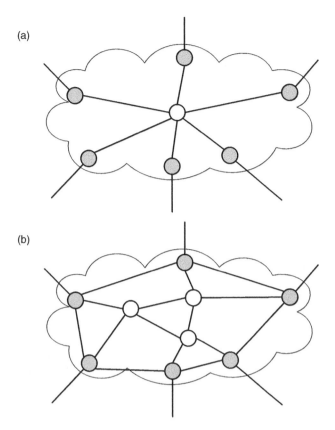

Figure 14.3 Dimensioning an over provisioned network in different network topologies. Filled circles are edge routers, open circles are interior routers

It is clear that many resources are wasted when a network is over provisioned to the extent that it can cope with any traffic situation however unlikely it may be, such as all traffic leaving the network via the same egress. Usually, network engineers expect statistical multiplexing and do not dimension that generously. Consequently, an over provisioned network does not give an absolute QoS guarantee. Rather, the QoS guarantee is probabilistic.

14.2.1.1.1 Admission Control
When network engineers would like to save on hardware resources in an over provisioned network, they must resort to greater sophistication by introducing **admission control**. Here, the ingress routers would allow or deny access to the network for new flows depending on the current load in the network.

The interesting question is how does the ingress router know the current load? A typical router having the "Internet spirit" only knows its own local traffic situation. It does not know the traffic situation at other routers. Picture the network in Figure 14.4a, where the dashed lines denote links close to congestion. A new flow desires access at ingress node **A** in order to travel via egress **B** to a destination node **C**. In the area of ingress **A** the traffic load is low, while at **B** the situation is becoming difficult. Ingress **A** thus says to the flow "Personally, I do not observe any

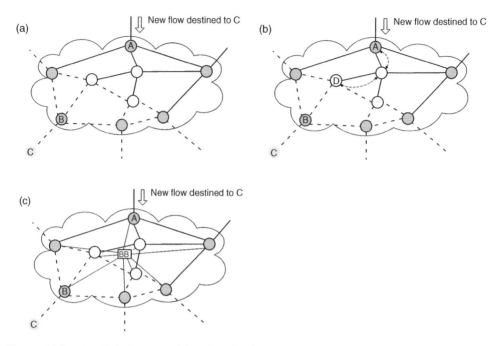

Figure 14.4 (a) Admission control based on local knowledge. Dashed lines denote links close to congestion. (b) Admission control based on signalling along the flow path (dashed arrows denote the signalling). (c) Admission control based on a Bandwidth Broker (BB) decision. The Bandwidth Broker communicates with network nodes along the thin lines

congestion, so please come in!" which will not result in the desired QoS to the new flow, and worsen overall the network situation.

A more informed admission control can be achieved in two ways, both adding overhead and complication. One possibility is for ingress **A** to collect the necessary information by QoS signalling towards the destination **C** along the flow path, as shown in Figure 14.4b. The signalling message carries information on the QoS requirements of the flow, e.g. what bandwidth it would need. Each router on the flow path receiving the QoS signalling message checks whether it has sufficient resources. Interior router **D** detects that it will not be able to accommodate the flow and returns the message to **A**. The new flow will be denied access.

The alternative, very popular with operators of Telecommunication Networks, is employing an omniscient **Bandwidth Broker**, see Figure 14.4c. The Bandwidth Broker has a complete picture of the network topology and network load. Ingress node **A** therefore always asks the Bandwidth Broker what to do. How does the Bandwidth Broker, however, acquire its knowledge? It needs to monitor the traffic situation in all routers closely. Moreover, it must know all of the routing tables in order to be able to determine whether a new flow will encounter any problems along its path. Those from the IETF community therefore typically regard a Bandwidth Broker as an unscalable entity.

14.2.1.2 Differentiated Services

Moving beyond overprovisioning, the IETF standardized **Differentiated Services**, also known as **DiffServ** [RFC 2475]. The idea is to assign packets to a limited number of traffic classes, called **Per-Hop Behaviour** (PHB). For example, the **Expedited Forwarding PHB** is used for packets derived from applications requiring low delay, low loss and low jitter—i.e. the typical multimedia applications. The **Default PHB** is suitable for packets with no particular QoS requirements. The IETF standardized a small number of PHBs, network operators can define more if desired.

The PHB which a packet should receive is coded in its IP header, in a tag in the **Type of Service** (TOS) field. This tag is called **DiffServ Code Point** (DSCP). The DSCP is a six bit field, allowing for 64 different PHBs. In practice, however, only a handful of different PHBs are ever used.

Routers are configured so that each packet receives the appropriate Per Hop Behaviour. For example, packets of the Expedited Forwarding PHB can be handled by **strict priority queuing**: they are placed in a queue dedicated to this PHB which is served with priority over all the other queues.

Of course, when too many packets are marked for Expedited Forwarding, a strict priority queue will also become congested, i.e. packets in the same PHB will compete for resources. Furthermore, a high load in the strict priority queue may lead to the other queues never being served, a phenomenon known as **queue starving**. Therefore, networks usually perform admission control at the ingress routers in order to limit the percentage of high-priority traffic. We already discussed above the different options for performing a more or a less informed admission control.

The QoS which a DiffServ network can guarantee is not absolute, because flows compete for resources in the same PHB. However, especially when combined with admission control, that guarantee is typically quite good.

14.2.1.3 Integrated Services

The most stringent method for providing QoS in a Computer Network is Integrated Services, also known as **IntServ** [RFC 1633]. With IntServ, the precise resources required for a flow are reserved in each node on the flow's path. This implies that the flow identifier and the per-flow resource requirements are signalled to each node on the flow's path, and that each node maintains per-flow resource state. Each packet arriving at an IntServ-enabled router runs through a header inspection process which allows for assigning the packet to one of the flows for which resources were reserved. Imagine the large backbone routers of the Internet maintaining state for millions of flows and classifying each packet. Clearly, this is not particularly scalable. In order to alleviate this problem, the IntServ concept was updated in order to describe how the per-flow reservations are **aggregated** into larger **reservations**, especially towards the core of the network, see Figure 14.5. However, IntServ is rarely considered to be an option today.

14.2.1.4 MPLS

The hearts of circuit-switched fans beat faster at the prospect of **Multiprotocol Label Switching** (MPLS) [RFC 3031]. Some say that MPLS is the most widely deployed QoS provisioning technique. Although, in fact, MPLS is not a QoS provisioning technique per se. MPLS was developed as a—at the time—fast alternative to IP routing: to this end, stable paths, the **Label Switched Paths** (LSPs), are established throughout the network. One can also think of LSPs as tunnels. When an IP-packet enters a network domain it is assigned to a particular

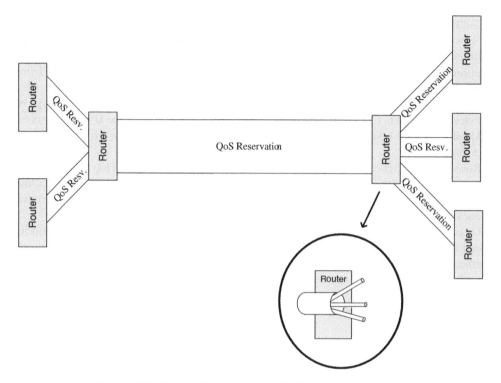

Figure 14.5   Reservation aggregation in the core of the network

LSP and pre-pended a label. The packet is then switched through the network on Layer 2.5, based on the label information. The IP address and routing tables are consulted again only at the egress of the LSP.

Today, the forwarding speed of IP routers is no longer an issue. MPLS, however, plays an important role because the LSPs are effectively circuits from an ingress node to an egress node, which can moreover be provisioned with a particular bandwidth. LSPs are more or less installed statically. In a basic MPLS configuration, each edge router maintains two LSPs to each other's edge router—one for each direction. Such a network is called **fully meshed**. Figure 14.6 illustrates an example of a mesh of LSPs for a network cloud.

Packets arriving at an ingress router are classified, assigned to an LSP and admitted if the LSP still has sufficient bandwidth. Note how the problem with admission control in other techniques—namely how does the ingress router know whether sufficient resources are available—is solved elegantly in MPLS: LSPs are basically ingress-to-egress tunnels for traffic aggregates and contain exactly those flows which the ingress node admitted previously. In other words, local knowledge of the ingress node is sufficient in order to guarantee QoS.

Compared to per-flow reservations in IntServ, MPLS has less of a scalability problem:. On the one hand, packet classification only happens on the ingress to an LSP rather than in each individual router. Furthermore, MPLS is foreseen to reserve resources for aggregate flows. The "reservations", i.e. the LSPs, are typically long-lived. They are dimensioned based on predicted bandwidth needs so that also the signalling overhead for maintaining the LSPs can be kept low.

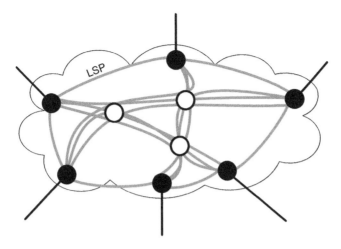

Figure 14.6   A full mesh of LSPs connecting edge routers (black circles). For the sake of clarity, only one LSP is shown per pair of edge routers, instead of two, one for each direction

However, in a network with n edge nodes, the "full mesh" of LSPs connecting each edge node with every other edge node in either direction requires **$2n$ (n-1) LSPs** which can result in a rather large number of LSPs.

14.2.2 QoS Signalling

QoS signalling can be performed by a variety of methods using a variety of protocols. We have already encountered one rather special example of QoS signalling, DSCPs. By tagging each IP packet with a DSCP, routers along the path know which QoS to give to that packet.

Generally, one can distinguish two different approaches: **on-path signalling** and **off-path signalling**. On-path signalling messages, illustrated in Figure 4.7a, travel along the path of the flow for which they are reserving resources and directly cause reservation state to be installed in the routers. They are addressed typically to the flow destination and thereby identify implicitly the correct path and the correct routers. Off-path signalling messages, illustrated in Figure 4.7b, travel from the network node initiating the signalling to dedicated network nodes that are not on the path of the flow. An example of such a node is a Bandwidth Broker (see Figure 14.4.c). The off-path node, in turn, configures the routers on the path of the flow. For off-path signalling, the address of the off-path node must be known and configured explicitly.

We now look at different QoS signalling protocols in more detail.

14.2.2.1 RSVP

The **Resource ReSerVation Protocol** (RSVP) [RFC 2205] is an on-path QoS signalling protocol. It is by and large the only QoS signalling protocol supported in commercial routers today. It was developed in the 1990s to signal for IntServ. In fact, it was not foreseen originally that it would signal for any other provisioning method. Later, it was extended to carry also DSCPs and thus support DiffServ [RFC 2996]. Another extension, **RSVP for Traffic Engineering** (RSVP-TE) was developed in order to signal for the installation of LSPs in

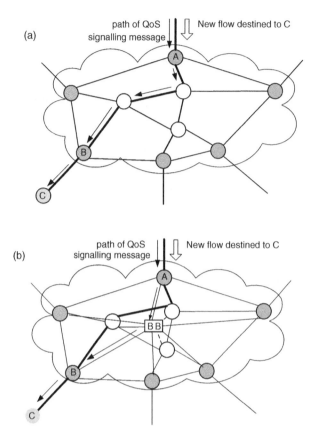

Figure 14.7 QoS signalling for a new flow entering the network domain at edge node A and leaving at edge node B. The bold line indicates the path the flow is expected to take. (a) on-path signalling and (b) off-path signalling to a Bandwidth Broker

MPLS [RFC 3209]. RSVP-TE has ultimately received wider deployment than the original RSVP.

RSVP is a very powerful QoS signalling protocol, however its strengths are in areas that have not attained the importance anticipated. For example, RSVP offers very versatile multicast support—at the cost of overhead even for unicast signalling. Multicast applications were expected to become commonplace at the time when RSVP was developed. This expectation was not realized. Instead, mobile applications became commonplace, and mobility, unfortunately, is something which RSVP does not support easily. For example, RSVP cannot deal with the double reservation problem (cf. Section 14.1.3): RSVP nodes identify signalling state by the IP address of the sender. Thus, if an MS's IP address changes because of a hand over, it has to completely renew an RSVP-signalled QoS reservation. Also, security requirements have increased since RSVP was developed and it turned out to be difficult to enhance RSVP accordingly.

A QoS reservation signalled by RSVP works as illustrated in Figure 14.8. The basic idea is that the actual reservation is initiated by the receiver of the flow rather than by the sender. This is

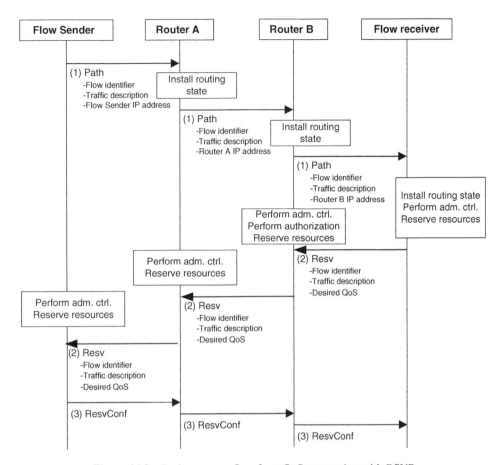

Figure 14.8   Basic message flow for a QoS reservation with RSVP

because multicast flows typically form a tree fanning out from a "root" at the sender. Reservations initiated by a flow receiver can simply be joined with the existing reservation at the node where it hits a branch of the "multicast reservation tree". They do not need to travel all the way to the sender, thereby saving on signalling traffic. Since, however, routing in a Computer Network is asymmetric, the path from sender to receiver must be found before the receiver can issue the actual reservation. The message flow therefore works as follows.

1. The flow sender sends a Path message to the flow receiver, containing its own IP address, the flow identifier and a description of the traffic generated by the flow, in terms of a token bucket and other parameters. RSVP runs directly on the IP layer. The Path message carries a **Router Alert Option** (RAO), which makes routers inspect it more deeply. Each RSVP-supporting router on the flow path thus inspects the packet and installs routing state which associates the flow identifier with the previous hop's IP address. This routing state later enables the router to route the actual reservation message correctly.
2. The flow receiver deduces the desired QoS from the traffic description. The flow receiver reserves the required resources, and sends a Resv message back upstream, addressed to the

previous RSVP hop. Each RSVP hop on the path now performs a local admission control, reserves the resources and sends on the message. If the RSVP hop is the ingress to a domain, an authorization of the flow sender is also performed. In the case of multicast flows, the Resv only proceeds up to the point where it hits an existing reservation for this flow.

3. The flow sender (in the case of multicast the branching point) acknowledges a successful reservation by sending a ResvConf message to the flow receiver.

RSVP state is soft-state, i.e. reservations have to be refreshed in regular intervals, typically in the order of minutes.

14.2.2.2 QoS NSLP

RSVP—except RSVP-TE—never received wide deployment. The reasons for this may be the inherent—but unused—multicast support which adds unnecessary overhead, the tight relation to IntServ, the absence of mobility support, scalability concerns, security problems, etc. There has been a multitude of amendments to RSVP and many were standardized. However, interest in a new QoS signalling protocol grew such that in 2001, the IETF chartered a new working group, **Next Steps in Signalling** (NSIS) to take care of this task. The scope of the NSIS was soon enlarged to that of a general protocol suite for controlling flows, e.g. it includes a protocol to open per-flow pinholes in NATs and firewalls.

The QoS signalling protocol of the NSIS family is called **QoS NSLP** [ID QoS NSLP]. Originally, it was an on-path signalling protocol; it was also possible to use it for off-path signalling [ID pds NSIS], however this idea did not catch on. QoS NSLP is more flexible than RSVP in that it collaborates with any QoS provisioning method and supports **sender-initiated** as well as **receiver-initiated reservations**. In sender-initiated reservations, the first message from sender to receiver is already a RESERVE message. The receiver replies with a RESPONSE. Sender-initiated reservations save half a round-trip as compared to receiver-initiated reservations, as depicted in Figure 14.8.

QoS NSLP is not expected to run into problems with respect to to mobility: It deals with the double reservation problem by identifying flows by random identifiers instead of by IP addresses. Partial reservations and double reservations are also possible, in principle. QoS NSLP security is designed to meet current expectations. The protocol also improves on a scalability problem with RSVP in that it usually only sends the first signalling message for a flow with the Router Alert Option. This message establishes associations between QoS NSLP-enabled routers, called **QoS NSLP Elements**, on the flow path and effectively installs a signalling overlay. Subsequent messages are addressed directly between QoS NSLP Elements.

At the time of writing, it is, however, unclear whether and how QoS NSLP will be adopted and deployed. It is up against an incumbent and deployed protocol, RSVP. Furthermore, in many scenarios, the flexibility which QoS NSLP provides is unnecessary, and the QoS signalling problem can be solved by more minimal means such as the SIP protocol introduced in the next subsection.

14.2.2.3 SIP

The **Session Initiation Protocol**, SIP, allows for the control of multimedia sessions. SIP messages are exchanged between the end points of the session, e.g. between two user terminals.

The messages are, however, intercepted by off-path network elements called SIP Proxies; the details will be discussed in Chapter 15.

As part of a session description, SIP can transport QoS information [RFC 3313], in order to allow the terminals to agree on the QoS they are going to reserve—SIP does not trigger reservations as RSVP or QoS NSLP do! The actual QoS reservation can subsequently be performed, e.g. by both terminals initiating unidirectional QoS signalling. However, the SIP Proxies can read the QoS information and forward it to a Bandwidth Broker. SIP becomes an off-path QoS signalling protocol and additional QoS signalling by the terminals becomes unnecessary.

Employing SIP for QoS signalling is especially popular in operator-controlled Telecommunication Networks which often feature a Bandwidth Broker. In these networks, the architecture and topology is usually sufficiently restricted to render the path-finding capabilities of on-path protocols such as RSVP or QoS NSLP unnecessary.

14.2.3 End-to-end QoS Signalling Scenarios

In this section we have encountered a considerable number of methods for QoS signalling and provisioning, which moreover can be combined. In Figure 14.9 we illustrate an end-to-end scenario employing several methods. We see a laptop (MS) and a Correspondent Node (CN) about to enter a multimedia session across several domains I to IV for which they need to reserve QoS. The MS initiates SIP signalling in order to coordinate the QoS to be reserved with CN. The SIP messages are intercepted by SIP Proxies (P). Once the two terminals have agreed on the QoS, the MS initiates QoS NSLP signalling in order to reserve the actual resources. The QoS NSLP message travels along the flow path through all domains I to IV.

- Domain (I) employs IntServ. QoS NSLP is interpreted by all routers on the flow path which reserve resources accordingly.
- Domain (II) is over provisioned. The first ingress router performs admission control on the basis of the QoS NSLP message and then passes it directly to the egress node.
- Domain (III) employs MPLS. LSPs are established by a separate signalling protocol, RSVP-TE. The QoS NSLP message is used for admission control and is then tunnelled directly to the egress node.
- Domain (IV) employs DiffServ and needs QoS information in order to perform admission control. It is, however, not interested in the QoS NSLP message and tunnels it to the egress.

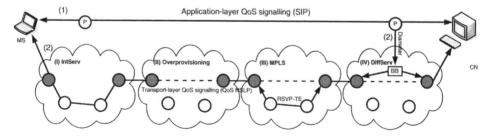

Figure 14.9   Combination of different QoS signalling and provisioning methods. Dashed lines indicate tunnelling of messages resp. direct forwarding edge-to-edge

Instead, it indirectly receives the QoS information from SIP signalling which was intercepted by the SIP Proxy (P) in domain IV. The SIP Proxy informs a Bandwidth Broker (BB), e.g. using the Diameter protocol. The Bandwidth Broker determines the path the flow will take and informs the ingress and egress router whether resources are available.

14.3 QoS in UMTS

The emphasis in UMTS is on *what* QoS shall be achieved. We therefore start this section with a description of the four different Traffic Classes defined by the UMTS standard. The QoS defined by the Traffic Classes must be delivered by each bearer to the bearer above (for an explanation of the bearer concept cf. Chapter 11, Section 11.2). Our focus is on how QoS is delivered by the UMTS Bearer extending from UE to GGSN. We deal with *how* the QoS defined by the Traffic Classes is achieved, with one section on QoS signalling and one section on QoS provisioning. Finally, we illustrate how it all collaborates in order to yield the End-to-End Bearer.

14.3.1 UMTS Traffic Classes

UMTS defines four **Traffic Classes** [3GPP 23.107]:

- **Conversational Class**, e.g. voice or video conferencing. This class has the highest real-time requirements, with a stringent limit on the values for delay and jitter. Services in this class need a guaranteed bit rate.
- **Streaming Class**, e.g. video streaming. Video streaming is different from video conferencing in that its real-time requirements are lower: Delay only plays a role when interacting with the server, e.g. for starting or stopping the service. However, services in this class need a guaranteed bit rate.
- **Interactive Class**, e.g. Web browsing or gaming. These are interactive services; however their real-time requirements are usually limited. They do not need a guaranteed bit rate.
- **Background Class**, e.g. emailing. This Traffic Class is usually not guaranteed any QoS.

The UMTS standard parameterizes QoS with a UMTS-specific set of more than a dozen parameters. Each Traffic Class must be given QoS in a standardized value range of these parameters. Table 14.1 provides an excerpt. We see that the Conversational Class requires a lower delay than the Streaming Class. Both the Conversational Class and Streaming Class are

Table 14.1 Range of QoS parameter values for UMTS Traffic Classes

	Conversational Class	Streaming Class	Interactive Class	Background Class
Maximum bit rate [kb/s]	up to 16 000	up to 16 000	up to 16 000	up to 16 000
Transfer Delay [ms]	100	300		
Guaranteed bit rate [kb/s]	up to 16 000	up to 16 000		
Traffic handling priority			1,2,3	

distinguished from the Interactive Class by offering a guaranteed bandwidth. Compared to traffic of the Background Class, however, the traffic of the Interactive Class can be given a higher priority.

14.3.2 QoS Signalling for the UMTS Bearer

The UMTS Bearer transports the data packets between the UE and its Correspondent Node. It extends between the UE and GGSN, cf. Figure 11.2. In a simple case, the QoS of the UMTS Bearer is determined by QoS signalling between the UE and PS Domain. This is the case discussed in the next subsection.

In a more advanced scenario, i.e. when an IMS service is used, the UE first of all performs a signalling exchange with the IMS using SIP, in order to determine which QoS should be used for the service—in Chapter 15 we will learn more about this. Only then does the UE signal to the PS Domain to set up a UMTS Bearer with the appropriate QoS in order to carry the service. Independently, the IMS and PS Domain coordinate via a policy interaction and make sure that the UE does what it is supposed to be doing. The details of this policy interaction will be covered in Chapter 17.

14.3.2.1 UMTS QoS Signalling to the PS Domain

QoS signalling with the PS Domain is performed between the UE and SGSN when the PDP Context is established, see Chapter 11, Section 11.6 and Figure 11.9. In the course of PDP Context establishment, the SGSN negotiates the QoS with both RNC and GGSN. This being UMTS, the signalling is not performed with a dedicated QoS signalling protocol. Instead, Traffic Class and QoS parameters are embedded in the GMM and GTP protocols along with other information on the PDP Context. RNC, SGSN and GGSN utilize the QoS information in order to perform admission control and authorization.

The advantage of embedding QoS information into PDP Context signalling is that, without any additional effort, seamless mobility is supported in the PS Domain. We have seen in Chapter 12 that the mobility of a UE can lead to a change of Node Bs, RNCs (SRNC relocation) or SGSN (Inter-SGSN Routing Area updates). Of these mobility events, change of Node Bs is not of interest to us since it is a radio interface topic only. We will look in more detail how seamless mobility is achieved in the case of SRNC or SGSN update. UMTS works with make-before-break: PDP Context information, including QoS information, is transferred between old SGSN (resp. RNC) and new SGSN (resp. RNC). Thus, resources along the new path can be reserved before the flow is rerouted.

We note that signalling QoS in the PDP Context only informs the SGSN, RNC and GGSN about the QoS that is to be reserved. These three nodes are, however, connected by an IP Network, see Figure 8.2b. The UMTS standard does not worry about whether and how QoS signalling is performed in this network as this is an intra-operator problem.

14.3.3 UMTS QoS Provisioning

Regarding provisioning, the UMTS standard is concerned mainly with the Radio Bearer, where QoS is provided by a dedicated channel whose bandwidth is regulated by the CDMA code (cf. Chapter 5, Section 5.2.4.3). In the case of the Background Class, a shared downlink channel can be utilized.

The UMTS standard determines that the GGSN, SGSN and RNC perform admission control based on information received via GTP signalling. The standard does not, however, describe how to provision the QoS in the IP-based part of the network, e.g. between GGSN and SGSN. This is an intra-domain problem, and its solution does not need to be standardized. DiffServ or MPLS are commonly used methods.

14.3.4 QoS of the End-to-End Bearer in UMTS

QoS of the End-to-End Bearer (cf. Chapter 11, Section 11.2) is comparatively simple to achieve if both UEs are in the same UMTS PLMN. In this case the End-to-End Bearer is the concatenation of two UMTS Bearers. However, what about scenarios involving more than one PLMN, e.g. a roaming UE, or a multimedia session between UEs homed in different PLMNs? In this case QoS must also be delivered between PLMNs, across the GRX (cf. Chapter 11, Section 11.8).

Inter-PLMN QoS is signalled and delivered on two levels. On a coarse-granular level, a particular QoS is reserved for the aggregate of all flows traversing network boundaries. On a fine-granular level, end-to-end QoS is given to each individual flow. We discuss these two levels in separate subsections.

14.3.4.1 Service Level Agreements

For QoS on the coarse-granular, per-aggregate level, PLMNs have static agreements, so-called **Service Level Agreements** (SLAs). An SLA describes the QoS per-Traffic Class provided to the aggregate of all flows between two networks. SLAs are negotiated and provisioned off-line between PLMNs and between each PLMN and the GRX. One reason for making SLAs static is that the amount of QoS-needy traffic between PLMNs is often constant, so that a dynamic adaptation is of moderate importance. Since there was no urgency, it has not yet been possible to agree on a protocol for signalling SLAs, so that today no standardized solution exists.

14.3.4.2 UMTS End-to-end QoS Scenarios

For QoS signalling on a fine-granular, per-flow level, 3GPP abandoned work on a standard but defined a number of—purely informational—end-to-end signalling scenarios in the Annex of [3GPP 23.207]. These scenarios do not presume the existence of a GRX, and also apply to the more general case of QoS between a UE and a Correspondent Node in an arbitrary IP-based network. The simplest scenario is illustrated in Figure 14.10.

The scenario in Figure 14.10 depicts a UE and a Correspondent Node (CN). The UE is attached to a PS Domain and the CN is attached to a remote Access Network. The PS Domain and remote Access Network are connected by an IP-based Core Network. By some offline procedure it is known that Core Network and remote Access Network use DiffServ as the QoS provisioning method—if the Core Network is a GRX, the existence of such an offline procedure is a realistic assumption.

The UE and CN agree on QoS via SIP signalling. Resources are reserved and provisioned as follows:

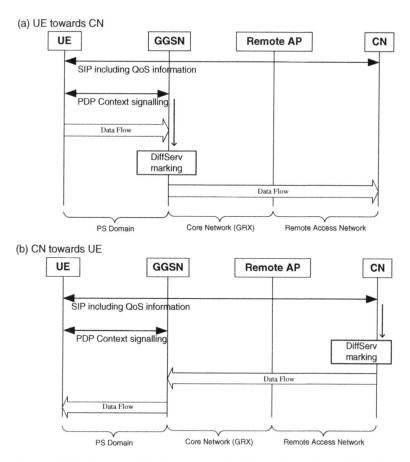

Figure 14.10   End-to-end QoS scenario in UMTS (a) UE to CN (b) CN to UE

- **PS Domain**: The UE reserves resources bidirectionally when it establishes the PDP Context. Admission control to the PS Domain is performed by the SGSN (uplink) and GGSN (downlink). Resources in the PS Domain are provisioned by a local method which is of no further interest. The IMS is not involved in this scenario so that coordination between QoS signalled via SIP and the QoS signalled in the PDP Context is not enforced.
- **Core Network and remote Access Network, in the direction GGSN to CN**: DiffServ marking is performed by the GGSN based on the information contained in the PDP Context. No resources are reserved explicitly, but admission control may be performed by the ingress node to the Core Network if it can obtain the necessary information from the GGSN.
- **Core Network and remote Access Network, in the direction CN to GGSN**: DiffServ marking is performed by the CN based on information obtained from SIP signalling. No resources are reserved explicitly, but admission control may be performed by the ingress node to the Access Network if it can obtain the necessary information from the CN.

14.4 Link-Layer QoS in a WLAN

The original 802.11 standard did not cover QoS. It was augmented to do so in 2005 with the 802.11e specification [IEEE 802.11e]. 802.11e, of course, only describes link-layer mechanisms for providing QoS on the radio interface. For end-to-end QoS, these mechanisms must be combined with the IP-based QoS signalling and provisioning mechanisms introduced in Section 14.2.

It is still too early to say whether the 802.11e amendment will become widely accepted. The upcoming udpate to the WLAN standard, 802.11n, which also promises a major increase in bandwidth (see Figure 2.4), integrates QoS support from the start.

The challenge in providing QoS for WLAN technology lies in the fact that the radio interface is a shared medium. Access to the radio resource is not coordinated centrally. Rather, in basic 802.11, each MS listens to the medium for a specified time $t = $ **DIFS** and only then is allowed to send frames. When a collision occurs, the MS waits for a random time t out of a back off window *CWmin* $< t <$ *CWmax* before attempting retransmission. With each attempt the back off window increases in size. This is called **exponential back off**.

One way in which 802.11e improves on this basic mechanism is by making the values of the parameters involved in the basic mechanism, e.g. *DIFS*, depend on the priority of a frame. Four priority classes are defined: voice, video, best effort and background –corresponding nicely to the four UMTS Traffic Classes. For sending high priority frames, a MS employs a smaller value of *DIFS* than that for sending low priority frames. Likewise, the back off window is decreased. These and a number of additional measures allow the MS to access the medium more quickly and more effectively so that a better QoS results for high priority frames. Clearly, however, the QoS guarantees are not absolute but depend upon the overall traffic volume.

14.5 Discussion

QoS support in UMTS to a large extent relies on tools developed by the IETF. 3GPP, however, stepped in where the telecommunications-style design could not be maintained with these tools and developed its own solutions, manifested in particular by the establishment of PDP Context.

QoS support in UMTS is, of course, designed with the complete system in mind. QoS control must collaborate seamlessly with other control functions such as mobility, security and session control. This is best achieved by integrating QoS parameters into the general control signalling for PDP Contexts. When the PDP Context is rerouted because of mobility, in one go the QoS is rerouted, also.

The complete-system approach of UMTS has another advantage. Whereas large-scale deployment of QoS in the Internet is somewhat difficult because of its patchwork nature, UMTS Networks, standardized so as to exhibit uniform features, can make sure that QoS is supported everywhere.

Another important Telecommunication Networks feature is operator control. However, the UE is indeed in control of the QoS which it requests from the network—when we discuss the future evolution of UMTS in Part II of this book we will see how this freedom is likely to be restricted. When a QoS reservation needs to be updated as a result of mobility events, the network is already in control, in current releases. In a Computer Network, the MS would typically be responsible for triggering this update.

Compared to UMTS, a WLAN network, as usual, pursues a different strategy. The original WLAN was specified without any QoS. It was of course always possible to integrate, in a toolbox fashion, QoS in a WLAN network on the IP layer, using the signalling and provisioning mechanisms defined by the IETF. This would, however, be only moderately successful without QoS support on the radio link. Such support was defined recently and awaits deployment.

14.6 Summary

QoS support plays an important role in UMTS because it is needed by multimedia applications, which are a main UMTS target. To a large extent, UMTS relies upon the QoS support developed by the IETF.

A universally agreed parameterization for QoS does not exist. An important parameter which is likely to appear in any QoS description is bandwidth, although different definitions are employed.

For parameterization, UMTS defines four Traffic Classes: conversational, streaming, interactive and background. Whereas the Conversational Class guarantees a specific bandwidth and a low delay, the Background Class does not guarantee QoS.

QoS support is based on two main ingredients: provisioning and signalling. QoS provisioning methods include overprovisioning, as well as the IETF-defined methods of DiffServ, IntServ and MPLS. UMTS can employ any of these methods, the details are not standardized. The PLMN operator only has to ensure that the QoS guarantees for each Traffic Class are met.

QoS signalling protocols include the IETF-defined RSVP, QoS NSLP and SIP. UMTS has integrated QoS signalling into the GMM and GTP protocols for PDP Context signalling in the PS Domain, and employs SIP for the IMS. QoS signalling across the IP Networks connecting RNC, SGSN and GGSN and inside the IMS is not standardized.

Aggregate inter-PLMN traffic as caused by roaming UEs, or sessions between UEs attached to different PLMNs, is guaranteed the necessary QoS by SLAs agreed off-line between operators. However, per-flow QoS needs to be signalled. 3GPP offers a number of informational scenarios as to how this can be achieved. For example, UEs exchange their QoS requirements with SIP signalling, and then signal for resources in their respective Access Networks with PDP Context signalling or by priority-marking packets with DSCPs.

15

Session Control

At this point, the reader knows how a UE establishes a bearer: secured IP connectivity with a certain QoS through the PS Domain which is maintained while the UE moves. The reader does not yet know, however, how multimedia services such as videoconferencing are provided.

The platform for multimedia services is the IMS (cf. Chapter 10). A multimedia session, providing multimedia services, is different from, e.g. a web-surfing session; we have already seen in the previous chapter one of the differentiating factors, namely the need for QoS. However, there is more. A web-surfing session involves shuffling data packets between a user terminal and a well-known server. A multimedia session, by contrast, involves sending several media (e.g. audio, video, whiteboard) between possibly a large number of participants. A multimedia session relies on a bouquet of network functions, such as locating the other participants, agreeing on QoS, authorizing service usage, coordinating the media (e.g. audio and video), etc. All of these functions are conveniently provided by *session control*, which is the topic of this chapter.

As we discussed in Chapter 10, the IMS differs from, e.g. the PS Domain in that it employs, Computer-Network-style, one protocol per functionality. Session control functionality is performed by the *Session Initiation Protocol* (SIP), which is a protocol standardized by the IETF—it is thus not 3GPP specific, although 3GPP solicited the IETF to define a number of SIP extensions. SIP seems the natural choice from today's perspective—the general trend is towards enabling networks to interwork on the basis of IETF protocols; we will hear more about this in Part II. However, at the time, it was a difficult decision to make, between SIP and the ITU Protocol H.323. An important argument in favour of SIP was that the protocol makes it easy to create new services.

We therefore start as usual with a general discussion of the problems, then introduce SIP and finally show how SIP is used in the IMS. Further reading is provided by the SIP specification [RFC 3261], the 3GPP standard on the IMS [3GPP 23.228] as well as [Camarillo 2005].

UMTS Networks and Beyond   Cornelia Kappler
© 2009 John Wiley & Sons, Ltd

Terminology discussed in Chapter 15:
Address-of-Record
Contact Address
Device mobility
IP Connectivity Access Network   IP-CAN
Location Server
Registrar
Service Profile
Session control
Session Description Protocol   SDP
Session Initiation Protocol   SIP
SIP Proxy
SIP Transaction
SIP trapezoid
User Agent
User mobility

15.1 Description of the Problems

We should classify multimedia services in order to better understand the problems. A multimedia session has two or more participants. The participants of a session can be users or Application Servers. An example of a session between two users is an IP telephony call. An example of a session between a user and an Application Server is a video streaming application. An example of a session with more than two participants is a telephone conference. In all of these examples, all of the participants are end points of the multimedia session.

Moreover, a multimedia session may be mediated by an Application Server, e.g. a video conference with a conference server providing floor control. This means that the chairperson of the conference can—by instructing the conference server accordingly—zoom the camera, control which video picture is transmitted where, switch microphones on and off, etc. Note that the Application Server in this scenario is not an end-point. Rather, it sits at the hub of a star-topology. An example of such a server is the Media Resource Function in the IMS (see Chapter 10, Section 10.2).

A multimedia session may also involve several Application Servers simultaneously or subsequently. For example, an Application Server could implement a "meeting service", which informs a user as to which of his friends are currently available—building on the Presence Service provided by the IMS (cf. Chapter 10, Section 10.1.2). The meeting service could also provide buttons that allow one to set up a phone conference with these friends, or to share an application. NetMeeting has provided just such a cascade of interworking services for many years (albeit in a proprietary, non-extensible way, and without QoS).

In this chapter, we only cover the simplest scenario of session control, i.e. sessions involving two participants as end points, without a mediator. The reader is encouraged to study more interesting cases in other sources such as [Camarillo 2005].

Let us now consider what it means to control a multimedia session. Controlling a session means creating, modifying and terminating it. We have already discussed in Chapter 10, Section 10.1 what it means to create a multimedia session: the initiating participant must locate and contact a server and/or other users. The participants must agree on the media to be used, codecs, ports, QoS, etc. At a later stage, the session may be modified, e.g. by removing a medium or by adding a participant. Finally, the session must be torn down and all the participants hang up.

Session control is complicated by the fact that a user, let us call him Bob, may own several devices and, depending on the circumstances, would like to handle a session from one device and not the other—for example, he would like to receive sessions on his laptop while at work, on his fixed phone while at home and, generally, all sessions initiated by his girlfriend on his mobile phone. Each device is reachable under a different address, e.g. the laptop answers on a temporary IP address and the fixed phone answers on an ISDN phone number. Furthermore, on the fixed phone only the medium "voice" is welcome, whereas on the laptop any medium, e.g. video, shared whiteboard, etc., is acceptable.

We can divide the problem of session control into the sub-problems of determining and locating the currently preferred device of the other participants, and then contacting the participants and negotiating session parameters. We will see how these sub-problems are solved in the following discussion on SIP.

15.2 SIP

SIP is a protocol for controlling multimedia sessions in an IP Network. In this chapter we only cover basic SIP functionality.

We begin this section with the introduction of an important SIP component, the SIP-specific identifier, and then give an overview of how session control is performed on the basis of SIP identifier and a dedicated SIP infrastructure. We continue with an overview of the SIP protocol structure. The SIP protocol itself only transports session descriptions. It does not understand them. We therefore also look at the language for describing multimedia sessions, the **Session Description Protocol** (SDP). We close this section by describing a number of example SIP flows.

15.2.1 SIP Identifiers

A SIP user has a public identifier (also called **Address-of-Record**) under which he can always be contacted, independent of his current location and the type of device he is currently using. The user must keep a dedicated SIP network infrastructure updated about the mapping of the public identifier to the currently active **Contact Address(es)**. The Contact Addresses are the IP addresses of one or more devices under which he can be reached. When a request for a session comes in, the network infrastructure determines the appropriate device, or possibly several devices, and directs the request to this device. The currently preferred device is thus located, and the user contacted.

Public identifier and Contact Addresses have the format of a Uniform Resource Identifier (URI) [RFC 2396], e.g. sip:arthur.dent@earth.org. The alert reader will recognize this as the format of the Public User Identity stored on the ISIM, i.e. the identification of users in the IMS, cf. Chapter 10, Section 10.2.1 and Chapter 9, Section 9.1.3.

It is instructive to compare the mobility problem solved by SIP to the mobility problem solved by Mobile IP (cf. Chapter 12, Section 12.4.2). Mobile IP addresses **device mobility**: Since a device's IP address changes because of mobility, Mobile IP introduces an immutable Home Address for each device which the Home Agent maps to a temporary Care of Address. At all times, the Home Agent maintains a single Home Address—to—Care of Address mapping. SIP, in contrast, addresses **user mobility**: Since a user may change devices, SIP introduces an immutable public identifier for each user, which the network infrastructure maps to temporary Contact Addresses. The network infrastructure can maintain a complicated rule set detailing under which conditions an incoming request should be mapped to which Contact Addresses.

15.2.2 SIP Infrastructure

We now look in more detail at the "network infrastructure" introduced in the previous section. The location information and the logic for mapping the public identifier to Contact Addresses are maintained in a **Location Server**. The exact information and the rules format in the Location Server are not detailed by the SIP standard. SIP only requires that the Location Server be able to take a public identifier as input and to provide one or more Contact Addresses as output.

Figure 15.1 illustrates how our user Bob updates his information in the Location Server, mediated by another infrastructure element called **Registrar**.

1. He sends a REGISTER message to the Registrar containing his current Contact Address.
2. The Registrar then updates the information in the Location Server. Note that the standard also allows other methods for updating the Location Server, e.g. by the network administrator.

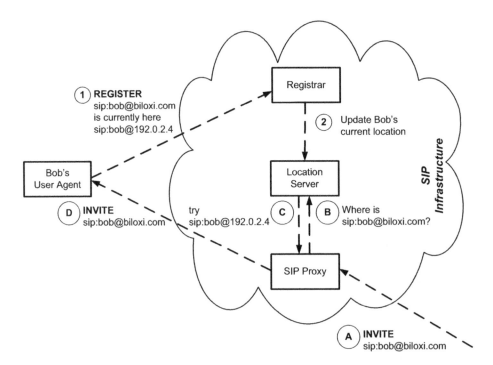

Figure 15.1 The SIP infrastructure mapping Bob's public identifier to his current contact address

Figure 15.1 also illustrates how another user invites Bob to a session using the SIP infrastructure.

(A)   A session request, an INVITE message, is addressed to Bob's public identifier and is received by his **SIP Proxy**.

(B,C) The proxy asks the Location Server to resolve the public identifier into Bob's current Contact Address,...

(D)   ...and then forwards the INVITE to this address.

Registrar, Location Server and SIP Proxy are logical roles which are often collocated in a single physical network element.

Another important SIP infrastructure element is the **User Agent**. All devices able to participate in SIP-controlled multimedia sessions, e.g. UEs or Application Servers, contain such a User Agent, whose job it is to handle SIP messaging on behalf of the device.

15.2.3 SIP Transactions

SIP messaging usually follows a simple logic: a SIP entity, e.g. Alice's User Agent, initiates a request, e.g. REGISTER, and the recipient of the message, e.g. Bob's User Agent, sends a final response. The response may be a success ("200 OK") or a failure ("400 Bad Request"). When message processing takes some time, the recipient may send provisional responses, e.g. "180 RINGING" in order to indicate that Bob is being alerted to the incoming session. We note that responses in SIP always carry a three-digit number. The message-response pattern is called a **SIP Transaction**. Table 15.1 describes how to build the most important SIP transactions in a UMTS context. The individual messages will be discussed in greater detail below.

Table 15.1 SIP requests, provisional responses and final responses

SIP Transaction:
Request + Provisional Response(s) + Final Response(s)
Exception to the rule: After a successful response to INVITE is received, an ACK is sent.

Requests:
  REGISTER
  INVITE
  PRACK (*provisionally acknowledge*)
  UPDATE
  BYE

Provisional Responses (*code 1xx*):
  100 TRYING
  180 RINGING
  183 Session Progress

Final Responses (*code 2xx for success, code 4xx for failure*):
  200 OK
  400 Bad Request
  400 Unauthorized

SIP is a somewhat typical IETF Protocol in that it provides rules for building message sequences, i.e. transactions. 3GPP-defined protocols usually do not exhibit this feature. Instead, in telecommunications-style protocols two entities exchange an arbitrary number of messages and close with acknowledgement messages when they are done.

15.2.4 Session Description

Finally, we look at the negotiation of session details. Above all, this requires a language for describing session details. This language is provided by the Session Description Protocol (SDP) [RFC 2327]—which is a textual convention rather than a protocol. It is noteworthy that from the perspective of SIP, the session description is an opaque container. SDP and SIP are independent and SIP would in fact also transport your private session description format. Session negotiation therefore consists of SIP transporting the opaque SDP container from the initiator to the receiver of the session in the INVITE message. The receiver can accept or reject the session based on this description.

Figure 15.2 provides an example of a session description where Alice invites Bob to participate in a video telephony session for a yoga class which she is teaching. The description contains a number of lines of the format "type = . . .", e.g. "v = . . .". SDP defines several types that are mandatory or optional for inclusion in a session description, only a subset is illustrated here.

v - (mandatory) protocol version.

o - (mandatory) the creator of the session and a session identifier.

s - (mandatory) session name.

c - (optional) connection information, i.e. the Internet (IN) IPv4 address where Alice is reachable.

t - (optional) the start and stop time of the session, with "0" coding "now" respectively "unbounded/until further notice".

m - (mandatory) the media type, the port under which Alice wants to receive the medium, the transport protocol (RTP with Audio Video Profile in this case) and a code for the codec to be used (e.g. 31 corresponds to the video codec H.261). When the session includes several media, one "m=" line is included for each medium.

a - (optional) is an attribute associated with the previous "m=" line and in our example states that both media are bidirectional.

Bob's User Agent analyses Alice's session description and determines whether his device supports the media and the offered codecs. If he does not support any media or codec, he

```
v=0
o=alice 2890844526 2890842807 IN IP4 126.16.64.4
s=Yoga class
c=IN IP4 126.16.64.4
t=0 0
m=audio 49170 RTP/AVP 0
a=sendrecv
m=video 51372 RTP/AVP 31
a=sendrecv
```

Figure 15.2 Session description with SDP

declines the session. Otherwise, he replies to Alice, including the description of his side of the session. For example, he includes his connection information and the port numbers under which he can receive the media.

The exchange so far cannot really be called a "negotiation" of session parameters. It is, however, possible for Alice to include more than one codec in each "m=" line, with her preferred codec listed first. If Bob does not like Alice's preferred codec—for example because he does not support the codec or because he cannot afford the bandwidth it requires—he could drop it from the list, and include those codecs he would rather use [RFC 3264].

15.2.5 SIP Example Message Flows

We have already encountered two important SIP messages, REGISTER and INVITE. In this sub-section we will present simple transactions with these messages that illustrate how SIP is used to establish a multimedia session. Note that the message flows presented in this subsection are not just simple but also simplified – only a subset of message content is shown.

We continue to observe Alice as she invites Bob to her yoga class. A prerequisite is that the User Agents of Alice and Bob, independently, register their current location with the respective Registrars, as illustrated in Figure 15.3. SIP messages always start with a status line containing the message name followed by a number of header fields and an optional message body. The REGISTER message contains—among others—the header fields "To" with the public identifier, and "Contact" featuring the current contact address. The REGISTER message is replied to with a 200 OK message and the transaction is closed.

Alice now invites Bob to the multimedia session. Since Alice doesn't know Bob's current Contact Address, nor the address of his SIP Proxy, she (respectively her User Agent) sends an INVITE message, with Bob's public identifier in the "To" header, to her own SIP Proxy. Her SIP Proxy determines Bob's SIP Proxy from Bob's public identifier and forwards the INVITE message there. Bob's SIP Proxy, finally, determines Bob's current Contact Address by querying the location server, and forwards the INVITE to this Contact Address. This four-hop signalling construction is also called the **SIP trapezoid**, as shown in Figure 15.4. Once the first SIP exchange is concluded, Bob and Alice know their respective Contact Addresses and can communicate directly.

Let us now look at the message flow in more detail, cf. Figure 15.5.

(i) Alice's User Agent sends an INVITE request to her SIP Proxy. The INVITE contains several header fields. "To" contains the public identifier of the user being invited to the

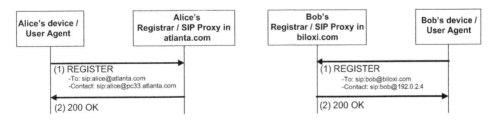

Figure 15.3 Message flow for SIP registration. In this example, Registrar, Location Server and SIP Proxy are co-located in one network element

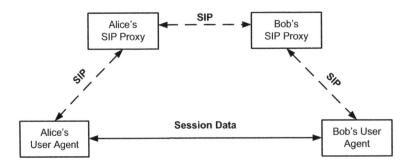

Figure 15.4 The SIP trapezoid

multimedia session. "From" contains the public identifier of the initiator of the session. "Via" contains the address to which the reply to the message should be sent and "Contact" contains the Contact Address of the initiator. The message body contains the actual description of the multimedia session, coded in SDP.

(ii) Alice's SIP Proxy determines the address of Bob's SIP Proxy, e.g. by a particular type of DNS look-up [RFC 3263]. It includes an additional "Via" header field, containing its own address, and forwards the INVITE message to Bob's SIP Proxy.

(iii) Alice's SIP Proxy sends a provisional response to Alice's User Agent that it is trying to forward the INVITE. We omit the header fields and possible body of the TRYING message.

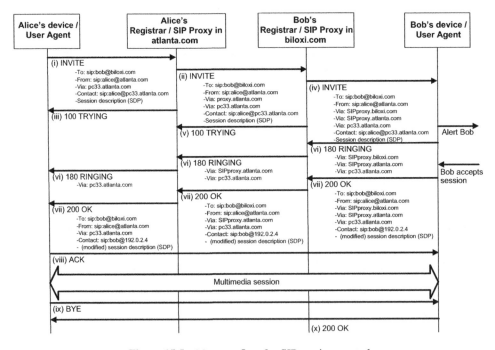

Figure 15.5 Message flow for SIP session control

(iv) Bob's SIP Proxy determines Bob's current contact address, resolves it into a routable address, e.g. by DNS look-up, includes an additional "Via" header field containing its own address and forwards the INVITE to Bob's current device.

(v)   Bob's SIP Proxy replies to Alice's SIP Proxy that it is trying to forward the INVITE.

(vi) Bob's User Agent can determine whether the current device can deal with the session description, e.g. whether it supports the suggested media and codecs. Since it does, it alerts Bob, e.g. by ringing the phone. It also informs the other nodes involved by generating a provisional response 180 RINGING, which is routed, via the SIP Proxies, to Alice's User Agent. Each proxy uses the Via header field in order to determine where to send the response and removes its own Via address from the header.

(vii) Bob accepts the invitation. His User Agent generates a 200 OK response which is routed backwards to Alice's User Agent, in the same way as the previous 180 RINGING message. The transaction initiated by the INVITE message is now closed.

(viii) Alice's User Agent acknowledges the successful conclusion of the signalling exchange. Now that contact addresses have been exchanged, the ACK is sent directly to Bob. Finally, the multimedia session can be held. Note that the ACK request is the exception to the transaction rule: it always follows a successful INVITE transaction and never generates a response.

(ix) Alice stops the session, her User Agent sends a BYE to Bob's User Agent (no header fields shown).

(x)   Bob's User Agent closes the transaction with a 200 OK message (no header fields shown).

It is instructive to compare the routing of SIP messages to the routing of RSVP or QoS NSLP messages (cf. Chapter 14, Section 14.2.2): Each of these protocols must solve the "backwards routing problem", i.e. the first signalling message travels from a sender to a receiver, thereby visiting a number of intermediate nodes (SIP Proxies in the case of SIP, RSVP routers in the case of RSVP, QoS NSLP Elements in the case of QoS NSLP). Subsequent messages may travel in the opposite direction, from the receiver to the initial sender of the message. These messages must visit the same intermediate nodes. RSVP and QoS NSLP solve this problem by installing backwards routing state in the intermediate nodes. SIP solves this problem by carrying the full set of intermediate routers in each backwards travelling message. The reason for this design decision appears to be that the number of intermediate routers in session control is smaller than in QoS signalling. Furthermore, SIP Proxies should contain as little state as possible.

15.3 SIP in the IMS

In this section we discuss how multimedia sessions are controlled by the IMS using SIP. IMS multmedia sessions normally feature a UE as one endpoint. The other endpoint may be one or more UEs, normal fixed-line phones, and generally any IP-based end-devices with a SIP User Agent. When the session is destined for a circuit-switched fixed-line phone, SIP signalling interworks with ISUP signalling in the MGCF (cf. Chapter 10, Section 10.2). Of course, the other endpoint can also be an Application Server, located in the IMS or outside the IMS. Furthermore, the IMS can provide server-mediated functions, e.g. multimedia conferencing or transcoding via the Media Resource Function (see also Chapter 10, Section 10.2).

The public identifier of a user is the Public User Identity stored on the ISIM, introduced in Chapter 9, Section 9.1.3. A User Agent is included in IMS-capable UEs. The SIP infrastructure is realized by the CSCFs introduced in Chapter 10, Section 10.2.

Keep in mind that the IMS needs a bearer in an **IP Connectivity Access Network** (IP-CAN) (cf. Chapter 10, Section 10.2.2), e.g. the PS Domain, in order to transport both the session signalling and the user-plane packets between UE and IMS. We will therefore also look at how the IMS collaborates with the PS Domain and how SIP signalling is intertwined with the PS Domain procedures "GPRS Attach" and "PDP Context Establishment" introduced in Chapter 11.

Of course, the PS Domain is just one possible IP-CAN for the IMS. As we will see in more detail in Part II of this book, 3GPP is currently integrating alternative IP-CANs into their network. Additionally, other standardization bodies such as 3GPP2, the WiMAX Forum, ETSI and CableLabs specify how to access the IMS (or a close variant) from their own networks (cf. Chapter 10, Section 10.2.2). Therefore, session control makes as few assumptions as possible on the IP-CAN. For example, it is assumed that the IP-CAN provides connectivity and hides UE mobility from the IMS—consequently, procedures related to UE mobility do not play a role in IMS session control.

In the following sub-sections, we first explain in more detail the how the SIP infrastructure maps onto the CSCF landscape.We then discuss the different steps in session control-user registration, session initiation, session release and user deregistration, and how they collaborate with PS Domain procedures, in particular.

15.3.1 SIP Infrastructure in the IMS

The CSCF is the main control node in the IMS. It takes on the SIP roles of Registrar, Location Server and SIP Proxy. At the same time, the CSCF appears in three different roles, Proxy CSCF (P-CSCF), Interrogating CSCF (I-CSCF) and Serving CSCF (S-CSCF), see Chapter 10, Section 10.2.1. These three CSCF roles are, however, independent of the three SIP roles. In fact, as we will see below, the three CSCF roles can transform the SIP Trapezoid into a SIP n-tagon with n = 8 or larger.

15.3.1.1 Proxy CSCF

The P-CSCF is the IMS contact point for the UE. From the UE's perspective, the P-CSCF is, among other things, a Registrar and a SIP Proxy. The UE discovers the P-CSCF as part of the registration procedure, and never changes the P-CSCF until deregistration.

As regards SIP signalling, the P-CSCF does, however, mostly act as a figurehead. It relays all SIP messages which it receives to a S-CSCF where they are processed. The S-CSCF, in turn, also communicates with the UE always via the P-CSCF. The S-CSCF may change but this is transparent to the UE.

When the UE is attached to the Home PLMN (HPLMN), the P-CSCF is, of course, also located in the HPLMN. When the UE is roaming, the P-CSCF may be located in the VPLMN or in the HPLMN. In any event, the P-CSCF is always in the same PLMN as the GGSN: As the reader may remember from Chapter 10, Section 10.1.1 and Chapter 14, Section 14.3.2, the P-CSCF, via the PCF, controls the QoS which a user may reserve by means of an interface with the GGSN, and operators do not appreciate network elements of foreign PLMNs controlling their GGSNs. Since the GGSN in current deployments is located in the HPLMN, the first IMS deployments are also expected to feature the P-CSCF in the HPLMN.

15.3.1.2 Serving CSCF

The S-CSCF is the node which actually performs the function of SIP Registrar and SIP Proxy, in addition to several other functions. For example, it obtains the **Service Profile** of users from the HSS after they have registered. The Service Profile is used for authorization and charging, and describes which services and which media a user may use, cf. Chapter 13, Section 13.3.3. Based on the Service Profile, the S-CSCF may involve additional Application Servers in the session.

The S-CSCF is always located in the HPLMN of the UE. An operator is expected to own several S-CSCFs with different capabilities. A S-CSCF is assigned to the UE based on the services to which the user is subscribed, the location of UE and S-CSCF, the current load distribution of S-CSCFs, etc. The UE is shielded from these decision algorithms by assigning a single P-CSCF as the UE's contact point.

15.3.1.3 Interrogating CSCF

The I-CSCF is introduced because operators typically like to hide the topology of their networks as well as the number and capability of their network elements. To this end, I-CSCFs are located at a PLMN border. When SIP signalling crosses a PLMN boundary, an I-CSCF is added to the signalling path. For example, a P-CSCF in a VPLMN receives a SIP message which it needs to relay to a S-CSCF in the HPLMN. Via a DNS query, it ascertains the IP address of a I-CSCF in the HPLMN and sends the SIP message there. The I-CSCF checks with the HSS in order to determine the S-CSCF and finally delivers the message.

The I-CSCF thus has the convenient capability of finding other CSCFs. It therefore ends up being employed even in single-operator scenarios, e.g. when the S-CSCF of one subscriber needs to find the current S-CSCF of another subscriber.

15.3.1.4 SIP Octagon

Let us now put the pieces together and determine how many CSCFs are involved in an INVITE transaction.

When the session is between a UE and an Application Server, and both are in the HPLMN of the UE, the INVITE would travel from User Agent to P-CSCF to S-CSCF and—because the S-CSCF knows the address of all Application Servers a user is entitled to use—from there directly to the Application Server.

When the session is between an originating UE (o) and a terminating UE (t), both with the same HPLMN and not roaming, the INVITE travels as follows: from User Agent (o)—P-SCSF (o)—S-CSCF (o)—I-CSCF (t)—S-CSCF (t)—P-CSCF (t)—User Agent (t). We see that S-CSCF (o) involves an I-CSCF in order to find the S-CSCF of UE (t).

We therefore see that the original SIP Trapezoid becomes a septagon if the initiator and receiver of the session have the same Home Network and if both are not roaming. The SIP Trapezoid can become an octagon or better when the initiator and receiver are subscribed to different PLMNs and roaming. A SIP Octagon is depicted in Figure 15.6.

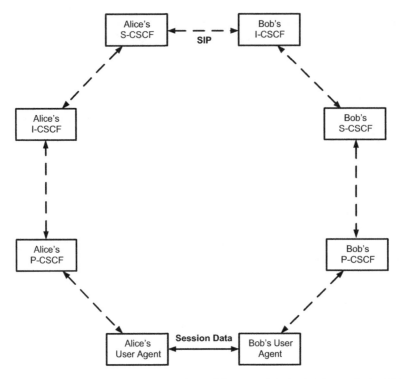

Figure 15.6 SIP Octagon when Alice and Bob have different HPLMNs. Bob is attached to his HPLMN whereas Alice is roaming

15.3.2 UE Registration in the IMS

When a user would like to use IMS services it first needs to register its presence using the SIP REGISTER message. Note that according to the SIP standard, the REGISTER message is just one way to update the Location Server about a user's presence. The IMS standard, however, mandates it as the first step. Only then can the user be invited to multimedia sessions by other users and can initiate its own sessions.

A prerequisite to being able to register with the IMS is that the UE has signalling connectivity over an IP-CAN. If the IP-CAN is the PS Domain this means that the UE has performed GPRS Attach and has established a PDP Context for signalling. Once a PDP Context is established the UE also has an IP address which it can use as Contact Address for IMS registration.

The UE sends the REGISTER message to the P-CSCF. To do so, it first needs the IP address of the P-CSCF. Generally, the UE can perform a DHCP query in order to obtain the domain name of the P-CSCF, and subsequently a DNS query to resolve the domain name into an IP address. When the IP-CAN is the PS Domain, the UE can already indicate in the Activate PDP Context Request message (cf. Figure 11.9) that it needs the P-CSCF's IP address. The GGSN gets the address and includes it in the Create PDP Context Response. This extension of the GTP protocol is a typical example of how 3GPP creates efficient, customized protocol solutions instead of using existing solutions out of a "toolbox".

The UE is now in a position to send the REGISTER message to the P-CSCF. The P-CSCF determines an I-CSCF in HPLMN, the I-CSCF determines S-CSCF and finally the user is registered with the S-CSCF. It is an important characteristic of this process that, from the UE's perspective, the procedure is always identical, independently of whether the UE is roaming or at home. In the next sub-section we provide an example message flow for UE registration.

15.3.2.1 Message Flow for UE Registration in the IMS

The message flow for UE registration is illustrated in Figure 15.7. We assume that the UE is roaming and the P-CSCF is in the Visited Network. Therefore, a I-CSCF is introduced between P-CSCF and S-CSCF.

(i) The UE sends a SIP REGISTER message to the P-CSCF including its Public User Identity, Private User Identity, Home Network domain name and its IP address. This information is included in the SIP header fields introduced in Section 15.2.4 as well as a number of SIP extension headers defined by the IETF based on requirements specified by 3GPP.

(ii) The P-CSCF performs a DNS query to find the IP address of an I-CSCF in the Home Network of the UE and then forwards the REGISTER to the I-CSCF in the HPLMN.

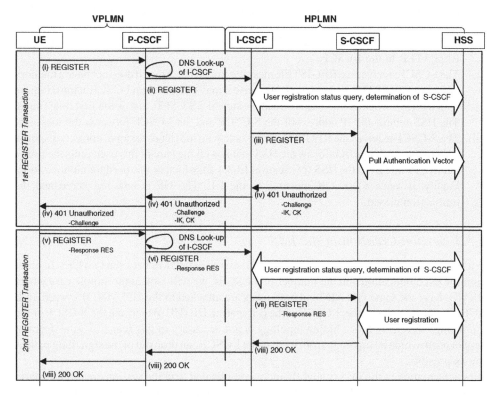

Figure 15.7 Message flow for UE registration in the IMS when the UE is roaming and the P-CSCF is in the Visited Network

(iii) The I-CSCF contacts the HSS and performs a basic authorization: is the user known to the HSS and is she allowed to roam to VPLMN? If yes, the HSS assigns a S-CSCF. This exchange is performed using the Diameter protocol, cf. Chapter 13, Section 13.4. Note that the I-CSCF is a node without memory for SIP messages. It handles a REGISTER message and forgets about it as soon as the transaction is closed—this design allows the DNS to perform load-balancing when returning the I-CSCFs address.

(iv) Before the UE can register with the S-CSCF it must be authenticated. We have already discussed the procedure in Chapter 13, Section 13.3.3 and see now how it fits into the overall message flow: The S-CSCF pulls the Authentication Vector including the challenge and temporary keys for integrity protection (IK) and encryption (CK) from the HSS. It then sends a 401 Unauthorized response to the UE, via I-CSCF and P-CSCF, in which the challenge is included—in much the same way as the https protocol can include a challenge-response handshake in its header. The S-CSCF also includes IK and CK for the benefit of the P-CSCF who will need them later. The transaction initiated by the REGISTER is closed. The P-CSCF removes and stores IK and CK when the message passes.

(v) The UE knows how to react to the 401 Unauthorized message: based on a secret key stored on the ISIM, it solves the challenge and generates a response RES. Unfailingly, it tries to register again, this time including RES in its REGISTER message. It also calculates the temporary keys IK and CK which are also known to the P-CSCF. UE and P-CSCF thus now have a Security Association (cf. Chapter 13).

(vi) The P-CSCF performs another DNS query to find an I-CSCF. It then forwards the REGISTER to the I-CSCF.

(vii) The I-CSCF receives the REGISTER message from a UE. Since it does not have a memory of former REGISTER messages, and because it may be a different I-CSCF, it must inquire with the HSS whether this UE is already assigned a S-CSCF, just as the first time round. The HSS returns the IP address of the S-CSCF and the I-CSCF forwards the message.

(viii) The S-CSCF receives the REGISTER message with the (let us assume: correct) response RES to the challenge. It informs the HSS that it is taking care of this user, pulls the user's Service Profile from the HSS (cf. Section 15.3.1.2), which is also used for authorization. Finally, it sends a 200 OK message to the UE. The UE is now registered and the transaction closed.

15.3.3 Session Creation in the IMS

We will now discuss the creation of a multimedia session in the IMS between two UEs. In order to minimize complication and the number of I-CSCFs, we will look at the simple case where both UEs have the same HPLMN and both users are attached to the HPLMN. However, an I-CSCF is involved: it helps the S-CSCF of the originating UE (UE o) to locate the S-CSCF or the terminating UE (UE t). Since hiding topology is not necessary in this scenario—the operator does not need to hide its topology from itself—the I-CSCF can drop out of the signalling path as soon as possible.

Session creation in the IMS is slightly more complicated than the session creation scenario discussed above (cf. Figure 15.5): in Telecommunication Networks, it is important that services are provided with the appropriate QoS. Therefore, the INVITE request comes with a **precondition**: It will only be successful if QoS has been reserved (see also Chapter 14,

Section 14.2.2.3). The desired, preconditional QoS is signalled in the SDP payload with an additional attribute "a = " for each medium [RFC 3313].

When the IP-CAN is the PS Domain, satisfying the QoS precondition means that both UEs are supposed to establish a PDP Context—with the desired QoS—that will carry the multimedia stream. The two UEs need to synchronize when they are done with their QoS reservation. Therefore, an additional transaction is embedded in the INVITE transaction: the originating UE sends an UPDATE request as soon as it has reserved QoS. The UPDATE carries an SDP payload describing the actual QoS reserved—which may be less than what was desired—and its consequences on codec choice, etc. Only after the terminating UE has sent the 200 OK response, can it also conclude the original INVITE transaction.

15.3.3.1 Message Flow for Session Creation in the IMS

The message flow for session creation—even in the simple scenario shown here—is so long that we need to split it into Figures 15.8a and 15.8b.

(i) UE-o generates a INVITE request. The SDP session description includes the envisaged media, codecs, ports and the desired QoS. The UE sends the INVITE to its P-CSCF-o.

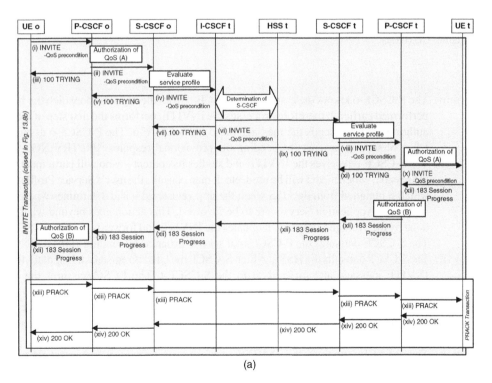

(a)

Figure 15.8a Part 1 of the message flow for session creation between an originating UE (UE o) and a terminating UE (UE t). Both users are subscribed to the same network and are not roaming

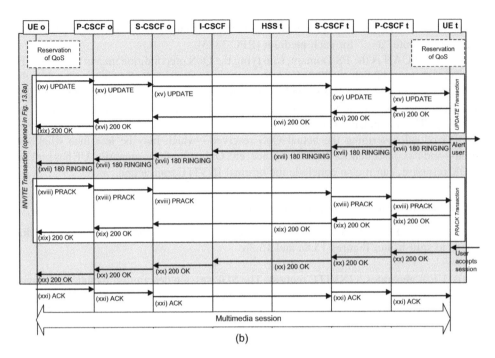

(b)

Figure 15.8b Part 2 of the message flow for session creation between an originating UE (UE o) and a terminating UE (UE t)

(ii,iii) The P-CSCF-o knows the associated S-CSCF-o from the registration which the UE performed earlier. It therefore processes the INVITE, performs the first step of QoS authorization and sends the message on to the S-CSCF-o. The P-CSCF-o assures UE-o of session progress by sending the provisional response 100 TRYING.

(iv,v) The S-CSCF-o receives the INVITE and studies its content—who will participate in the session, which media will be used, etc. It then consults the user's Service Profile—which it obtained from the HSS when the user registered—and determines whether additional Application Servers are to be involved. This action goes beyond what is done by the average SIP Proxy. It updates the INVITE and forwards it to the I-CSCF. The S-CSCF-o assures the P-CSCF-o of session progress by sending 100 TRYING.

(vi,vii) The I-CSCF consults the HSS at which S-CSCF the UE-t is registered. It updates the INVITE message and forwards it to the S-CSCF-t. The I-CSCF assures the S-CSCF-o of session progress by sending 100 TRYING.

(viii,ix) The S-CSCF-t receives the INVITE and evaluates the Service Profile of the user of UE-t. It updates the INVITE and forwards it to the P-CSCF. The S-CSCF-t assures the I-CSCF of session progress by sending 100 TRYING.

(x,xi) The P-CSCF-t updates the INVITE, performs its own QoS authorization and forwards the message to the UE-t. It sends 100 TRYING to the S-CSCF-t.

(xii) UE-t recognizes from the INVITE that it needs to reserve QoS before proceeding with the transaction. Prior to the reservation, however, UE-o and UE-t need to agree on which media to use and which codecs, in order to fix the QoS parameters. Therefore, UE-t sends a provisional response 183 Session Progress back to UE-o, including an SDP description of the session characteristics it can accept. The provisional response travels via the same CSCFs as the INVITE message. When it passes the P-CSCFs, they trigger the final authorization of the QoS that is currently envisaged. Chapter 17 will explain the QoS authorization procedures in more detail. When the message passes the S-CSCF-o, it learns the address of S-CSCF-t.

(xiii) 3GPP would like UE-o to acknowledge receipt of the provisional response in order to increase the reliability of the message exchange. As we saw above, however, only a sucessful response to an INVITE message triggers an ACK. Therefore, a new request message was defined, the provisional acknowledgement (PRACK). The PRACK defines a transaction in its own right, and, unlike ACK, requires a response. We therefore see in Figure 15.8a a PRACK transaction being embedded in the overall INVITE transaction. At the same time, the PRACK carries the final set of media and codecs being agreed on in this session. The PRACK travels from UE-o to UE-b. Note that S-CSCF-o sends it directly to S-CSCF-t, bypassing the I-CSCF, which has served its purpose of finding the address of S-CSCF-t.

(xiv) In response to the PRACK, UE-t sends a 200 OK back to UE-o. This message follows the same path as the PRACK, i.e. it bypasses the I-CSCF.

(xv) Both UEs can now reserve QoS. If the IP-CANs are PS Domains, they set-up a PDP Context. When this is done, UE-o initiates a new transaction in order to tell UE-t about this deed. It sends an UPDATE request, which may contain updated QoS and codec information in case UE-o could not reserve QoS as necessary. The UPDATE transaction bypasses the I-CSCF.

(xvi) If it has also successfully reserved QoS, UE-t sends a 200 OK response back to UE-o. The response may again contain an updated SDP description.

(xvii) After sending the 200 OK response, UE-t alerts the user of the incoming session. It informs UE-o with a provisional response 180 RINGING. Note that this response is part of the original INVITE transaction and thus travels via the I-CSCF.

(xviii,xix) UE-o acknowledges the receipt of the 180 RINGING message with a PRACK transaction. UE-t replies with 200 OK.

(xx) We are lucky, the user of UE-t accepted the session, all of the signalling effort was worthwhile! UE-t informs UE-o with a 200 OK message and concludes the INVITE transaction.

(xxi) As required by the SIP standard, UE-o sends a final ACK. The multimedia session can now be started.

15.3.4 Session Release and UE Deregistration in the IMS

When a UE wants to discontinue a multimedia session, it stops the media stream. It issues a BYE request to its P-CSCF and tears down the QoS reservation, e.g. the PDP Context associated with the session. The BYE travels via various CSCFs to the other participant, i.e. a UE or an Application Server, where a 200 OK is produced, and QoS is torn down as well. When a UE wants to stop using IMS services, e.g. because it detaches from the UMTS Network, it deregisters with the P-CSCF with a straightforward procedure.

15.4 Discussion

Session control in the IMS is a rather interesting area for studying the convergence of Telecommunication Networks and Computer Networks. The IMS, unlike the other UMTS architectural components, employs solely IETF Protocols in the "one-protocol-per-functionality" style. And yet, the resulting system does not quite look like a Computer Network. Compare the original SIP trapezoid (Figure 15.4) with the IMS's SIP octagon (Figure 15.6), or compare the original SIP call flow for session creation (Figure 15.5) with the corresponding IMS SIP call flow (Figure 15.8). 3GPP has surely managed to introduce greater complication. Why did they do this?

Again, we must consider that the business model of a UMTS operator is delivering high-quality services to a user, with minimum user involvement.

- 3GPP specifies a complete system. They must go beyond rules of the type "if A then B", e.g. "if you are a SIP Proxy and receive an INVITE, then reply with a TRYING and forward the INVITE"—which is what a typical IETF standard provides. Rather, 3GPP needs to standardize the complete machinery, i.e. they also need to standardize under which circumstances "A" comes into existence, e.g. when and how an INVITE is issued.

- 3GPP must describe how session control collaborates with other control functions, e.g. QoS and security. Many of the additional messages in user registration and session creation can be attributed to this.

- Minimizing user involvement—and, by the same token, optimizing operator control—leads to the division of the SIP Proxy entity into P-CSCF, I-CSCF and S-CSCF. The P-CSCF is the constant contact point of the UE into the IMS; the operator gains freedom in assigning the S-CSCF depending upon the service requested and the load situation. The I-CSCF allows for shielding the network from unwelcome inquisitiveness.

We therefore see that it is also possible to build a Telecommunication Network on the basis of IETF Protocols.

15.5 Summary

In this chapter we have discussed the control of multimedia sessions in the IMS. Session control in the IMS is based on the SIP protocol which is standardized by the IETF. The original SIP protocol was extended to take into account the 3GPP's requirements.

Session control addresses the problems of mobility of the session participants (i.e. user mobility, not to be confused with device mobility), contacting the participants and, finally, negotiating session parameters such as which media are used, which codecs, which ports, which QoS, etc.

SIP solves the participants' mobility problem by introducing an immutable public identifier for SIP participants. In an IMS context, this is the Public User Identifier stored on the ISIM. The mapping of public identifer to the current Contact Address is provided by the SIP infrastructure, particulary Registrar, Location Server and SIP Proxy – often collocated on a single network element. Users can REGISTER their Contact Address with the REGISTRAR who then saves it to the Location Server. Participants are invited to a multimedia session by an INVITE request to their SIP Proxy. The SIP Proxy consults the Location Server and forwards the INVITE to the current Contact Address.

The actual session parameters are described in a format called SDP and are included, e.g. in the SIP INVITE message.

The IMS realizes the SIP infrastructure in its CSCFs. Three types of CSCFs exist, P-CSCF, I-CSCF and S-CSCF. The P-CSCF is the contact point for UEs in the IMS. It never changes for the duration of a registration. The I-CSCF is the contact point for SIP requests from other PLMNs, and for a CSCF searching another CSCF. It also serves to hide operator topology. The S-CSCF performs the actual session-control. It is always located in the Home Network.

For each SIP transaction, the SIP request travels from the UE to P-CSCF to S-CSCF, possibly via an I-CSCF. When the transaction involves another UE, e.g. an INVITE transaction, the request continues from the S-CSCF of the initiating party to an I-CSCF in the PLMN of the called party, and from there via S-CSCF and P-CSCF to the UE. Of course, several additional I-CSCFs could be included.

IMS must collaborate with an IP-CAN which provides connectivity and QoS and which hides UE mobility from the IMS. When the IP-CAN is the PS Domain, the interworking with PS Domain procedures such as PDP Context set-up is described in the 3GPP standards. The collaboration with other IP-CANs has not yet been specified in detail.

16

Charging

Charging is a key function in any Telecommunication Network, and with the advent of UMTS its complexity has grown considerably.

Charging in GSM is a comparatively simple affair: since all calls occupy the same bandwidth and receive the same QoS, there is a single most important charging parameter: the duration of a call. Variation is introduced by making charging depend also on the time of day, the day of the week and the destination PLMN. All parameters are collected by a single network element, the MSC.

Charging in Computer Networks is also relatively simple. In a modular fashion it is usually performed, if at all, by the Access Network and by application servers. The Access Network charges on the basis of traffic volume or usage duration. Application servers charge on the basis of content. Charging by Access Network and application servers is not coordinated and separate bills are delivered to the user. Generally, the IETF community is more interested in accounting, i.e. in the collection of traffic data, the results of which may be used to bill the user, for auditing or just for statistical purposes.

Charging in UMTS is comparatively complex because it can depend on many parameters beyond usage duration and traffic volume, e.g. QoS, service used, radio access technology, etc. The data necessary for charging is collected by numerous network elements, e.g. SGSN, GGSN, P-CSCF and application servers, as the case may be. Complications arise because UMTS charging is aimed at delivering a single bill to the subscriber in which all consumption parameters are included.

This chapter follows the approach of the other control chapters. We start with a description of the problems, then introduce charging in UMTS, and finally discuss charging in Computer Networks. Further information on charging in UMTS is available from [3GPP 32.240].

Terminology discussed in Chapter 16:
AAA Server
Accounting
Accounting Data

UMTS Networks and Beyond   Cornelia Kappler
© 2009 John Wiley & Sons, Ltd

Application Function	AF
Bearer-level charging	
Billing	
Billing Server	
Billing System	
Chargeable Event	
Charging Characteristics	
Charging Data Function	CDF
Charging Data Record	CDR
Charging Gateway Function	CGF
Charging Information	
Charging Rule	
Charging Trigger Function	CTF
Correlation	
Correlation identifier	
Flow-based Charging	
Metering	
Metering Data	
Offline charging	
Online charging	
Online Charging Function	OCF
Online Charging System	OCS
Policy and Charging Rules Enforcement Function	PCEF
Policy and Charging Rules Function	PCRF
Proxy AAA Server	
Rating	
Service Flow Template	SFT
Service-level charging	
Subsystem-level charging	

16.1 Description of the Problems

The goal of the operator is receiving payment from the user for service usage. The operator thus needs to perform charging. The charging process can be broken down in several steps, as shown in Figure 16.1. We will illustrate a very simple process through the example of user Bob watching a streaming video.

- **Metering**
  The service usage resulting from the video stream is metered in a number of network elements. The metering can be performed both on the control-plane and on the user-plane:
    — on the control-plane, the flow-related signalling can be analysed, in order to determine, e.g. the QoS reserved or the start and termination time of the service
    — on the user-plane, the actual flow can be analysed in order to determine the traffic volume, packet delay, etc.

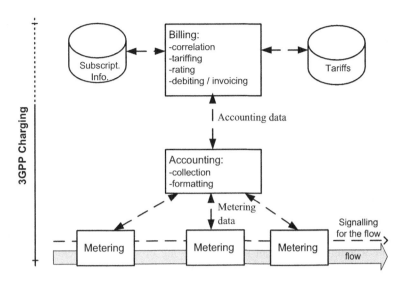

Figure 16.1   The charging process

In the case of Bob's video streaming service, the Access Point or the SGSN meter the traffic volume and the Application Server reports an identifier for the video and the QoS reserved. As a result, **Metering Data** is produced.

- **Accounting:**
The Metering Data from all meters is **collected** in one or more central network elements. The accounting elements, furthermore, **format** the sets of Metering Data in some uniform way. The result is called **Accounting Data**.

- **Billing:**
The Accounting Data is sent to a network entity performing **billing**. Billing is the most complex of the three steps. First of all, the Accounting Data must be **correlated**. This means that from all the Accounting Data arriving at the billing entity, data that relate to the same user and to the same service usage are assigned to one record, in our case the traffic volume; the video's identifier and the QoS reserved go into one record for Bob. Redundant or irrelevant measurements are now discarded. For example, if Bob's subscription allows him to receive videos for a fixed price, independent of their size, the data on traffic volume or QoS is dismissed.

It is an interesting question how the billing entity recognizes data that should be assigned to one record: the metering entitites (e.g. Access Point/SGSN and Application Server) must include a **correlation identifier** in the Metering Data. This correlation identifier needs to be generated for each service usage and must then be distributed to all metering entities. This can be an important task, especially when many metering entities are involved.

After correlation, the billing entity **rates** the Accounting Data, i.e. the monetary value of the service usage is determined based both on subscription information and tariff tables. The result of the calculation is assigned to Bob's account. At some point in time, all of the entries in Bob's account are summarized in an invoice which is delivered to the user. Alternatively, if Bob uses a pre-paid card or a voucher, the result of the calculation is fed back in real-time to the meters so they can stop service delivery if need be, and the pre-paid card or voucher is debited.

The process just described is, in principle, independent of the nature of the network, be it GSM, UMTS, cdma2000 or a Computer Network. The reader should be warned, however, that variations in emphasis exist. This is reflected in that there is no uniform terminology for describing the charging process. For example, what is called charging in 3GPP (and in this book) is defined so as to consist of the steps described above of metering, accounting and of most of the billing process, up to the point where the bill is sent (offline charging) or the prepaid card or voucher is debited (online charging). In the terminology of other communities, by contrast, charging is just one step in the billing process.

16.2 Charging in Computer Networks and WLAN

Charging in WLANs is normally performed using proprietary solutions. The IEEE has not defined a specification for this topic.

For Computer Networks, the IETF has described a set of informational architectures for metering and accounting. They can support charging or other management tasks such as network capacity analysis, QoS monitoring or intrusion detection.

In [RFC 2722], a flow measurement architecture is described consisting of a meter performing metering and a meter reader performing accounting. This architecture thus takes care of the first two steps of the charging process as defined above. The meter writes the Metering Data in its local **Management Information Base** (MIB). The meter reader polls this MIB using the **Simple Network Management Protocol** (SNMP) [RFC 2570]. The drawback with this approach is that the meter reader needs to poll all meters periodically independent of whether they have collected any Metering Data. In networks with a large number of meters especially this is not scalable. In more recent work on the topic [RFC 3955], the metering element therefore actively pushes the Metering Data to the accounting element. To this end, several protocols could be used, e.g. Diameter or, as a more specialized solution, the NetFlow protocol [RFC 3954].

In [RFC 2975], an "accounting architecture" is defined consisting of network devices, **Proxy AAA Servers**, **AAA Servers** and **Billing Servers**, see Figure 16.2. This is a good opportunity to bear in mind the differences in terminology within the different communities, because this accounting architecture would be called a charging architecture by 3GPP. In the IETF accounting architecture, the network devices perform what we have, in Section 16.1, defined as metering and accounting. In the simple intra-domain case, the Accounting Data is passed to a AAA Server (cf. Chapter 13, Section 13.5) for correlation using, e.g. the Radius or Diameter protocol. The AAA Server sends the correlated Accounting Data to a Billing Server for billing, e.g. via ftp or http. When the user is roaming, the scenario is slightly more complicated because the billing is, ultimately, performed by the home domain. The correlated Accounting Data is thus proxied by the AAA Server in the visited domain and sent to the AAA Server in the home domain. This is the same set-up as for authentication in roaming scenarios, see Chapter 13, Section 13.5.1.3 — which is not surprising since the same AAA infrastructure is used.

16.3 Charging in UMTS

UMTS, just as GSM, features two different methods for charging: **offline charging** and **online charging**. We will look at these two methods, then discuss the general UMTS charging

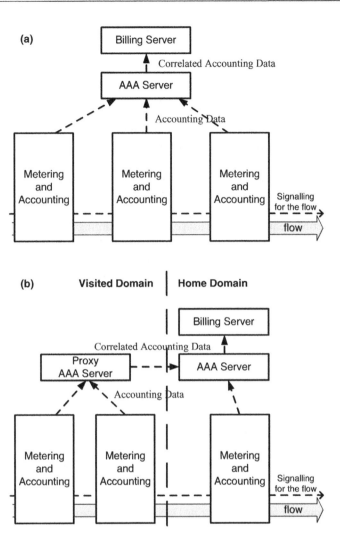

Figure 16.2 The IETF Accounting Management architecture (a) intra-domain accounting (b) inter-domain accounting

architecture and proceed finally to charging in the PS Domain and charging in the IMS. Note that the charging architecture was substantially re-organized and extended in Rel-6. We will therefore only describe charging from Rel-6 onwards.

16.3.1 Offline Charging and Online Charging

Traditional charging is **offline charging**: a subscriber uses a service and later, off-line, is sent the bill for the accumulated costs. In offline charging, it is only the metering which is performed in real time. The other steps of the charging process can be performed later. Furthermore, the information flow between the network entities involved in charging is simple and unidirectional: the metering elements pass data to the accounting element, and the accounting element passes data to the billing entity.

In some scenarios, however, offline charging is not sufficient. For example, support of pre-paid cards requires greater sophistication. With a prepaid card, the user has an account with the operator in which he deposits a certain sum of money. The user can use network services until the account is empty. The operator thus performs the entire charging process in real-time in order to detect when the account is drained. Such charging is known as **online charging**. Online charging is also necessary in order to be able to inform the user about the costs before or during service usage.

Online charging, as opposed to offline charging, requires bidirectional signalling between the network entities involved in the charging: the billing, accounting and metering entities collaborate in order to determine how long the subscriber may use a service.

When the subscriber requests service usage in online charging the metering entities prompt the billing entity to check the balance of the account. Only when the balance allows, is the user authorized. The billing entity reserves credit on the user's account and passes a particular usage quota to the metering entities, e.g. "ten minutes of video streaming", or "2 Mbyte of data". When the quota is consumed, the metering element prompts the billing entity to debit the subscriber's account and to assign new quota. When the account is emptied no quota is generated and service delivery is terminated.

Clearly, online charging requires more functionality than offline charging. In fact, in UMTS, the online charging architecture is distinct from the offline charging architecture, as will be discussed in the next subsection.

16.3.2 UMTS Charging Architecture

A single service, e.g. the delivery of a video stream, can invoke metering and reporting on very different levels: the usage of network resources (bandwidth and QoS) and the usage of an application (the video). Correspondingly, in UMTS charging on several levels may be distinguished:

- **Bearer-level charging** provides Metering Data from the usage of a bearer. The bearer can be provided by the PS Domain, or generally by any IP-CAN (cf. Chapter 15, Section 15.3) Metering Data can be collected both on the user-plane by metering the actual bearer, e.g. the volume of the traffic transported in the bearer, and on the control-plane by metering the bearer-related signalling, e.g. the QoS signalled with PDP Context establishment.
- **Subsystem-level charging** provides Metering Data derived from usage of the IMS, e.g. the start and stop time of service delivery. Subsystem-level charging, as it is specified today, does derive all Metering Data on the control-plane, e.g. from SIP signalling. The user-plane data is not metered.
- Finally, **service-level charging** provides Metering Data resulting from usage of services, e.g. the delivery of a particular video. Service-level charging can be performed both on the control and on the user-plane.

Historically, for each level and each domain the charging architecture was specified independently. With Rel-6, the number of charging architectures was reduced to two generic ones: one architecture for offline charging and one for online charging. They are illustrated in

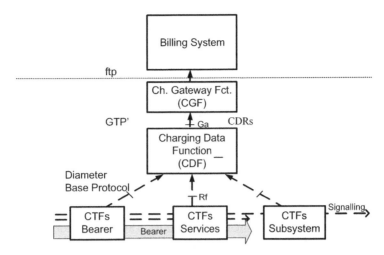

Figure 16.3 The UMTS offline charging architecture. CTF — Charging Trigger Function

Figures 16.3 and 16.4, respectively, and are discussed below. Furthermore, in Rel-6 and Rel-7 a number of additional charging features were introduced. The most important of these is **Flow-based Charging** allowing for better coordination of bearer-level charging with subsystem-level charging and service level charging, as we will see in more detail below.

16.3.2.1 Offline Charging Architecture

The offline charging architecture of Rel-6 defines one or more network elements for each step of the charging process in Figure 16.1:

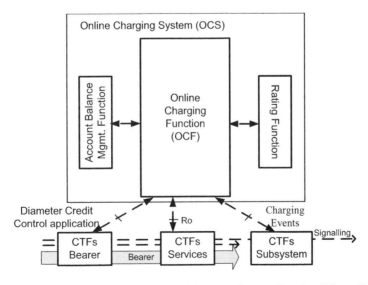

Figure 16.4 The UMTS online charging architecture. CTF — Charging Trigger Function

- Metering is performed by elements called **Charging Trigger Function** (CTF). They meter what is known as **Chargeable Events**, e.g. PDP Context Activation and Deactivation. The parameters they meter are called **Charging Information**, e.g. the QoS and APN of a particular PDP Context, as well as the volume of traffic which passes through it.

  CTFs are integrated into the network elements which we have already encountered, on all of the three charging levels: For example, CTFs for bearer-level charging are SGSN, GGSN and MSC. A CTF for subsystem charging is the P-CSCF and a CTF for service-level charging is an Application Server.

- When a Chargeable Event occurs, a CTF sends corresponding Charging Information to the first stage of accounting, the **Charging Data Function** (CDF). A CDF accumulates the Charging Information and for each CTF generates **Charging Data Records** (CDRs). A UMTS Network may feature several CDFs. For example, in the PS Domain CDFs are normally integrated into SGSN and GGSN.

  The interface between CTF and CDF is called Rf. The protocol for the Rf interface is the Diameter base protocol, in particular the ACR and ACA messages (cf. Chapter 13, Section 13.5.1.4.1) plus vendor-specific extensions.

- The second stage of accounting is performed in the **Charging Gateway Function** (CGF). The CGF collects and reformats CDRs and stores them until further usage. At some point it sends the collected information as CDR files to the **Billing System**. The CGF can be integrated into existing network elements such as GSNs, or it can be a separate physical box.

  The interface between CDF and CGF is called Ga, and the protocol is a variant of GTP called GTP'.

- The Billing System performs billing, as described in Section 16.1. It receives the CDR files from CGFs via simple ftp. The specification of the Billing System is outside the scope of 3GPP and is usually not standardized. The literature available on this topic is somewhat limited.

16.3.2.2 Online Charging Architecture

The online charging process is less linear than the offline charging process. This is reflected in the architecture. Again, we will introduce the generic architecture of Rel-6.

- CTFs meter the Chargeable Events, just as in offline charging. The CTFs in online charging have, however, additional functionality: when a request for service usage comes in, they must delay service delivery and first obtain quota and authorization from the **Online Charging Function** (OCF). When no authorization is possible or the quota is exhausted, they must deny access or terminate service delivery, respectively. The interface between CTF and OCF is called Ro, and the protocol is the Diameter Credit Control application [RFC 4006]. It provides the messages Credit Control Request (CCR) and Credit Control Answer (CCA).

- The accounting and billing entity is summarized in one functional block, the **Online Charging System** (OCS). Inside, the OCF issues authorization and quota to the CTFs. The OCF collaborates with the Rating Function, performing rating, and the Account Balance Management Function, where a user's account information is kept.

The number of CTFs in online charging must be as low as possible, because Chargeable Events must be metered, accounted, correlated and rated in real-time. Also the amount of Charging information must be reduced to the essential. As we will see in more detail below, a considerable proportion of Charging information is, however, collected for statistical purposes only. For example, operators are interested in the Routing Area Identifier in order to evaluate where network usage is high and where it is low. Such statistical information does not need to be available in real-time. Therefore, in online charging, CTFs are a subset of CTFs employed in offline charging — for example, the I-CSCF is a CTF in offline charging, but not in online charging; CTFs for online charging are selected so that they can provide all of the Charging information that is actually used for charging. The "statistical" Charging information for the same Chargeable Events can be collected independently by offline charging.

16.3.2.3 Flow-based Charging

The charging architecture described so far has a shortcoming: for bearer-level charging in the PS Domain and other IP-CANs it is only possible to charge with the resolution of a bearer, e.g. a PDP Context. Service-specific charging is difficult. For example, a particular video stream, in order to apply a special, reduced, tariff on the bearer level, cannot be identified. Another way of looking at the problem is that charging on the bearer level is not sufficiently coordinated with charging on subsystem and service level.

A PDP Context is characterized by its end point — also known as the APN (cf. Chapter 11, Section 11.6.1) — which translates into a particular GGSN. A single PDP Context can carry flows belonging to several services. If the IMS is accessible via a single APN, all IMS sessions requiring the same QoS use the same PDP Context. The PS Domain does not understand the packet semantics inside the PDP Context. Operators today often work around the problem by assigning service-specific APNs, e.g. the APN for sending an MMS is different from the APN for surfing the Internet.

Diversity of APNs, however, does not yet solve the problem of PDP Contexts used for signalling: for example, a PDP Context for registration in the IMS or for initializing a session obviously has the same APN as the subsequent session. However, it should not be charged. The somewhat inelegant solution to this problem is the introduction of extra Charging information collected by the GGSN (see Table 16.1): "Is it a PDP Context for signalling?"

In Rel-6, the problem was addressed more fundamentally by introducing **Flow-based Charging** (FBC) [3GPP 23.125]. Once Rel-6-conformant equipment is employed, it will be possible to perform service-specific charging with resolution finer than a bearer. The underlying idea is that CTFs in an IP-CAN request rules dynamically determining which packets belong to the same flow and should thus be charged jointly.

More generally, the charging rules can be regarded as policies, and thus the Flow-based Charging architecture is a form of policy architecture. In fact, in Rel-7, Flow-based Charging [3GPP 23.125] is being integrated into a more general concept, **Policy and Charging Control** (PCC) [3GPP 23.203], which additionally provides, e.g. access control policies.[1]

In Figure 16.5, the Flow-based Charging architecture is illustrated. We see a **Policy and Charging Rules Function** (PCRF) that contains the **charging rules**. To paraphrase, a rule could say "all packets addressed to a DNS server are not metered" — because the DNS service shall not be charged. The charging rule contains a filter criterion, the **Service Flow Template** (SFT),

---

[1] This generalization brought with it a change in terminology for network elements and interfaces. In this book we adopt the Rel-7 terminology and skip the Rel-6 terminology.

Table 16.1 Charging information in the PS Domain collected for the Chargeable Events PDP Context Activation, Modification and Deactivation

Charging Information
IMSI, MSISDN, PDP address (IP address of UE)
PDP Context Identifier
SGSN identifer(s) and their PLMNs
Access Point Name (APN) and IP address of GGSN
Time of activation
PDP context duration
Traffic Volume, QoS
Radio Access Technology (UMTS? GSM?)
Is it an IMS Signalling PDP Context?
Cell Identifier, Routing Area Identifier
External Charging Identifier (e.g. from IMS)

which allows the PCEF/CTF to recognize the respective packets, e.g. the IP address and port number of the DNS server. Other rules could say "meter the traffic volume for flows destined to the general Internet", and "for our IMS-supported Voice over IP service, meter only the QoS"—the Charging information for the Voice over IP service would be completed by the S-CSCF recording from SIP signalling the time of activation and the duration. More formally, a rule contains a filter criterion that allows for recognizing the packets of the flow that is to be charged. Furthermore, the rules also indicate the charging method—i.e. online or offline charging—and the Charging information to be reported, e.g. traffic volume, or duration.

When a CTF detects the set-up of a new bearer, it wonders how to meter this bearer and requests a corresponding charging rule from the PCRF over the Gx interface. The CTF thus

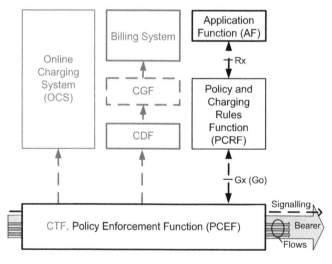

Figure 16.5 The Flow-based Charging architecture (using Rel-7 terminology)

enforces the policies provisioned by the PCRF, and assumes the role of a **Policy and Charging Rule Enforcement Function** (PCEF).

The charging rules stored in the PCRF may contain some wild-cards, especially in the filter criterion that serves to identify a flow: the IP address and port number of services can change, and in particular the IP address of the UE is known only at the time of service delivery. In case the IMS is involved in the set-up of the session for the service, the P-CSCF is able to provide this real-time information. In other cases it may be an Application Server. The abstract term for a network element able to provide such information is **Application Function** (AF).

16.3.2.3.1 Message Flow for Flow-based Charging
In Figure 16.6 we illustrate how a PCEF, alias a CTF in an IP-CAN, obtains charging rules from the PCRF in the case of offline charging. The charging process is initiated by the UE negotiating session details with the AF.

(1) The AF sends charging-specific information, e.g. the IP address and MSISDN of the UE to the PCRF. The PCRF acknowledges receipt of the information. The Diameter NAS application (cf. Chapter 13, Section 13.5.1.4.1) is used as a protocol.

(2.i) The UE requests bearer establishment. If the IPCAN is the PS Domain, the PCEF is located in the GGSN. The UE sends an Activate PDP Context Request to the SGSN, and

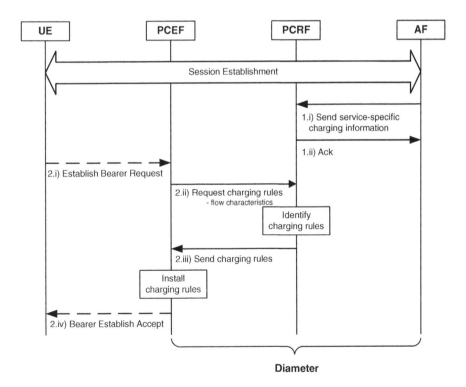

Figure 16.6   Message flow for offline Flow-based Charging

the SGSN translates the message into a Create PDP Context Request message which it sends to the GGSN, cf. Figure 11.9.

(2.ii) The PCEF requests a suitable charging rule from the PCRF, by providing flow information, e.g. the MSISDN or IP address of the UE. The protocol between PCEF and PCRF is the Diameter Credit Control application (cf. Chapter 13, Section 13.5.1.4.1).

(2.iii) The PCRF determines the appropriate charging rule(s) and sends them to the PCEF.

(2.iiv) Only now that it knows how to perform metering for the bearer, can the PCEF accept the request for bearer establishment. If the IP-CAN is the PS Domain, the GGSN will send a positive Create PDP Context Response to the SGSN.

For online charging, the message flow in Figure 16.6 is slightly extended: after step 2.iii, the PCRF (alias CTF) performs a signalling exchange with the OCF in order to obtain quota and authorization. Only when this can be obtained, is the request for bearer establishment admitted.

16.3.3 Charging in the PS Domain

The PS Domain performs bearer-level charging. In this section we will look at how the generic UMTS charging architectures and concepts, including Flow-based Charging, are applied in the PS Domain: Where are the architectural elements located and how do they collaborate? Which Chargeable Events are defined? What Charging information is metered? Which complications arise in roaming scenarios? PS Domain charging is described in [3GPP 32.251].

Generally, CTFs in the PS Domain are SGSN and GGSN, see Figure 16.7. The SGSN collects Charging information related to Radio Access Network and mobility management, whereas the GGSN collects Charging information related to the usage of external networks. Both nodes collect Charging information related to the PDP Contexts.

When a UE performs the GPRS Attach procedure (cf. Chapter 11, Section 11.5, Figure 11.7), the SGSN pulls GPRS Subscription Data, including **Charging Characteristics**, from the HLR. The Charging Characteristics define the Chargeable Events of the subscriber, e.g. a Chargeable Event is produced after a particular traffic volume or a specific time has passed. When a PDP

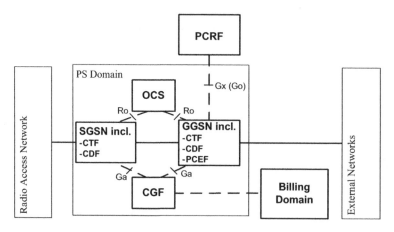

Figure 16.7 The charging architecture of the PS Domain

Context is established, the SGSN passes on the Charging Characteristics to the GGSN (cf. Figure 11.9). The GGSN pulls a rule from the PCRF on whether online or offline charging is to be applied. It furthermore generates a correlation identifier for this PDP Context, the **GPRS Charging Identifier** (GCID), and sends all this information back to the SGSN in its response.

16.3.3.1 Offline Charging in the PS Domain

Together with the CTFs, the CDFs in the PS Domain are normally integrated into SGSN and GGSN. The CGF is, however, a separate node. The CGF's address can be preconfigured into SGSN and GGSN, or it can be provided from the HLR together with the Charging Characteristics.

The main chargeable events in the PS Domain are PDP Context activation, modification and deactivation. For each Chargeable Event, a multitude of Charging information is collected. Table 16.1 provides a simplified overview; the full list is much longer. It is mandatory to collect only a fraction of Charging information, and operators customize CTFs in order to collect what is of interest to them.

In Table 16.1, we recognize one subset of Charging information which serves to identify the subscriber in the Billing System, e.g. IMSI, MSISDN (cf. Chapter 9) and PDP Context identifier and PDP Address (cf. Chapter 11). We recognize another subset of Charging information which is used to determine tariffs and to calculate rates, e.g. time of activation, duration, QoS, traffic volume, PLMN identifier (to identify roaming UEs) or Radio Access Technology. Yet another subset of the Charging information such as Cell Identifier or Routing Area Identifier is only collected for statistical purposes. Finally, the Charging information includes external correlation identifiers that relate to the same service usage; e.g. the IMS correlation identifier if the IMS was used: PS Domain and IMS exchange their correlation identifiers over the Go and the Gx interface, respectively, cf. Chapter 10, Section 10.2.1.

16.3.3.1.1 Offline Flow-based Charging in the PS Domain

In addition to the charging just described, Flow-based Charging can be performed in the PS Domain. As discussed in Section 16.3.2.3, Flow-based Charging allows for a more fine-granular, service-specific charging. The PCEF for Flow-based Charging is located in the GGSN. The PCRF is a network element located in the IMS.[2] Charging information for Flow-based Charging is a subset of the Charging information for general PS Domain charging: the charging rule can to some extent restrict the Charging information and this way reduce overhead—for example the duration of a PDP Context will not be reported when volume-based charging is applied.

16.3.3.2 Online Charging in the PS Domain

Up to Rel-6, online charging in the PS Domain was performed with only the SGSN as CTF, communicating with the OCS via a protocol from the SS7 family. In Rel-6, however, the online charging architecture was reorganized by introducing flow based charging. The GGSN, as PCEF, is now the only CTF. The SGSN, however, continues to collect Charging information that is not available to the GGSN, e.g. the radio access technology used, and sends it to the

---

[2] The reader should be warned that this changes in Rel-8 when the UMTS architecture undergoes a major reorganization, see Part II of this book.

GGSN, embedded in GTP. As for offline charging, the address of the OCS is either preconfigured, or provided by the HLR as part of the Charging Characteristics.

The chargeable events and the Charging information in PS Domain online charging are almost the same as in offline charging, see Table 16.1.

16.3.3.3 Roaming Scenario

We must also look at how PS-Domain charging is performed in roaming scenarios. In a roaming scenario, at least the SGSN and possibly also the GGSN are located in the VPLMN (cf. Chapter 11, Section 11.8)—in other words, one or more CTF, CDF and a CGF are in the VPLMN. The issue we therefore face is that a CGF in the VPLMN must interface with the security-sensitive Billing System in the HPLMN. Two cases can arise:

- When the GGSN is located in the HPLMN, the Charging information collected by the SGSN in the VPLMN does not enter the user's bill. It is only used by the operators in order to determine how many resources they spent on each other's subscribers and to settle their inter-operator bill.
- When the GGSN in located in the VPLMN, all bearer-level charging is collected in the VPLMN. Therefore an interface between the CGF in the respective VPLMN and the Billing System in the HPLMN must be established, and the Charging information collected in the VPLMN enters the user's bill.

16.3.4 Charging in the IMS

The IMS performs subsystem-level charging. It is described in [3GPP 32.260] and Camarillo, 2005. A concise overview is given in Kühne, 2007.

Chargeable events in the IMS are related to SIP signalling messages, e.g. the registration of a user with REGISTER, or the initiating of a session with INVITE. Correspondingly, CTFs are the network elements processing SIP messages, e.g. CSCFs and Applications Servers, as illustrated in Figure 16.8.

When a UE performs a SIP registration, the S-CSCF pulls the subscriber's Service Profile from the HSS (cf. Chapter 15, Section 15.3.1.2). The Service Profile includes the type and address of the network element to which Charging information is to be sent, i.e. the address of the CDF or OCS. The S-CSCF distributes this address to other CTFs by including it in the subsequent SIP messaging, using a private SIP header, the P-Charging Function Addresses header [RFC 3455]. In this way CTFs also learn whether they should perform offline and online charging.

The correlation identifier for SIP sessions is called *IMS Charging Identity* (ICID) and is generated by the P-CSCF. The P-CSCF distributes the ICID to other CTFs with the SIP messaging which it performs, using another private SIP header, the P-Charging-Vector header [RFC 3455].

16.3.4.1 Offline Charging in the IMS

The CTFs for offline charging in the IMS are manifold: all CSCFs, SIP Application Servers, the BGCF, MGCF, MRCF and the MRFP (cf. Chapter 10, Section 10.2). Via Diameter, CTFs send the Charging information to the CDF. The CDF and CGF are normally integrated into one network node.

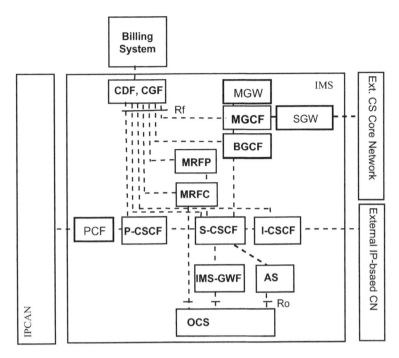

Figure 16.8   The charging architecture of the IMS

The reception of particular SIP messages by a CTF is a Chargeable Event, e.g. a 200 OK message indicating the acceptance of an INVITE—respectively the reception of an ISUP message in case of the MGCF. Table 16.2 presents a subset of the Charging information collected. As for PS Domain charging, one subset of Charging information serves to identify the subscriber in the Billing System, e.g. subscriber identifier and session identifier. Another subset of Charging information is used to determine tariffs and to calculate rates, e.g. type of event, type of SIP message, time stamps, SDP session description, authorized QoS and inter-operator identifier (to identify roaming UEs). Finally, the Charging information includes the

Table 16.2   Selected Charging information in the IMS

Charging Information
Subscriber Identifier
Session Identifier
Inter-Operator Identifier
Event Type, SIP method
Role of CTF (originating/terminating)
Time Stamps
SDP Session Description
Authorized QoS
Access Correlation ID (e.g. from PS Domain)

corresponding correlation identifier of the bearer-level charging, e.g. the GCID, which is exchanged over the Go or Gx interface, respectively.

16.3.4.2 Online Charging in the IMS

For online charging, the number of CTFs is much smaller. Only those network elements collecting Charging information necessary for credit control are included. For example, in offline charging, the I-CSCF collects a variety of Charging information, e.g. the address of the S-CSCF and the **SIP method** i.e. the function evoked on the I-CSCF by the SIP message. However, the identical Charging information is also collected by other CTFs. The I-CSCF collects this information only so the operator can provide statistics as to how often and in what manner the I-CSCF is involved in the signalling path. Thus the I-CSCF is not an online charging CTF. IMS online charging CTFs are the AS, the MRFC and, indirectly, the S-CSCF.

The S-CSCF is not an online-charging CTF per se. Instead, it sends its Charging information using SIP to the **IMS-Gateway Function** (IMS-GWF). From the perspective of the S-CSCF, the IMS-GWF is an Application Server, to which it is instructed to send Charging information. The IMS-GWF, in turn, sends the Charging information to the OCS. The reason for this approach—breaking with the general charging architecture—is better backwards compatibility with existing implementations: the S-CSCFs design does not need to be changed. Of course, the backwards compatibility argument would also hold for other online charging CTFs, e.g. the MRFC. However, the result of a standardization process is often a balance of the interests of the different parties involved. It is not necessarily the technically obvious solution.

16.3.4.3 Roaming Scenario

In a roaming case, the S-CSCF is also located in the HPLMN. The P-CSCF may be located in VPLMN or HPLMN, depending on the location of the GGSN (cf. Chapter 15, Section 15.3.1.1). The other CTFs may also be located in either of the PLMNs. This may mean that the Billing System has to correlate the CDRs received from different PLMNs. However, the details of IMS charging in roaming scenarios are not yet standardized.

16.4 Discussion

Charging provides a good example for illustrating the difference in emphasis between UMTS and Computer Networks.

In UMTS, charging is a central topic. Two charging architectures are defined, one for online charging and one for offline charging. These charging architectures introduce a number of new network elements. As usual, the architectures are agreed first and the protocols are designed as the final step.

For Computer Networks, by contrast, a standardized solution for charging scarcely exists. In fact, commercial WLANs normally employ proprietary solutions. Note that this is possible because charging functionality can be added to a WLAN in a modular fashion, whereas, of course, charging is an integral part of a UMTS Network. The IETF defines a high-level charging architecture. It focuses mainly, however, on the collection of Metering Data. For metering a number of different solutions are described, and a variety of protocols can be employed. The data thus collected can be used for charging, but the details are not specified. Along the same lines, the accounting entity is not a dedicated network element but is integrated conceptually into the AAA Server. Compare this with UMTS, where the CDF and CGFs are, of course, not part of the HLR or HSS.

16.5 Summary

This chapter covered the control functionality of charging. The term "charging" is defined differently by different communities; in this book, we regard charging as being composed of the steps of metering and accounting and as including some parts of the billing process. A particular service usage may cause several metering entities to produce Metering Data. This Metering Data is collected centrally during the accounting step. For billing, Charging Data related to the same service usage is correlated and tariffs are assigned. Correlation requires that a correlation identifier be included by the metering entities together with the metered data. This correlation identifier needs to be generated and distributed for each service usage.

For Computer Networks, the IETF describes a general charging architecture; the details, however, are worked out only for metering and accounting.

For UMTS, 3GPP distinguishes offline charging and online charging. Separate charging architectures are defined for each case.

In offline charging, CTFs perform metering, CDFs and CGFs perform accounting and a Billing System performs billing. Only the metering is a real-time process. In the PS Domain, SGSN and GGSN function as CTFs. The most important Chargeable Events are PDP Context manipulations. For offline charging in the IMS, almost all of the network elements processing SIP or ISUP messages act as CTFs. Chargeable Events relate to the reception of SIP or ISUP messages. The specification of the Billing System is outside the scope of 3GPP.

In online charging, CTFs perform metering as above; additionally, however, they have the task of asking the OCS for quota before a new service usage is admitted. Furthermore, they must interrupt service delivery when the quota is exhausted.

The PS Domain and generally an IP-CAN perform bearer-level charging, whereas the IMS performs subsystem-level charging. Bearer-level charging was originally only possible with the resolution of a bearer. Flow-based Charging, however, allows service-specific charging on the bearer-level of any resolution. Flow-based Charging is based on policies and is realized on the basis of a policy architecture featuring a PCRF in the IMS and a PCEF in the IP-CAN, e.g. in the GGSN. Flow-based Charging thus also allows for coordination of the bearer-level, subsystem-level and service-level charging.

17

Policy Control

In Chapters 11–16 we covered the most important UMTS control functions. In the next two chapters we will look at the additional functionality introduced in the last two UMTS releases. Generally, we observe that the recent releases have developed UMTS so as to become more flexible. In this chapter, we will introduce the policy architecture **Policy and Charging Control** (PCC) of Rel-7, which is of a more general application than the Flow-based Charging of Rel-6 (cf. Chapter 16, Section 16.3.2.3). In Chapter 18, we will look at how a WLAN can be attached to a UMTS Network.

Traditional networks do not need to be flexible. The number of external events is quite low. For example, in a fixed-line circuit switched network, the events are grouped around "a user picks up the phone to make a call", "a user picks up the phone to receive a call" and "a user hangs up". Also, in traditional networks, the possible reactions of a network element to events are of a low complexity. For example, in a simple Computer Network, a router receives a packet, looks up the destination address in the routing table, determines the outgoing interface and sends out the packet. Under such circumstances it is feasible to hard-code inflexibly the reaction of network elements to external events.

In recent years, however, the variety of external events has increased dramatically. When a user has switched on a UMTS phone, events can be "user is roaming", "user is moving", "user places a circuit-switched call", "user wants to surf the Internet", "user wants to establish an IMS session", etc. The complexity of network elements, and respectively their reaction, has also increased. For example in a Computer Network, users are authenticated and authorized, and mobility can be supported. In such a complicated environment, hard-coding the mapping of events to reactions becomes too inflexible. Instead, it is desirable to separate the mapping engine from the rules governing its behaviour. The rules, known as **policies**, should be exchangeable; in fact they should be exchangeable "on the fly".

The IETF started work on policy control related to AAA services in the 1990s. In the meantime, generic policy architecture and a number of protocols have been developed. In UMTS, policy architecture, PCC, is being worked on for Rel-7. PCC comprises control of some aspects of charging, control of the QoS delivered to a flow and control of which packets pass a gateway to an IP-CAN, e.g. a GGSN.

UMTS Networks and Beyond   Cornelia Kappler
© 2009 John Wiley & Sons, Ltd

As usual, this chapter starts with a description of the problems and a general approach. We then describe the policy work for Computer Networks by the IETF and continue with UMTS' PCC.

Terminology discussed in Chapter 17:	
Application Function	AF
Authorization Token	
Policy	
Policy and Charging Control	PCC
Policy Decision Function	PCF
Policy Decision Point	PDP
Policy Enforcement Point	PEP
Service-based Local Policy	SBLC

17.1 Description of the Problems

The problem solved by a policy infrastructure is the following: a network element receives an event and must decide how to react to it. The event is described by a number of parameters, e.g. originator of the event, time of day, etc. The mapping of event to reaction, taking into account some of the parameters, is described by a policy. The network element must therefore find the right policy to apply and then enforce this policy.

Policies can be applied in very diverse scenarios. For example, policies can be used to control the access to a network, for configuring network elements upon installation, for describing how to react to errors or unexpected events, for prescribing how to treat particular flows, etc. In fact, a large body of recent research and development work is focused on basing the entire network management system on policies.

An important property of policies is that they can be exchanged easily in order to modify the behaviour of the network. A policy change must be possible during the run time of the network without having to recode network elements or stop network operation. Policies are therefore stored in a **policy repository** where they are accessible centrally to both the network elements applying policies and the operator of the network.

Policies generally consist of a **condition** triggering usage of the policy, and one or more **actions** which are to be executed when the condition is met. Put more formally,

IF (condition) THEN (action) ELSE (other_action).

Thereby, condition is a Boolean statement evaluating to TRUE or FALSE. Of course, it is allowed to AND or OR several conditions. For example, let us consider the event of a user requesting a particular QoS for a flow. For this event, a policy may apply that says, in pseudocode:

IF ((subscriber_is_authorized) AND (bandwidth_is_available)) THEN admit ELSE reject

Another example is a policy for Flow-based Charging that could say:

IF (packet_destined_to_DNS) THEN (do_not_meter)

Policies can be attached with a priority. In this way it is possible to express a general policy which is always applicable, and to handle exceptions to the general policy by adding "subpolicies" with a higher priority. For example, generally, the volume of all flows should be metered. However, if a flow relates to signalling, then it will not be metered at all.

For applying policies, a policy language must be defined that is sufficiently powerful to describe the conditions and actions for the scenario in question. Policy languages in use range today from quite simple, e.g. when they are applied in limited environments such as Flow-based Charging, to highly complex, so that their usage requires a full-blown expert system.

17.2 Policy Control in Computer Networks

In this section we will look at the application scenarios for policy control in Computer Networks and at the different solutions developed by the IETF. The motivation for applying policies in the IETF arises from the areas of roaming and admission control, discussed in the following two subsections.

17.2.1 Policy Control in Roaming

A simple network access scenario consists of a subscriber desiring access to her home domain, as illustrated in Figure 17.1a. The subscriber's Authentication Client is authenticated by a NAS

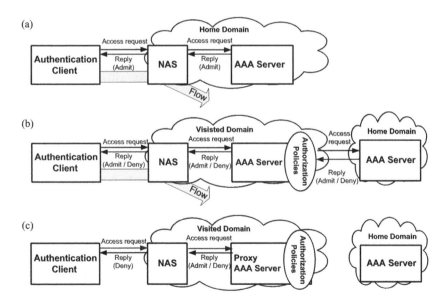

Figure 17.1 Network access scenario with policy controlled authorization. (a) Subscriber desires access in home domain, AAA Server in home domain. (b) Subscriber desires access in visited domain (c) Subscriber desires access in visited domain, policy control in visited domain denies access

on the basis of decisions made by an AAA Server (cf. Chapter 13, Section 13.5.1.2). The AAA Server bases its decisions on a simple, hard-coded, rule: if and only if the subscriber can authenticate herself will she be admitted. In other words, authentication normally does not need updatable policies.

When the subscriber is roaming, the NAS is located in the visited domain and the authentication decision is proxied by a Proxy AAA Server in the visited domain to the AAA Server in the home domain. The Proxy AAA Server does not participate in the authentication decision. However, it may contribute an authorization decision: for example, the visited domain may admit roaming subscribers from a particular home domain only at certain times of day. Such authorization rules should be simple to update, because, e.g. the list of home domains or the time of day may change. In other words, authorization should be implemented as policies. The scenario is illustrated in Figure 17.1b/c: an access request is passed from the Authentication Client via the NAS to the Proxy AAA Server. In Figure 17.1b, the Proxy AAA Server authorizes the access request, and it is passed to the AAA Server for authentication. By contrast, in Figure 17.1c, the Proxy AAA Server decides that it cannot accept the user and instructs the NAS to deny access. In this case, the AAA Server in the home domain never receives the access request. Alternatively, the Proxy AAA Server may first pass on the access request but then dismiss the acceptance decision by the AAA Server and reject the subscriber after all (not shown in Figure 17.1) [RFC 2607].

17.2.2 Policy Control in QoS Authorization

Some subscribers may be more equal than others when it comes to QoS reservations. This preferential treatment is also expressed by authorization policies. Thus, when a subscriber performs QoS signalling in order to reserve resources for a flow, the network, e.g. the Access Router, determines whether the subscriber is authorized to receive the QoS desired (cf. Chapter 14, Section 14.1.2, Figure 14.7) by querying a **Policy Decision Point** (PDP). This PDP could be an AAA Server. If the user is not authorized—or if resources are not available—the user is denied access. We illustrate this situation again in Figure 17.2. The reader will notice the architectural similarity to Figure 17.1a.

17.2.3 The IETF Policy Architecture

Motivated by the QoS authorization problem, the IETF devised a general policy architecture [RFC 2753], depicted in Figure 17.3. It consists of a **Policy Enforcement Point** (PEP) receiving a request that requires a policy decision—e.g. a QoS request or an access request. The PEP formulates a policy decision query and sends it to the Policy Decision Point (PDP). The query message contains information such as who issued the request or what QoS is desired. The PDP consults its policies, which may be stored in a policy directory, and returns a decision. It is the responsibility of the PEP to enforce the decision.

When we map this policy architecture to the QoS authorization problem, we recognize that the PEP is the Access Router in Figure 17.2. Of course, it can be also a Bandwidth Broker (cf. Chapter 14). More generally, each router receiving a QoS reservation request may be a PEP and perform authorization, so that a QoS request is authorized more than once. The IETF also developed a protocol that the PEP uses in order to pull policy decisions

(a)

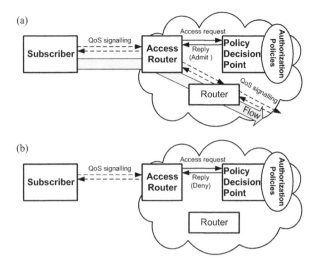

(b)

Figure 17.2 QoS authorization with policy control. (a) Resource request is admitted. (b) Resource request is denied

from the PDP, it is the COPS protocol [RFC 2748] which we have already encountered in Chapter 10.

Mapping the policy architecture to the roaming problem, we will recognize that the NAS acts as PEP and the AAA Server acts as PDP. The protocol between PEP and PDP in this case is, of course, Diameter or RADIUS, which therefore provides an alternative to COPS.

17.2.4 Policy Push

The policy examples which we have encountered are still fairly restricted: The policy decision process is always triggered by a request to a PEP. The PEP subsequently **pulls** a policy decision from the PDP in real-time. In a more general setting, the PDP would proactively **push** policies to the PEP with a flexible timing. This model is especially applicable to decisions that relate to network configuration rather than to authorization decisions. For example, the PDP could push a policy to the Access Routers which prescribes that 50% of the inbound bandwidth is reserved for a real-time services. The IETF policy architecture in Figure 13.3 can also accommodate

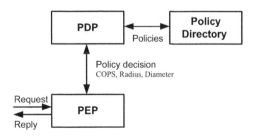

Figure 17.3 IETF policy architecture

such push-policies. The corresponding protocol support is provided in [RFC 3084] with a COPS protocol extension.

17.3 Policy Control in UMTS

Policy control was first introduced in UMTS in Rel-5: the P-CSCF performs policy-based QoS authorization decisions indirectly which are enforced at the GGSN, as PEP (see Chapter 10, Section 10.2 and Figure 10.1). This mechanism is called **Service-based Local Policy** (SBLC) [3GPP 23.207]. Independently, in Rel-6, Flow-based Charging was introduced (cf. Chapter 16, Section 16.3.2.3 and Figure 16.5). Here, the CTF is a PEP querying decisions on how to meter flows from the PCRF. In Rel-7, the two policy mechanisms were streamlined and united in the **Policy and Charging Control** (PCC) architecture [3GPP 23.203]. In the following we will first describe SBLC and then show how it has evolved and, together with Flow-based Charging, becomes PCC. Note that at the time of writing, PCC is only just being specified. Some details, therefore, are still to come.

17.3.1 Service-based Local Policy

When a subscriber wants to use an IMS service, the UE performs SIP signalling with the CSCFs for service invocation and control (cf. Chapter 15, Section 15.3). This SIP signalling includes the QoS necessary for the service as a precondition. The actual delivery of the service is then via the IP-CAN, e.g. the PS Domain. The UE therefore needs to set up a bearer, e.g. a PDP Context, with QoS suitable for the desired service. In other words, the UE performs two independent signalling procedures.

From an operator's perspective, this independence is a problem: the UE could signal a particular QoS via SIP in the IMS, and then signal a different QoS for the bearer in the IP-CAN. How would this be accounted for in charging? Clearly, the operator needs better control. A link was therefore introduced between P-CSCF and GGSN which allows one to ensure that the QoS signalling performed by the UE in IMS and PS Domain is consistent—for other IP-CANs such a link was introduced later, starting with Rel 7.

The link between IMS and PS Domain consists of the GGSN querying a policy decision—an authorization—from a **Policy Decision Function** (PDF) in the IMS whenever it receives a request for PDP Context setup that is related to an IMS service. Only when the PDF gives its ok, does the GGSN allow the PDP Context establishment to continue. We recognize the policy architecture: the GGSN is a PEP, querying policy decisions from the PDF which is a PDP (Policy Decision Point). Actually, it was debated whether to name the PDF in an IETF-conformant way, but the idea was discarded because the acronym "PDP" already has another meaning in the 3GPP world.

On what basis does the PDF make its decisions? It learns which is the QoS signalled via SIP (the "SIP QoS") from the P-CSCF and this QoS is what it authorizes. The GGSN compares the authorized QoS to the QoS signalled with the PDP Context setup (the "bearer QoS") and only allows the PDP Context setup to continue if the two are aligned. The corresponding architecture is shown in Figure 17.4.

Now imagine that the PDF receiving numerous data sets from the P-CSCF on "SIP QoS" arriving in SIP messages. At the same time, the PDF receives numerous queries from the GGSN for authorizing "bearer QoS" requests from a variety of UEs. How does the PDF know which

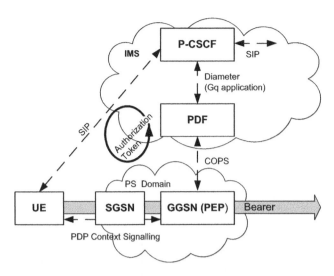

Figure 17.4 SBLC architecture. The curved arrow denotes the itinerary of the Authorization Token

belongs to which? We need an identifier! Note that a UE identifier, e.g. the current IP address, is not sufficient to bind the "SIP QoS" with the "bearer QoS", because a single UE may request several IMS services simultaneously. Furthermore, a UE may setup a single bearer containing several flows, e.g. one flow carrying an IMS service and another flow carrying a video stream from a server outside the IMS. In this case the PDF would be unable to map the "bearer QoS" to any "SIP QoS" it knows.

The **Authorization Token** was therefore introduced as binding information. It is generated by the PDF for each "SIP QoS" data set, and passed to the UE by the P-CSCF. The UE includes the Authorization Token in its PDP Context establishment signalling, it is extracted by the GGSN and passed back to the PDF. The PDF is now in a position to inform the GGSN of the "SIP QoS" corresponding to the "bearer QoS". If the two do not match, the GGSN blocks the PDP Context setup. In other words, the UE is prevented from setting up PDP Contexts carrying IMS-related flows and non-IMS related flows simultaneously. The itinerary of the Authorization Token is sketched in Figure 17.4. In the following sub-section, the message flow is described in detail.

17.3.1.1 Message Flow for Service-based Local Policy

The message flow for Service-based Local Policy illustrated in Figure 17.5 which we are now going to study binds the message flow for PDP Context establishment (Figure 11.9) and the message flow for IMS session creation (Figure 15.8). In particular, we will see how the "black boxes" in Figure 15.8 which are labelled "Authorization of QoS A/B" and "Reservation of QoS" are filled in.

A note on numbering of the messages in Figure 17.5: the Figure depicts several interweaved protocol exchanges: a SIP exchange between UEs and CSCFs, a Diameter exchange between P-CSCF and PCF, a COPS exchange between GGSN and PCF, etc. In order to help the reader follow the structure, each such message exchange is indexed with an arabic number, and the

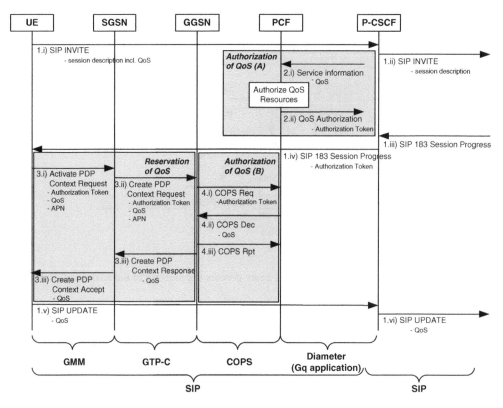

Figure 17.5   Message flow for SBLC

individual messages inside each exchange are sub-numbered with roman numerals. For
example, the first SIP message between UE and P-CSCF is numbered 1.i.

(1.i,ii)   The UE generates an INVITE request and sends it to the P-CSCF. The SDP session
           description in the INVITE includes the desired QoS. The P-CSCF forwards the
           INVITE to the S-CSCF. This message sequence corresponds to the messages (i, ii) in
           Figure 15.8a. In the following we omit some intermediate SIP messages which not
           pertinent to this subject matter, such as the 100 TRYING or PRACK messages.
  (2.i)   The P-CSCF sends service information, including the QoS description, to the PDF
           and requests that an Authorization Token be generated. For protocol support, 3GPP
           defined the Diameter Gq application for interaction of P-CSCF and PDF.
 (2.ii)   The PDF consults its policies and authorizes (or not) the QoS for this service. Note that this
           decision is not subscriber-specific, the PDF does not consult the HSS. The decision
           depends only upon which QoS the operator considers suitable for the requested service.
           If the QoS can be authorized, the PDF generates the Authorization Token and includes it
           in its reply message to the P-CSCF.
(1.iii,iv) P-CSCF receives the 183 Session Progress message from the S-CSCF. It includes the
           Authorization Token in a private header extension to SIP which was defined by the

IETF in collaboration with 3GPP and other standardization bodies from the tele-communication world [RFC 3313]. The augmented message is forwarded to the UE.

(3.i) Upon reception of the 183 Session Progress message, the UE starts the setup of the bearer for transporting the session. In other words, the UE sends an Activate PDP Context Request to the SGSN, containing the APN for the service, the desired QoS and the Authorization Token.

(3.ii) The SGSN determines the GGSN, authorizes the QoS, checks whether sufficient resources are available and sends a Create PDP Context Request message to the GGSN. The Authorization Token, the APN and the desired (possibly downscaled) QoS are included in the message.

(4.i) The GGSN recognizes from the APN that it should query the PDF to authorize the desired QoS. It sends a COPS request (Req) message to the PDF with the Authoriza-tion Token.

(4.ii,iii) The PDF uses the Authorization Token to identify the corresponding authorization status and the QoS requested originally via SIP. It sends a COPS decision (Dec) message to the GGSN with the authorized QoS. The GGSN acknowledges with a COPS report (Rpt) message.

(3.iii,iv) The GGSN compares whether the QoS authorized by the PDF is equal to the QoS contained in the PDP Context Setup message. If this is the case, the GGSN is allowed to continue with the PDP Context Setup. It acknowledges this good news with a Create PDP Context Response message to the SGSN and ultimately to the UE. Note that the PDP Context just established is a secondary PDP Context—the UE obviously already has a PDP Context, including an IP address, to perform the SIP signalling.

(1.v,vi) The UE has successfully established a bearer with suitable QoS. It continues the SIP signalling with an UPDATE message in order to inform its communication partner of the possibly updated QoS. Please consult Figure 15.8b for the continuation of this exchange.

17.3.2 Policy and Charging Control

Policy and Charging Control (PCC) unites and generalizes the applicability of SBLC and Flow-based Charging. Let us start by asking ourselves what are the communalities of the two procedures. Both SBLC and Flow-based Charging aim to coordinate IMS service and bearer: SBLC coordinates the resource requests, and Flow-based Charging coordinates the charging.

Both SBLC and Flow-based Charging install—in IETF language—a PEP in the GGSN. When the GGSN receives a request for PDP Context setup, it queries a policy decision from—in IETF language—a PDP. In the case of SBLC, the PDP makes a policy decision relating to the authorization of QoS. In the case of Flow-based Charging, the PDP makes a policy decision relating to the charging rule that shall be applied. The PDP includes in its decision information it received from the P-CSCF on the IMS session for which the PDP Context is to be installed. Once the GGSN receives a positive decision from the PDP, it proceeds with the PDP Context setup.

We may see that SBLC and Flow-based Charging only differ in the character of the policies being applied. The basic mechanisms and the corresponding architectures are the same. PCC unifies the architectures in that the PDP for SBLC and Flow-based Charging becomes one network element, called PCRF, as illustrated in Figure 17.6.

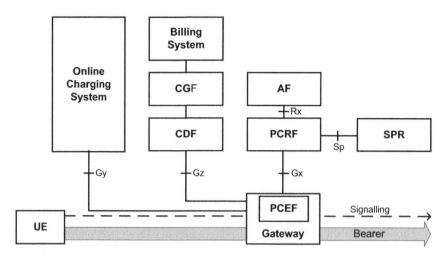

Figure 17.6   PCC Architecture

PCC, however, satisfies more goals. We have already learned in Chapter 10 that the IMS will be accessible from any IP-CAN. In particular, other telecommunication standardization bodies such as ETSI, ITU-T and 3GPP2 are currently specifying how to use IMS services from their networks. Just as with 3GPP, these standardization bodies appreciate operator control, and they need especially to bind the IMS service to the bearer in their IP-CAN. PCC thus expands the binding concept to general IP-CANs beyond the PS Domain. The PEP, called PCEF, is located in a network element called generically a **Gateway**. It pulls both QoS authorization and charging rules from the PCRF. In the case of the PS Domain, the Gateway with the PCEF is the GGSN. In the Telecommunication Networks defined by ETSI, ITU-T and 3GPP2, the PCC concept is adopted, and such Gateways are also identified. In Chapter 18, Section 18.1.8 we discuss how a WLAN can become an IP-CAN, as yet another example.

As a final generalization, in PCC the service for which the bearer in the IP-CAN is established does not need to be an IMS service. Service-specific information is delivered by a network element called the **Application Function** (AF) (cf. Chapter 16, Section 16.3.2.3). The AF can be the P-CSCF or, for example, an Application Server. We therefore see how PCC is positioned as a generic binding and control mechanism between an IP-CAN and a Service Network, with a potential existence quite independent of UMTS. In Part II we will see how PCC is evolved further.

Together with the introduction of PCC, the COPS protocol between GGSN and PCF/PCRF was abandoned and replaced by Diameter: for communication between AF and PCRF, the Diameter Rx application, and for communication between PCRF and PCEF, the Diameter Gx application.

Comparing the PCC architecture in Figure 17.6 to the SBLC architecture in Figure 17.4 and the Flow-based Charging architecture in Figure 16.5, we notice an additional, interesting network element, the **Subscriber Profile Repository** (SPR). The SPR contains subscriber information and bears considerable resemblance to an HSS. Presumably, the SPR includes a subset of the information in the HSS plus some additional information, but at this point in time the details are not yet clear; remember that the PCC specification is still new. As the 3GPP authors in their deadpan prose put it: "The SPR's relation to existing subscriber databases is not specified in this release".

The main reason, however, that the SPR is interesting is because neither SBLC nor Flow-based Charging derive policy decisions from subscriber-specific data. This possibility is newly introduced with PCC. For example, this enables the PCRF to authorize QoS not only on the basis of the service requested via the AF. Instead, the PCRF can also determine whether the subscriber is authorized for such a QoS in the first place. Note that authorization is also performed by the SGSN and the S-CSCF. PCC should, however, work with any IP-CAN and also without IMS. This functionality is therefore pulled also into the policy-control architecture.

The combination of SBLC with Flow-based Charging in PCC, furthermore, allows for simplification of the authorization procedure, particularly of how the service ("SIP QoS") is bound to the bearer ("bearer QoS"): In SBLC, the binding is done by the Authorization Token. The Authorization Token is abolished in PCC and replaced by a filter criterion, the **Service Flow Template** (SFT), which allows for identification of a particular flow within a bearer, cf. Chapter 16, Section 16.3.2.3. When the GGSN queries a QoS authorization— presenting a UE identifier and the QoS desired—the PCRF consults its database, pulls out a suitable SFT and charging rule, and passes both to the GGSN [3GPP 29.213]. A binding is possible because the SFT permits a unique identification of a single service-related flow within a bearer. The GGSN can apply the SFT and allot the authorized QoS to the packets thus identified. With PCC it is therefore also allowed to mix flows that are authorized via PCC with flows that do not need to be authorized within the same bearer. With SBLC, by contrast, authorization could only be performed with per-bearer resolution, leading to the difficulty of mapping bearer to service.

17.3.2.1 Policy and Charging Control in Roaming Scenarios

How does PCC work in a roaming scenario? We expect the Gateway, i.e. the PCEF, to be located in the VPLMN. The PCRF needs access to subscriber-specific data and HPLMN-specific charging rules; therefore it is best located in the HPLMN. 3GPP is currently investigating proxy architecture with a Visited-PCRF in the VPLMN proxying authorization requests to the Home-PCRF in the HPLMN, see Figure 17.7. The AF, e.g. the P-CSCF, is envisaged to be located in the HPLMN. With this architecture it is possible for the V-PCRF to add its own policy decisions to the policy decisions of the H-PCRF. For example, the V-PCRF may reject a QoS authorization decision by the H-PCRF and block a particular flow.

The PCC roaming architecture for Rel-7 is located in the purely informative annex of [3GPP 23.203]. It will thus not become part of Rel-7. It will, however, be detailed and included in Rel-8.

17.4 Discussion

Both Computer Networks and UMTS employ policies. Compared to the general applicability of policies, e.g. in the areas of network management or network configuration, the policy usage in Computer Networks and UMTS is still restricted to access control and configuration of

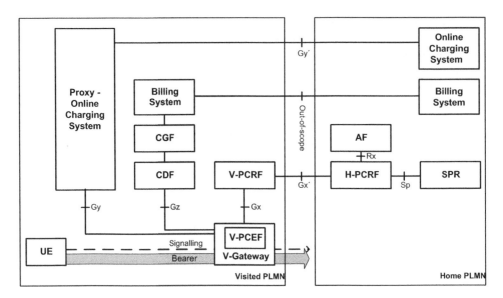

Figure 17.7   PCC Architecture in a roaming scenario

charging rules. We will see in Part II of this book how policies may take on a more important role in the future.

The 3GPP policy architecture is intended to conform as far as possible to IETF standards [3GPP 23.207]. Comparing the IETF policy architecture in Figure 17.3 to the UMTS SBLC and PCC architectures in Figures 17.4 and 17.6, however, we may observe that they are not identical: there are some minor differences in that the IETF locates the policy directory explicitly outside the PDP, and that 3GPP has the PDP-equivalent, the PCRF, interact with a subscriber database. The more interesting difference is the existence of the AF in PCC, which has no equivalent in the IETF policy architecture.

What is the role of the AF? An important purpose of PCC is binding a service, e.g. an IMS service, to a bearer, e.g. a PS Domain bearer, and ensuring their consistency. The AF supplies service information to the PCRF which is used to police the bearer passing the PCEF. This kind of distributed information gathering is not considered in the IETF policy architecture. In the cases of IETF policy use—QoS authorization and roaming—all information relevant to the event or flow being policed is available locally in the PEP. The IETF does not think in terms of all-encompassing architectures with coordinated control functions inside the network.

When we compare the envisaged PCC roaming architecture (Figure 17.7) to an IETF roaming architecture (Figure 17.1), we may observe, however, commonalities: PCC places a proxy PCRF in the VPLMN which may interfere with the authorization decisions reached in the HPLMN. Likewise, the IETF locates a Proxy AAA Server in the visited domain which may reject or update decisions from the AAA Server in the home domain. By contrast, the older PS Domain roaming architecture does not show a proxy: the SGSN in the VPLMN communicates directly with the HLR in the HPLMN in order to authenticate and authorize the subscriber (cf. Figure 11.10).

17.5 Summary

Policies allow for a real-time configuration of control functions such as authorization or charging. Applications for policies range from comparatively simple policies for authorization to highly complex policy-based network management systems.

For Computer Networks, the IETF standardized policy usage related to AAA services in roaming scenarios and for QoS authorization. Generic policy architecture was developed featuring a PDP making policy-based decisions that are enforced by a PEP.

3GPP first employed policies for QoS authorization; in particular for enforcing that the QoS requested for the bearer in the PS Domain conforms to the QoS requested for the IMS service. In Rel-5 the corresponding mechanism is called SBLC. The SBLC architecture involves a PDF as—in IETF language—PDP, making policy decisions that are enforced at the bearer by the GGSN as—in IETF language—the PEP. The PDF additionally receives service-specific information from the P-CSCF. The binding of "bearer QoS" to "SIP QoS" is realized by means of an Authorization Token generated in the PDF, which is passed to UE and, via the GGSN, back to the PDF.

In Rel-6, 3GPP broadened its policy portfolio by employing charging policies in Flow-based Charging, which allowed for the charging of individual flows rather than of entire bearers. Additionally, Flow-based Charging allows for coordination of IMS charging and bearer charging.

In Rel-7, PCC was introduced as generic policy infrastructure for both QoS authorization policies and charging-related policies. PCC thus unites and generalizes SBLC and Flow-based Charging. In PCC, policy decisions are made by the PCRF and are enforced by a PCEF. The PCRF receives additional service-related information from the AF. Compared to SBLC and Flow-based Charging, PCC is more general as it works for any IP-CAN and any service.

18

WLAN and Other Alternative Access Methods

Traditionally, mobile Telecommunication Networks have had their own, specific, Radio Access Network. The early releases of UMTS already exhibited some flexibility in that a 3GPP Core Network can modularly connect to any 3GPP-standardized RAN, i.e. UTRAN and GERAN. The result is called a **3GPP System** rather than UMTS (cf. Chapter 4, Section 4.6). With Rel-6, 3GPP is breaking new ground: it introduces greater flexibility by specifying how a UE can access the 3GPP Core Network via a WLAN Access Network.

The rational for integrating WLAN as an Access Network into the 3GPP System is, of course, the potential for creating operator revenue.

- On the one hand, by 2006 WLAN hotspots had become commonplace. In fact, many 3GPP operators that run a 3GPP System have themselves entered the WLAN hotspot market. However, they usually operate WLAN and UMTS as independent networks. It is of course desirable to create synergies by sharing the subscriber base and by allowing subscribers to move between the two network types.
- 3GPP operators can also benefit from interworking with WLAN hotspots operated by third parties. For example, new customers can be gained when users of a WLAN hotspot can access the 3GPP System directly, e.g. for using IMS services, without additional bureaucratic overhead.
- Furthermore, WLAN equipment costs less than UMTS equipment, both in purchase price and in operation. It is therefore advantageous for 3GPP operators to extend their coverage with WLAN Access Networks, in areas where the virtues of a UTRAN—seamless handover and QoS support—are not so important or too expensive.

Interestingly, 3GPP developed two independent solutions for WLAN integration: **Interworking WLAN** (I-WLAN) and **Generic Access Network** (GAN), the latter also known as **Unlicensed Mobile Access** (UMA). In fact, GAN is not only applicable to WLAN but also to any wireless broadband IP-based Access Network or Radio Access Network, e.g. Bluetooth

UMTS Networks and Beyond   Cornelia Kappler
© 2009 John Wiley & Sons, Ltd

or even the **Digital Enhanced Cordless Telecommunications** (DECT), the technology to connect your cordless phone at home to a fixed-line base station.

As an aside, GAN thus allows mobile operators to address the fixed-line market. However, as we will discover in Part II, the I-WLAN approach is more important for the evolution of 3GPP Systems.

Independently of I-WLAN and GAN, another approach for accessing a 3G System seems to be gaining momentum: **femtocells**. Here, the operator sells miniature Node B's or even entire UTRANs to its customers. This customer-premises equipment is connected to the 3GPP System via a fixed line, e.g. **Digital Subscriber Line** (DSL). Femtocells are only now being standardized, and a large variety of solutions exists.

We therefore divide this chapter into a section on I-WLAN, a section on GAN, a comparison of these two approaches, and finally a description of the femtocell idea.

Terminology discussed in Chapter 18:	
Enhanced GAN	EGAN
Femtocell	
Generic Access Network	GAN
Interworking WLAN	I-WLAN
Packet Data Gateway	PDG
Unlicensed Mobile Access	UMA
WLAN Access Gateway	WAG

18.1 Interworking WLAN

We start this section with an overview of I-WLAN usage scenarios. We then describe the I-WLAN architecture defined in Rel-6 in detail, and cover the various control aspects—mobility, security, QoS, charging and policy control—in the same sequence as we did for traditional UMTS. An overview of the issues is also given in [Ahmavaara 2003].

18.1.1 I-WLAN Scenarios

Work on I-WLAN started in the 3GPP SA1 working group with a feasibility study [3GPP 22.934]. Six consecutive interworking scenarios were defined, each building on the one preceding it and each being more powerful than the one before. The idea was to define an evolutionary path from the isolated co-existence of WLAN and 3GPP technology to fully seamless system interoperation. The technical solution for interworking should not affect the WLAN specification; it is only a 3GPP issue.

- 1st Scenario—Common billing and customer care
  In this scenario, the user has a single subscription that enables her to access independently both the 3GPP PLMN and a (set of) WLAN hotspot(s)—she can access the Internet from the WLAN, but she cannot access the 3GPP network. However, she receives a single bill covering both 3GPP and WLAN usage. In other words, in this scenario the only relation

between the 3GPP System and the WLAN is that they share subscribers. Since billing is outside the scope of 3GPP, the support of this scenario does not require any update of a 3GPP specification.

- 2nd Scenario—3GPP-based access control and charging
  In this scenario, the subscriber also has a single subscription for both a 3GPP PLMN and WLAN hotspots. As in the 1st scenario, she accesses the 3GPP PLMN and the WLAN hotspots independently and it is not possible to use 3GPP services when being attached to a WLAN. However, subscriber information is stored exclusively in the 3GPP Core Network. Consequently, when the subscriber accesses a WLAN hotspot, the authentication and authorization information is pulled from the 3GPP network. Charging is also performed in the 3GPP network. For the WLAN operator and the 3GPP operator, the advantage of this scenario is that subscriber handling is simplified.
- 3rd Scenario—Access to 3GPP packet-switched services
  This scenario builds on the first two scenarios, and also allows the UE to set-up a bearer to the PS Domain for accessing packet-based services in the 3GPP System, e.g. IMS services. It is, however, not yet possible to handover from the WLAN access to the UTRAN.
- 4th Scenario—Service continuity
  This scenario additionally supports handover between the WLAN and UTRAN: services, e.g. a streaming video service, continue to work after the handover; they do not need to be set-up again. However, the handover does not need to be seamless and might be noticeable to the subscriber.
- 5th Scenario—Seamless service
  With this scenario, the handover between WLAN and UTRAN becomes seamless for packet-based services.
- 6th Scenario—Access to 3GPP circuit-switched services
  This last scenario adds access to CS-Domain supported services from the WLAN—without, of course, implementing circuit-switched technology in the WLAN. For circuit-switched services, handover between WLAN and UTRAN will also be seamless.

We thus see how each scenario offers more functionality than the previous one. The document [3GPP 22.934] describes the scenarios and is of course just a feasibility study. For I-WLAN, 3GPP has provided technical specifications for comparatively simple scenarios only: in Rel-6, the 2nd Scenario and 3rd Scenario are included. Further scenarios, e.g. mobility between heterogeneous access systems, are only included in Rel-8 when the topic is approached in a much more general fashion, see Chapter 21.

18.1.2 I-WLAN Architecture

Let us look at the relation between I-WLAN hotspot and the 3GPP System in more detail. The I-WLAN hotspot accepts all users that are authenticated by the 3GPP System. Users are charged by the 3GPP System, and the I-WLAN operator is reimbursed by the 3GPP operator: one might say that the I-WLAN hotspot and the 3GPP PLMN have a kind of roaming relationship, see Figure 11.1, with the I-WLAN hotspot in the role of a VPLMN. However, this analogy has a shortcoming: the I-WLAN is an Access Network; it is not a full PLMN. Therefore, when an I-WLAN is interworking with a 3GPP System, some core network functions must be performed in the 3GPP System. In any event, we may safely assume that

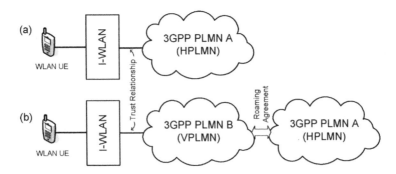

Figure 18.1 I-WLAN high-level architecture (a) non-roaming case, the I-WLAN is attached to the
HPLMN of the UE. (b) roaming case, the I-WLAN is attached to a VPLMN

an I-WLAN has a long-term relationship of trust with the 3GPP System involving a shared
secret and a secure tunnel.

Of course, the I-WLAN has a relationship of trust only with one or two 3GPP PLMNs—we
exclude the case that has no relationship of trust at all as not being relevant for our discourse.
This means, from the UE's perspective, that we can have either of the two situations, illustrated
in Figure 18.1:

- Non-roaming case
  A UE has a subscription with PLMN *A*, its HPLMN. The I-WLAN happens to have a
  relationship of trust with the same PLMN *A*. Thus, when the UE accesses the I-WLAN,
  authentication information is pulled directly from the HPLMN.
- Roaming case
  The UE has a subscription with PLMN *A* as before. The I-WLAN, however, has a relationship
  of trust with another PLMN *B*. When the UE accesses the I-WLAN, the authentication
  information must thus be pulled from the PLMN *A* via PLMN *B*. Of course, this is only
  possible if the two PLMNs have a Roaming Agreement.

The general problem that needs to be solved when designing the I-WLAN solution is that the
WLAN specification shall not be touched: a WLAN does not understand 3GPP-specific protocols,
nor can it make sense of 3GPP-specific network elements. Consequently, the solution is to shield
the 3GPP-specifics from the I-WLAN. With this in mind we now examine the I-WLAN
architecture specified in [3GPP 23.234] and discuss how the different scenarios from
Section 18.1.1 are realized. We start with the simpler non-roaming case illustrated in Figure 18.2a.

18.1.2.1 Non-roaming Case

For **the 2nd Scenario**, a UE accesses the I-WLAN and is authenticated in the 3GPP System: The
UE has a UICC with an I-WLAN subscription. It is authenticated and authorized following the
usual WLAN procedure (cf. Chapter 13, Section 13.5.1.2) by an AAA Server in its 3GPP
HPLMN. The AAA Server, in turn, is synchronized with the HLR or HSS. Once authenticated
and authorized, the UE can access the Internet; it cannot send user-plane packets via its
HPLMN. The I-WLAN performs accounting and sends the Accounting Data to the AAA
Server. From there it is passed to the Offline Charging System or the OCS.

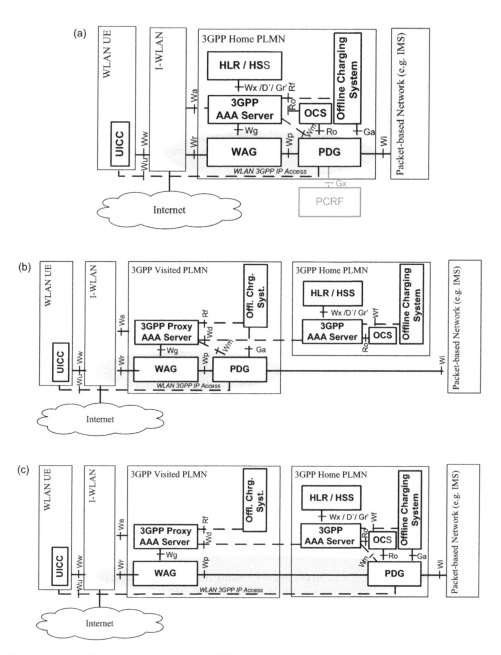

Figure 18.2   I-WLAN architecture (a) the I-WLAN is attached to the HPLMN of the UE. (b) and (c) the I-WLAN is attached to a VPLMN. As usual, dashed lines indicate signalling interfaces, solid lines indicate a data path

The 3rd Scenario extends the 2nd Scenario in that it allows the UE to access packet-based services in the 3GPP System. This is facilitated by two additional user-plane nodes, the **WLAN Access Gateway** (WAG) and the **Packet Data Gateway** (PDG). Once the UE is authenticated, it performs a DNS look-up of the PDG and establishes an IPsec tunnel, via the WAG, to the PDG. All user-plane packets are routed through this tunnel. 3GPP packet-based services, e.g. in the IMS, can be accessed through this tunnel. The role of the PDG with respect to an I-WLAN access is thus analogous to the role of the GGSN with respect to a UTRAN. In fact, it is possible to extend a GGSN to include a PDG. The task of the WAG, as gateway, is making sure the tunnel from the UE is indeed routed to the PDG. Its role becomes more important in the roaming case, as we will see below.

Let us examine the role of the IPsec tunnel. As noted above, the I-WLAN and the 3GPP System have a long-term relationship of trust involving a shared secret and a secured tunnel between I-WLAN and WAG. However, the I-WLAN is not fully trusted! UE and PDG communicate via their own secured tunnel (so we have a tunnelled tunnel between WAG and PDG) so that the I-WLAN cannot tamper with the packets exchanged between the 3GPP system and UE.

18.1.2.2 Roaming Case

We have learned in Chapter 11, Section 11.8 and Figure 11.10 how a roaming scenario results in the network elements of the PS Domain becoming distributed between HPLMN and VPLMN: The HLR is always located in the HPLMN, the SGSN is always located in the VPLMN, and the GGSN can be located in HPLMN or VPLMN.

The situation in the roaming case for the I-WLAN is analogous, cf. Figure 18.2.b/c: The AAA Server is always located in the HPLMN, the WAG is always located in the VPLMN, and the PDG can be located either in the HPLMN or in the VPLMN. However, one additional network element, a Proxy AAA Server, is introduced in the VPLMN: in keeping with the AAA roaming architecture (cf. Chapter 13, Section 13.5.1.3), the NAS in the I-WLAN contacts the Proxy AAA˜Server in the VPLMN instead of the AAA Server in the HPLMN.

18.1.3 I-WLAN Basic Functionality

The steps for a UE from choosing to attach to an I-WLAN to being able to send the first packet are analogous to the steps a UE performs when connecting to a UMTS network (cf. Chapter 11, Sections 11.4, 11.5, 11.6):

- The UE establishes WLAN radio connectivity using the WLAN-specific procedures that we have already studied in Chapter 11, Section 11.9.2—this step is analogous to RRC Connection set-up.
- The UE discovers that it can use its 3GPP Systems subscription to access the WLAN. While 3GPP does not yet provide a mechanism for this, the IEEE is working on this problem and expects to publish the corresponding amendment as 802.11u in 2009.
- The UE authenticates with the AAA Server in its HPLMN using an IETF-standardized procedure, along the lines discussed in Chapter 13, Section 13.5.1.2. The UE then receives an IP address from the I-WLAN. This IP address is called the UE's **local IP address**. The UE can now access the Internet via the WLAN according to 2nd Scenario—this step is analogous to GPRS Attach, except that a GPRS-attached UE does not yet have an IP address and cannot access the Internet

- For the 3rd Scenario, the UE must also set-up a bearer to the 3GPP Core Network: the UE locates the PDG via a DNS query. It then uses IKE to establish a Security Association with the PDG (cf. Chapter 13, Section 13.5) in order to establish the IPsec tunnel between UE and PDG. In the course of this procedure, the UE and PDG also authenticate each other. The PDG additionally pulls authorization information from the AAA Server, and pulls an IP address from, e.g., a DHCP server. This IP address it allocates to the UE as the **remote IP address**. The remote IP address is used for the packets inside the IPsec tunnel, to access packet switched services. This final step is analogous to PDP Context establishment.

We may observe how the I-WLAN mechanisms defined by the IEEE and IETF Protocols are combined to rebuild UMTS functionality.

18.1.4 I-WLAN Mobility

The mobility of the UE—beyond WLAN link-layer mobility (cf. Chapter 12, Section 12.3)—is not covered up to Rel-7. For the upcoming Rel-8, mobility between IWLAN and a 3GPP System is specified for pre-Rel-8 3GPP architectures [3GPP 23.327]. Indeed, Rel-8, which already prepares 4G, includes major architectural updates and solves the mobility problem more fundamentally, for the Access Networks of any technology. We will therefore postpone the mobility discussion until Chapter 21 (Part II).

18.1.5 I-WLAN Security

The authentication scenario follows the usual WLAN procedure described in Chapter 13, Section 13.4.2: the authentication is between the UE (the Supplicant), the I-WLAN Access Point (the front end) and the AAA Server (the back end), with possibly an AAA Proxy Server in-between. The identifier which the UE presents to the AAA Server is a **Network Access Identifier** (NAI) just as for the IMS (cf. Chapter 10, Section 10.4). A NAI has the format user@realm, with "user" containing the IMSI of the subscription. Since "realm" identifies the subscriber's HPLMN, an AAA Proxy Server is able to find the appropriate AAA Server.

The back end protocol between I-WLAN Access Point and AAA Server is RADIUS or Diameter, encapsulating EAP with the EAP-AKA authentication method. In other words, the actual authentication method between UE and 3GPP System is always AKA, independent of whether the UE attaches via UTRAN or via I-WLAN.

18.1.6 I-WLAN QoS

The original access via I-WLAN in Rel-6 is without QoS support. However, for Rel-7, 3GPP gave some thought to how to guarantee QoS in the 3rd Scenario; by contrast, the 2nd Scenario only provides Internet access, in which case QoS support is not considered important.

The proposal is to signal QoS between PDG and UE using **DiffServ Code Points** (DSCPs) in the tunnel's external IP header (cf. Chapter 14, Sections 14.2.1.2 and 14.2.2). Presumably, the routers along the path are DiffServ-enabled. The DSCP information can also be used to provide link-layer QoS on the WLAN air interface (cf. Chapter 14, Section 14.4).

18.1.7 I-WLAN Charging

The charging solution for the I-WLAN [3GPP 32.353] follows the same principles and architectures as charging for 3GPP Systems in general (cf. Chapter 16, Section 16.3). In particular, both offline and online charging is supported. We discuss only the location of the CTFs, the entities collecting Charging information and the choice of correlation identifiers.

In the 1st Scenario, the I-WLAN sends CDRs to the Billing System. In the 2nd Scenario, the I-WLAN collects per-user Charging information. In the 3rd Scenario, bearer-level charging is performed.

CTFs are located as follows:

- One CTF is assumed to be in the I-WLAN itself, but 3GPP does not determine where exactly the CTF resides in the I-WLAN. As Charging information, it collects uplink and downlink data volumes as well as—in the case of the 3rd Scenario—the duration of a bearer.
  Since 3GPP aims to specify the I-WLAN so that all 3GPP-specifics are located in the 3GPP System, the IETF-specified architecture (cf. Figure 16.2) and IETF Protocols are employed to report Charging information: The I-WLAN CTF reports to the AAA Server, or AAA Proxy Server in roaming scenarios, using the RADIUS or Diameter protocol. The AAA Server is located in the 3GPP System, and acts as a proxy for the I-WLAN CTF, interfacing with the online and offline charging systems, cf. Figure 18.2. Note that the support of online charging— with its credit-control features—requires a special RADIUS extension, the RADIUS prepaid extension, to be implemented between the I-WLAN CTF and the (Proxy) AAA Server.
- Another CTF is located in the PDG—applicable only in the 3rd Scenario. The PDG can also meter the traffic volume and duration of a bearer. Having an additional CTF in the PDG allows for differentiating the 2nd Scenario from the 3rd Scenario traffic: only the 3rd Scenario, traffic passes the PDG. In fact, the PDG supports charging not only with bearer resolution it also supports Flow-based Charging, i.e. it contains a PCEF that receives its charging rules from the PCRF, cf. Figure 18.2.
  It is still a subject of debate whether an additional CTF is assigned to the WAG, e.g. in order to verify the amount of traffic relayed between VPLMN and HPLMN.

Correlation identifiers are necessary for the Billing System to recognize that Charging information generated by different CTFs belongs to the same service usage and the same user. In the I-WLAN case, both the I-WLAN CTF and the PDG generate a correlation identifier. In the 3rd Scenario, they exchange the identifiers via the AAA Server when the PDG requests authorization of a bearer.

18.1.8 I-WLAN Policy Control

Last but not least we look at policy control in the I-WLAN. The general idea of policy control is binding services to bearer; particularly for charging and QoS requests. The general 3GPP policy architecture from Figure 17.6 is applied straightforwardly to the 3rd Scenario of I-WLAN [3GPP 23.203] and is shown in Figure 18.2a.

The PDG contains a PCEF pulling policies from the PCRF for Flow-based Charging and for QoS control. Note that the PDG thus receives two types of QoS policies: per-subscription QoS authorization policies from the AAA Server, and per-bearer QoS policies from the PCRF. On the basis of these policies, the PDG filters the UE's packets. For example, it may remark the

DiffServ markings, i.e. the DSCPs (cf. Chapter 14, Section 14.2.1.2) or even drop the packet. The details for roaming scenarios have not yet been worked out.

18.2 Generic Access Network

The work on GAN originated from a non-standard initiative of mobile operators and vendors. The results were brought into 3GPP and were standardized for Rel-6 [3GPP 43.901, 3GPP 43.318]. This partly explains why 3GPP offers two different solutions for accessing a 3GPP System via WLAN.

The goal of the GAN work is to connect a generic IP-based Access Network—e.g. a WLAN, DSL, Bluetooth—to a 3GPP Core Network over the A/Gb interface—remember that A/Gb connects a GERAN to 2.5G network elements in a 3GPP Core Network (cf. Figure 4.6). In other words, neither the WLAN specification nor the existing A/Gb interface shall be changed because of GAN. This tells us two things:

- CS Domain and PS Domain perceive a UE accessing over A/Gb. They are not aware of whether a UE is accessing via a GERAN or via an IP-based Access Network. Thus, in GAN the IP-based Access Network is indeed an alternative Radio Access Network—note how the limit between the terms *Access Network* and *Radio Access Network* is indeed blurred. By contrast, I-WLAN connects to the 3GPP core through a back-door.
- So far, GAN applies only to the "2.5G part" of a 3GPP System, see Chapter 4, Section 4.6, Figure 4.6 and Figure 18.3a. At the time of writing, work is ongoing on technical options for

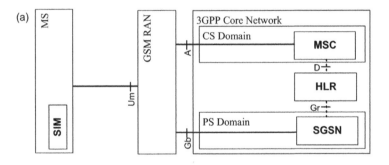

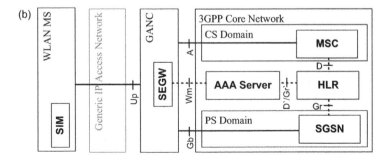

Figure 18.3 3GPP System architecture with (a) GERAN and (b) GAN. Note that only the network elements relevant for GAN are shown

an **enhanced GAN** (EGAN) that provides generic access also over the Iu interface and thus to the "3G part" of the 3GPP System.

Figure 18.3b depicts the GAN architecture: the generic IP-based Access Network is only used as a relay. The GERAN is replaced by a **GAN Controller** (GANC). The GANC terminates the A/Gb interface towards the 3GPP Core Network, and communicates with the MS.

It is interesting to consider the problem from a protocol perspective, see Figures 18.4 and 18.5: in GAN, the upper layer of the protocol stack as well as the A/Gb interface are not touched. Between MS and GANC they are tunnelled over the IP-based Access Network. The lower layers of the GERAN radio interface are replaced by a set of GAN-specific protocols running over a secured IP-stack. These GAN-specific protocols take care of 3GPP-style mobility, security and charging so that the A/Gb interface can interact with GAN the same way as with a GERAN.

Now let us go through the various control functions:

- GAN Mobility

  Handover between different GAN cells as well as between GAN and GERAN, or UTRAN, is supported by means of the GAN-specific protocols that reproduce the normal 3GPP

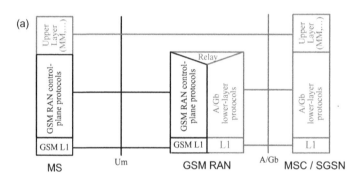

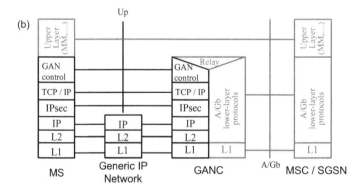

Figure 18.4 Control-plane comparison (a) GERAN attached to the 3GPP Core Network via A/Gb interface (b) GAN attached to the 3GPP Core Network via A/Gb interface

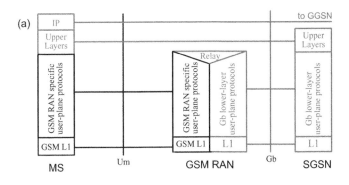

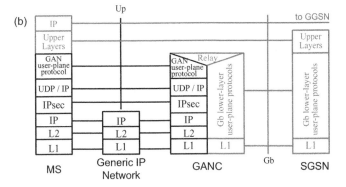

Figure 18.5 User-plane comparison (a) GERAN attached to PS Domain via Gb interface (b) GAN attached to PS Domain via Gb interface

functionality. It is also necessary to transfer the concept of "cell" to GAN: The coverage area of a GAN Access Point is called a cell and has a cell ID. The 3GPP System must know the neighbourhood relation of the GAN cells with UTRAN or GERAN cells in order to be able to control handovers meaningfully.

- GAN Security
  A MS accessing a 3GPP Core Network via GAN utilizes the same procedures for authentication, authorization, encryption and integrity protection as when accessing via GERAN. However, it additionally, and upfront, needs to authenticate with a security gateway (SEGW, see Figure 18.3) in the GANC in order to establish an IPsec tunnel over which subsequent traffic is transported securely. The SEGW pulls the subscriber information from an AAA Server which in turn is in contact with the HLR.

- GAN QoS
  No QoS is assured between MS and GANC.

- GAN Charging
  Charging-relevant functions, e.g. CTFs, are always located in the Core Network. Therefore, existing charging mechanisms can continue to be used for GAN. However, minor adaptations are necessary; e.g. the SGSN/MSC report the radio access technology as Charging information and must be informed when the MS accesses via GAN.

- GAN Policy control
  Policy control is not applicable as it operates in the Core Network rather than in the RAN.

18.2.1 Enhanced GAN

For Rel-8, 3GPP is working on developing GAN for use with a 3G PS Domain—access to the CS Domain will remain via the A interface. The result is called enhanced GAN (EGAN). At this point in time, a feasibility study is being drafted [3GPP 43.902].

One enhancement provided by EGAN will be the support of QoS. The EGAN feasibility study is, however, especially interesting because it disregards the SGSN. In particular, the **enhanced GANC** (EGANC) will take on some of the functions of the SGSN and interface directly with the GGSN via the Gi interface (i.e. the usual SGSN-GGSN interface). By removing one node from the path latency overhead will be reduced. In Part II of this book, when covering the future evolution of 3GPP Systems, we will encounter the same idea.

18.3 Comparison and Discussion

As we saw from the descriptions above, 3GPP offers two quite different solutions for WLAN access to 3GPP Systems. In this subsection we will compare the two approaches—a summary is provided in Table 18.1.

The most obvious difference is that GAN is indeed an alternative access, on a par with UTRAN and GERAN. The Core Network cannot tell whether a UE is using GAN or GERAN. By contrast, for I-WLAN, new network elements were defined in the PS Domain so that I-WLAN traffic is handled in parallel to UTRAN-originated traffic. The underlying reason is that a GAN is assumed to be operated by the entity operating the 3GPP System. The GANC is thus bestowed with a large amount of system-internal information. By contrast, the I-WLAN was conceived such that it can also be operated by third parties.

As a consequence of its architectural role as alternative access, GAN must support mobility in the same way as UTRAN and GERAN—otherwise the Core Network would indeed notice a difference. Note that this argument does not hold for QoS: while the Core Network signals to the GANC via the A/Gb interface which QoS should be supported, the GANC does not provision the QoS required and just ignores the request. Compared to this, the I-WLAN can take a much more relaxed attitude and even introduce mobility in some later release.

Another consequence of the architectural role is that for GAN specific, new protocols had to be developed that fit the A/Gb interface. For I-WLAN, in contrast, IETF Protocols could just be re-used.

Table 18.1 Comparison of I-WLAN and GAN

	I-WLAN	GAN
Architectural role	Alternative RAN plus new network elements in the Core Network	Alternative RAN on a par with UTRAN and GERAN
Protocols	IETF protocols	GAN-specific protocols
Control functions	AAA	AAA, mobility
Technologies	Restricted to WLAN	Any IP-based access technology

Last but not least we should mention that GAN is applicable to generic IP-based access technologies including WLAN, DSL or Bluetooth, whereas I-WLAN was developed for WLAN specifically—although the I-WLAN specification is not too hard to generalize.

18.4 Femtocells

A femtocell is a piece of equipment that produces a small private cell with a 3GPP air interface in homes or offices—strictly speaking "femto" stands for the factor 10^{-15} (just as "milli" stands for 10^{-3}) although here it should be understood to mean just "miniscule". The end-user buys femtocell equipment off-the-shelf and connects it to his fixed broadband access. By means of an operator-owned gateway, the femtocell is integrated into the 3GPP System. A femtocell is only accessible to a restricted number of authorized subscribers, e.g. the owner of the femtocell or all employees of a company. Otherwise, a femtocell is thought to offer the same functions as a normal 3GPP cell, including mobility support and QoS. However, it is likely that there is a broad spectrum of manufacturer-specific differences.

The envisaged advantage of femtocells for the end-user is that his fixed-line services, beyond connectivity, become obsolete. The mobile operators, on the other hand, can extend their business at the expense of fixed-line operators. In mid-2008, femtocells were being trialled, some have already been deployed in the US.

Femtocell technology developed independently of 3GPP; the first moves towards standardization happened only recently [3GPP 25.820]. Therefore, a number of approaches are utilized for attaching a femtocell to a 3GPP System, for example:

- The femtocell equipment is a Node B, a Node B plus RNC, or a Node B plus RNC plus some SGSN functionality.
- The femtocell is a variety of GAN.
- The femtocell utilizes the I-WLAN means of connecting to the network. In this case, for example, mobility and QoS are not supported.
- The femtocell provides access to the IMS only on the basis of SIP.

In all cases, however, the femtocell equipment is controlled by the operators as an integral part of the operator network. Upon being connected to the broadband access, the femtocell contacts its operator's gateway, authenticates, and configures. The support of femtocells poses a number of interesting technical challenges:

- A femtocell operates in the licensed band of the operator and may interfere with the surrounding cells, upsetting the carefully crafted spectrum allocation. To solve this problem, the operator may set aside a part of the precious spectrum for femtocells—a rather wasteful solution. Alternatively, the femtocell may offer sophisticated algorithms for interference minimization.
- In order to support handover, the network needs to localize the femtocell and determine its neighbourhood relations. If the user enjoys taking the femtocell with him on travels and plugs it in at the new location, the network must be able to recognize this. If the user travels abroad out of the realm of the operator and his license, the femtocell must stop working.

More information on femtocells is available from the web page of the Femto Forum (Femto Forum), an organization promoting femtocell deployment.

Comparing femtocells to I-WLAN and GAN, we may note that they offer an alternative access to a 3GPP System, and aim to enhance 3GPP coverage. However, they achieve this goal in different ways: Femtocells employ native 3GPP radio technology, whereas I-WLAN and GAN employ non-3GPP radio technology, in particular WLAN. Furthermore, femtocells target a different market than I-WLAN and GAN: whereas a femtocell is a private cell owned by a subscriber extending mobile coverage in his home or office, I-WLAN and GAN are owned by operators and extend coverage in hotspot areas. In any event, we may notice that 3GPP Systems evolve so as to allow access by a variety of technologies beyond UTRAN.

18.5 Summary

This chapter presented how the 3GPP System evolves in order to allow access via alternative RANs, in particular WLAN. 3GPP developed two alternative solutions to the problem, I-WLAN and GAN. Both solutions do not affect the WLAN standard. Only the 3GPP specifications need to be modified.

The I-WLAN work is based on six scenarios that build on each other and exhibit sequentially greater functionality. The I-WLAN specification of Rel-6 supports the first three scenarios: a UE accessing via I-WLAN is authenticated, authorized and charged by an AAA Server in the 3GPP System and can then access 3GPP services. Mobility, however, is not supported. Architecturally, the I-WLAN access is attached to the PDG in the PS Domain, whose role is similar to that of the GGSN. For support of I-WLAN, only IETF Protocols are employed.

For upcoming releases, 3GPP works on expanding the I-WLAN concept to any IP-based Access Network, and to include seamless mobility support and QoS.

The GAN work originated from a non-standard initiative of mobile operators and vendors, and was later standardized, for Rel-6. From the start, GAN is applicable to any IP-based technology and supports mobility. The IP-based Access Network is indeed a fully-fledged RAN, architecturally and functionally equivalent to UTRAN and GERAN. For accessing the 3GPP Core Network, it uses the same A/Gb interface as GERAN. Consequently, GAN employs protocols developed specifically for the purpose by 3GPP.

Femtocells are yet another alternative access to the 3GPP System, albeit not yet standardized. Femtocells are very small private cells operated in homes or offices. They are private in that they can only be accessed by particular subscribers. Femtocells are connected to the 3GPP System via a fixed broadband access, and are managed by the operator via a gateway. Femtocells offer a 3GPP air interface and operate within the licensed band of the operator.

19

UMTS Releases Summary

In this final chapter of Part I we will review the evolution of UMTS through the different releases: which feature was added when, and what is the current deployment status? In other words, this chapter provides an overview of the content presented in previous chapters, and orders this content on the timeline; some features are also introduced that have so far not been mentioned. The reader should be aware that the description of the releases provided in this chapter is by no means complete. A comprehensive overview is provided on the 3GPP web pages [3GPP].

Terminology discussed in Chapter 19:	
Evolved Packet System	EPS
High Speed Downlink Packet Access	HSDPA
High Speed Uplink Packet Access	HSUPA
Long Term Evolution	LTE
Multimedia Broadcast Multicast Service	MBMS
Session Traversal Untilites for NAT	STUN
System Architecture Evolution	SAE

19.1 Release 99

R-99 was released in the year 2000. It is the first release by UMTS. GPRS had already introduced the PS Domain and supports IP-based communication. UMTS R99 differs mainly from GPRS in two respects:

- The GSM RAN of GPRS is replaced by the UTRAN. The UTRAN employs W-CDMA on the radio interface, and offers a higher bandwidth than GPRS, namely up to a theoretical limit of 2 Mb/s. Furthermore, macrodiversity and soft handover are introduced.
- Whereas GPRS only supports best-effort service on the basis of IP, UMTS introduces four Traffic Classes—conversational, streaming, interactive and background—each offering a

different QoS. The reason is that GPRS is intended as mobile access to the Internet, whereas UMTS is meant to deliver multimedia services.

The first UMTS Networks based on R99 were deployed in 2003. The bandwidth offered by these networks is typically 384 kb/s per cell.

19.2 Release 4

Rel-4 was released in 2001. It is a so-called "small release", containing mostly minor updates. The most important enhancements, compared with R99, are the division of the (G)MSC into a transport node, the Media Gateway and a control node, the (G)MSC Server. Furthermore, the standard made the CS Domain bearer-independent: it can be circuit-switched natively, ATM-based or even IP-based (cf. Chapter 7, Section 7.1). Rel-4 conformant equipment has also been deployed.

19.3 Release 5

Rel-5 was published in 2002. It is a major release, adding the following main features:
- IMS (cf. Chapter 4, Section 4.5)
  With the introduction of IMS, UMTS now has a standardized means of offering the multimedia services for which it was built.
- High Speed Downlink Packet Access
  High Speed Downlink Packet Access (HSDPA) is the first instalment of HSPA (cf. Chapters 2 and 5). It enhances the UMTS air interface on the basis of 16-QAM and other techniques in order to introduce a downlink shared channel with packet rates up to a theoretical value of 14.4 Mb/s. The bandwidth offered by HSDPA is higher than the 2 Mb/s required from a 3G Network. A UMTS Network with HSDPA therefore already qualifies as "3.5G".
- 3GPP System (cf. Chapter 4, Section 4.6)
  The UMTS specification was extended in order to allow not just a UTRAN but also a GERAN to be attached to PS Domain and CS Domain. The result is called a 3GPP System.
- Iu Flex (cf. Chapter 8, Section 8.1)
  In earlier releases, the relation of RNCs and SGSNs/MSCs is strictly treelike and hierarchical: each SGSN serves several RNCs, and each RNC is attached to exactly one SGSN and MSC (cf. Figure 8a). This restriction was lifted, allowing many-to-many relations between SGSNs/MSCs and RNCs.

At the time of writing, in 2008, Rel-5 is being rolled out. While the first standard-conformant IMS's are just being deployed, HSDPA is already widely available and has a share of one fifth of all 3GPP subscriptions.

19.4 Release 6

Rel-6 was published in 2005. It is also a major release featuring, among others, the following:
- IMS supporting services (cf. Chapter 10, Section 10.1)
  The specification of services for mobile networks is the responsibility of the Open Mobile Alliance (OMA). 3GPP therefore standardized only a small number of supporting services, e. g. Push Service, Instant Message Service and Presence Service.
- I-WLAN (cf. Chapter 18, Section 18.1)
  A trusted WLAN Access Network is defined as alternative access to the 3GPP Core Network.

The WLAN and 3GPP Core Network communicate with each other using standard IETF Protocols. The I-WLAN, however, is neither architecturally nor functionally equivalent to a UTRAN or GERAN. For example, the I-WLAN does not support mobility.

- GAN (cf. Chapter 18, Section 18.2)
  The goal of GAN is to attach a generic IP-based Access Network to the 3GPP Core Network via the A/Gb interface. This is achieved by using an additional network element, masking the GAN to the Core Network. A number of 3GPP-specific protocols is employed to this end. Functionally and architecturally, a GAN is equivalent to a GERAN.

- Update of the charging architecture (cf. Chapter 16, Section 16.3.2)
  A harmonized charging architecture was introduced for bearer-, subsystem- and service-level. Pre-Rel-6, for each level and each domain an independent charging architecture had been developed.

- Flow-based Charging (cf. Chapter 16, Section 16.3.2.3)
  Until the advent of Flow-based Charging, charging is only possible with the resolution of a bearer: when two or more service-level flows utilize the same bearer, they cannot be differentiated. Flow-based Charging supports service-specific charging on the bearer-level of any resolution and, furthermore, allows for the coordination of charging on the bearer- subsystem- and service-levels.

- High Speed Uplink Packet Access
  High Speed Uplink Packet Access (HSUPA) extends the downlink HSPA from Rel-5 by adding a dedicated uplink channel with bandwidth up to 5.76 Mb/s—since uplink a sufficient number of codes has been available (cf. Chapter 5, Section 5.2.4.3), there is no need to introduce shared channels. HSUPA in Rel-6 is, however, specified only for FDD.

- Multimedia Broadcast Multicast Service
  As we already know, UMTS designers are somewhat concerned about saving resources on the radio interface. In scenarios such as multicast of sport events, multi-party conferencing or broadcast of emergency information it may happen that several UEs in a cell receive the same data stream. The **Multimedia Broadcast Multicast Service** (MBMS) [GPP 23.246] enables a resource efficient data transfer to many UEs in parallel: by means of an additional network element in the Core Network and new protocols, the data is sent only once to each cell and all UEs access the same channel for receiving it.

Most of Rel-6 is currently in the trial phase. The exception is HSUPA of which a number of networks have already been deployed.

19.5 Release 7

Rel-7 was published at the end of 2007. Among others, the following features are part of Rel-7:

- PCC (cf. Chapter 17, Section 17.3.2)
  Whereas in previous releases, the authorization of QoS in the PS Domain by the IMS (via SBLC) and Flow-based Charging relies on an independent mechanism, Policy and Charging Control (PCC) provides a single policy infrastructure in order to handle both kinds of policies. Furthermore, PCC is applicable to any IP-CAN, and to any service network. Together with the introduction of PCC, the COPS protocol between GGSN and PCF/PCRF was abandoned and replaced by Diameter.

- HSPA+
  This is an enhancement of the HSPA technology, in particular towards higher bandwidth—28 Mb/s downlink and 11 Mb/s uplink. This is achieved by employing 16-QAM and MIMO.

- HSUPA for TDD

  The specification of HSUPA has been extended to also work with TDD.
- IMS support of UEs behind NATs

  The original 3GPP idea is that of the UE as a single piece of equipment connected directly to the RAN. Of course, it is also possible for the subscriber to set-up a home network of devices, e.g. laptop and mobile phone, and to facilitate UMTS Network access for all devices via the mobile phone. In this case the UE consists of several pieces of equipment and one UICC, see Figure 19.1.

  What is more, the subscriber may install a Network Address Translator (NAT) and a firewall between the UE and the UMTS Network. The NAT separates the address spaces of the subscriber's home network and UMTS Network: In the home network a private address space is used, while towards the outside, the IP address assigned by the GGSN is employed. In other words, the NAT translates the IP address and may be even the port number of any packet traversing it.

  NATs are known to create considerable problems in cases where the IP address, respectively the port number, appears in "unexpected" places, i.e. somewhere else than in the IP header, respectively the transport header: the NAT cannot translate them. For example, the payload of a SIP signalling message can contain the IP address of the originator of a session (cf. Chapter 15, Section 15.2.4). A UE behind a NAT is not aware of its outside IP address and would of course include its local IP address. As a result, the recipient of the SIP message is unable to contact the UE.

  3GPP specifies two solutions for this problem. One solution is employing an **Application Level Gateway** (ALG), as defined in [RFC 2663], co-located with the P-CSCF. The ALG performs deep packet inspection and has application-specific knowledge that allows it to translate appropriately, e.g. IP addresses in session descriptions, possibly in interaction with the NAT.

  Alternatively, the **Session Traversal Utilities for NAT** (STUN) [ID STUN] is employed. The STUN specification includes a STUN Server in the external network—the IMS in our case—which the entity behind a NAT—the UE in our case—can query in order to learn its external address. This allows the UE to include right away the correct address in its session description. Of course, the UE must be aware that it is supposed to contact the STUN Server.

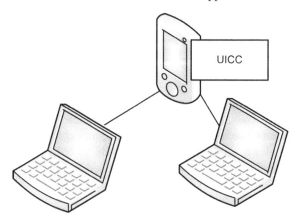

Figure 19.1 A home network of laptops connecting to the Internet via the mobile phone

19.6 Outlook

Concurrently with the work on Rel-7, 3GPP is concentrating on feasibility studies for its next Rel-8. Rel-8 will contain a major update of the 3GPP architecture and the radio interface, known as **Evolved Packet System** (EPS) or **Long Term Evolution/System Architecture Evolution** (LTE/SAE). Rel-8 will move the 3GPP System towards 4G.

The EPS architecture will support multiple Access Networks, of both the 3GPP and the non-3GPP variety, including mobility between these Access Networks. The EPS air interface will support peak data rates of 100 Mb/s downlink, and 50 Mb/s uplink. Simultaneously, the architecture and protocols will be simplified, and IETF Protocols will play a major role. Of course, backwards compatibility with the existing 3GPP System must be maintained. We can thus look forward to an interesting balancing act. The details are provided in Part II of this book.

12.6 Outlook

Part I Epilogue—Convergence

In Chapter 1, the introduction to Part I, we described how Telecommunication Networks, in particular UMTS and Computer Networks, have different roots and different design traditions. At the same time, Telecommunication Networks and Computer Networks appear to converge—they increasingly offer the same services and employ the same technology. What, however, does this convergence mean? To what extent do they converge?

Part I of this book thus covered UMTS, and for each of its technical features discussed how it has been realized in telecommunication-style in UMTS, how the same feature would be realized in a Computer Network, how these two solutions differ and what they have in common. In other words, we examined to what extent UMTS and Computer Networks exhibit a technological convergence that goes beyond employing the IP protocol.

What did we find? UMTS uses IP; it features a packet-switched domain and the IMS where IP-based multimedia services are supported. We also found that UMTS and in particular the IMS employ many IETF protocols. Indeed, the IMS leverages the modularity of its "IP heritage" in order to enable the ad-hoc creation of new services by operators. Furthermore, any service offered on the Internet can, in principle, be accessed via UMTS—depending upon the firewall settings of the network operator.

However, UMTS operators are Telecommunication Network operators with a corresponding business model. UMTS is therefore clearly designed in telecommunication-style as we can see when revisiting the design principles laid out in Chapter 1, Section 1.2:

- The cathedral and the bazaar
  UMTS is designed cathedral-style as a complete system. UMTS includes the Radio Access Network, a Core Network and a subsystem for the support of service delivery. Care is taken for all control functions to interwork. For example, as we saw in Chapter 12, a handover in UMTS always includes authentication at the new Node B, context transfer between SGSNs and set-up of QoS on the new path. In Chapter 15 on session control we saw how the original SIP protocol is extended for use in the IMS in order to add coordination with security, QoS and charging, and how additional network elements are added on the signalling path. Furthermore, the IMS specification includes details on when and where to issue protocol messages; something not considered necessary in an IETF specification.

UMTS Networks and Beyond   Cornelia Kappler
© 2009 John Wiley & Sons, Ltd

This also implies that UMTS—unlike a Computer Network—is not modular on a functional level: It is, for example, impossible to set-up a UMTS Network without mobility support as a "fixed network". With a Computer Network, this is entirely straightforward.

- Operator control and user control
  Operators of Telecommunication Networks sell an end-to-end solution to users whom they shield from the technical details; they do not sell connectivity to "techies". The IMS was therefore designed originally for services offered by operators, not for services offered by users. However, we may observe some movement as some operators have declared recently an open IMS interface that allows third party developers to create new services.

  However, the operator needs comprehensive control of the network and the services it offers. Technically, this manifests itself in UMTS' numerous elaborated control functions, e.g. for mobility control, QoS, charging, etc.

  UMTS also is a typical Telecommunication Network in that control is exerted in a centralized fashion by omniscient entities. For example, the RNC controls the Node Bs. The central control entity in the PS Domain is the SGSN. The current evolution of UMTS in the context of policies introduced yet another centralized, omniscient entity, the Application Function (AF), cf. Chapter 17, Section 17.6. PDPs query the AF about the flows for which they must make a policy decision. Generally, intelligence is located in the network instead of in the UE. For example, the handover decisions are made by the network.

  Computer Networks today can also offer many control functions, the IETF has standardized numerous protocols for network control in recent years, e.g. Mobile IP and other mobility protocols, RSVP, NSIS, SIP, Diameter, etc. However, IETF Protocols traditionally place decision power and intelligence in the MS. It should be noted, however, that one of the more recently founded Working Groups of the IETF is investigating a mobility protocol, **Proxy Mobile IPv6** (PMIPv6) that locates decision power in the network. We will hear more about this in Part II.

- "In the beginning is the architecture" and "In the beginning is the protocol"
  UMTS clearly follows telecommunication-style design principles in that it is founded on an architecture that distributes functionality to network elements. Protocols are added later. The UMTS protocols developed by 3GPP are typical telecommunication-style protocols: for example, GTP is specified for communication between SGSN and GGSN. It carries all the information which SGSN and GGSN need to exchange, relating to QoS, mobility, charging, etc. By contrast, IETF Protocols typically are a modular one-protocol-per-functionality. UMTS also employs a number of IETF protocols. Indeed, the IMS uses IETF protocols exclusively, for example Diameter and SIP. We have observed, however, that 3GPP had to augment, e.g. SIP for coordinating it with QoS signalling and security signalling.

We will therefore conclude as follows: when looking at UMTS as it is standardized up to Rel-7, 3G Telecommunication Networks and Computer Networks have converged at a high level regarding the functionality and the services they can provide. They have, however, not converged on a deeper level, regarding how they provide the functionality and services. Operators of UMTS Networks and Computer Networks continue to build their network based on different design principles. As we will see in Part II, however, the business models of operators appear to change slowly, and as a consequence network design is adapted. We will see what this means for convergence in the Epilogue to Part II.

Part II

Beyond UMTS
Networks

In Part I of this book we examined one particular 3G Telecommunication Network, UMTS. Indeed, UMTS evolved in order to allow Radio Access Networks other than the UTRAN and was henceforth called the **3GPP System**. In Part II, we will look at the future evolution of the 3GPP System. To add interest, we will take a somewhat broader approach, looking not only at the evolution of the 3GPP System itself, but at the future of mobile Telecommunication Networks in general. Obviously, in the process we will have to leave the firm ground of facts and standards, and perform a certain amount of guesswork.

Which user services will be supported by the next generation of mobile Telecommunication Networks? What will be the design principles and the distinguishing technical characteristics? Which business models will be successful?

While research on the topic has been ongoing for several years, a concensus is only just emerging. For a long time there was not even an agreement on the name of the next generation of mobile Telecommunication Networks, with **Beyond 3G** (B3G) and **4G** as the top runners. Recently, the term 4G appears to be emerging as the winner, so we will use this term in the book. The ITU is currently finalizing a set of requirements on 4G Networks with their **IMT-Advanced** concept, analogous to their earlier definition of 3G Networks with the IMT-2000 concept (cf. Chapter 2, Section 2.3).

In Part II of this book we will review the current development towards 4G Networks. In Chapter 20, we take the 3G status quo as a starting point, discuss upcoming short-term evolution and finally provide an overview of 4G capabilities which the ITU is likely to require. In Chapter 21, we provide a detailed technical discussion of 3GPP System Rel-8, i.e. the short-term evolution of 3GPP Systems towards 4G. In Chapter 22, we discuss the short-term evolution of other technologies, in particular cdma2000, WiMAX and fixed networks. In Chapter 23, finally, we present the additional features and ideas that have been worked on in the context of 4G Networks—and future Communication Networks in general—which, however, have not yet become part of 4G.

20

4G Motivation and Context

One of the goals of this chapter is to provide as clear an idea as possible of 4G Networks. The other is to depict the likely path of evolution towards 4G. To this end, we will take the following approach:

- We start with an evaluation of today's mobile Telecommunication Networks regarding services, technology and business models; in particular how these facets have evolved since the conception of 3G in the 1990s.
- Subsequently, we investigate the likely short-term evolution—which may be within the next five years. Which services are likely to be offered, which technology trends are being incorporated in standardization, and how will business models evolve further?
- Finally, we discuss the upcoming official definition of 4G Networks by the ITU.

Indeed, when this book went to print in mid-2008, the definition of 4G has not yet been released by the ITU and therefore any description of 4G is preliminary. The reader may, however, have seen some technology or other being advertised as "4G". One could share the view expressed in [3G Americas 2007]: "Any claim that a particular technology is a 4G technology or system today is, in reality, simply a market positioning statement by the respective technology advocate".

Terminology discussed in Chapter 20:	
4th Generation networks	4G
Access Network	AN
Always-on	
Converged Access Network	CAN
Digital Subscriber Line	DSL
Evolved UTRAN	E-UTRAN
Evolved Packet Core	EPC

UMTS Networks and Beyond   Cornelia Kappler
© 2009 John Wiley & Sons, Ltd

Fixed-Mobile Convergence	FMC
Generic Authentication Architecture	
IMT-Advanced	
Mobile WiMAX	
Moving Network	
Multi-homed	
Multi-hop	
Personal Area Network	PAN
Personal Network	PN
Proxied Roaming Agreement	
Quadruple Play	
Radio Frequency Identification	RFID
Service Network	
Software Defined Radio	SDR
Ultra Mobile Broadband	UMB
Vehicular Ad-hoc Network	VANET
Wireless Access in Vehicular Environments	WAVE
Wireless Sensor Network	WSN

20.1 Today's Mobile Telecommunication Networks

20.1.1 Today's Services and Technology Trends

The services to be offered by 3G Networks were listed by the ITU-T in their IMT-2000 concept (cf. Chapter 2, Section 2.3.2): Based on high-quality, anytime anywhere communication, support would be offered for data transport, Internet and Intranet access and, in particular, multimedia services. It is instructive to compare this vision with the reality of 2007.

The first thing to observe is that "anytime anywhere communication" has not quite become reality. While *a* mobile network is available almost anywhere—UMTS, cdma2000, WLAN hotspots, WiMAX, etc.—the user is not necessarily able to access it: technologies are heterogeneous and incompatible, inter-technology roaming is not possible. Users need a variety of radio interface equipment, a multitude of subscriptions and some technical expertise in order to always be connected. On the other hand, devices are increasingly **multi-homed**: they support several radio interfaces, e.g. HSPA, WLAN and Bluetooth, in order to always be able to connect. It is noteworthy however, that these radio interfaces are operated independently. The user has to actively select which interface shall be used and it is not possible to handover a session between them.

How are the mobile networks being used today? WLAN and WiMAX offer nomadic or mobile Internet and Intranet access. They are mobile Computer Networks in that they offer connectivity (cf. Chapter 1, Section 1.2.2). Of course, these networks are often used together with services such as a **Virtual Private Network** (VPN) or Voice over IP. These services, however, are usually configured by the user. They are not offered by the operator of the mobile network.

By contrast, we would expect the "traditional" 3GPP System,[1] as a Telecommunication Network, to offer actual services. Indeed, the 3GPP System was developed in order to provide multimedia services. Remote banking, telemedicine, remote monitoring and control of the home and location-based services were also envisaged (cf. Chapter 2, Section 2.1.2), although for the latter no standardized provisioning methods have been developed.

The 3GPP Systems deployed today do not fully satisfy these expectations. They are often used for voice services via the CS Domain, and for mobile Internet and Intranet access—web surfing, file downloading, email checking, etc.—via the PS Domain. In most parts of the world, other usage has not yet taken off. In this book we do not attempt to evaluate the reasons for this. We content ourselves with enumerating the reasons which others have brought forward:

- The quality of multimedia services is unsatisfactory because 3GPP System bandwidth is not sufficient for this purpose.
- Today's mobile phones are unsuitable for multimedia services or to play games, i.e. the screen is too small, processing power and memory are not sufficient and the keyboard-equivalent is too inconvenient to operate.
- Users do not sufficiently trust the security measures for performing mobile banking.
- The service offerings (multimedia, mobile banking, mobile gaming, location-based services) are still limited.
- The costs are too high, and just as importantly, cost structures are often not transparent, i.e. the user cannot determine how much she has to pay in total. It is thought that a flat-rate can alleviate this problem.
- User habits are difficult to change. A mobile phone is viewed as a telephone and not as a TV, credit card or play station.

On the other hand, in specific contexts, mobile services delivered via Telecommunication Networks are in fact quite popular: the successful Apple iPhone integrates a large variety of services, and even offers a **Software Development Kit** (SDK) that allows users to develop their own services, so-called Web Applications, which other users can even download and install in a simple way. Another example is the Amazon electronic book, Kindle, which includes a subscription for downloading book files via cdma2000.

For supporting the service of "remote monitoring and control of the home" it may also be true that 3GPP technology, including IMS services, is not sufficient: hardware support in the home is also necessary, e.g. in the form of controllers embedded in all electric devices, as described in [digitalstrom].

Location-based services are not yet widely used, but this is likely to change. The original idea was to employ a network-based positioning method, using the ID of the cells in which the mobile phone is located. The accuracy of this method is, however, only in the range of hundred metres to several kilometres which is not sufficient for applications such as navigation. Furthermore, privacy issues may prevent the network operator from using location information freely. Meanwhile, terminal-based positioning methods using **Global Positions System** (GPS) capability built into the mobile phone has become available. GPS is more accurate—in the range of 10 metres—making it possible to use the mobile phone for route planning and navigation. With terminal-based methods, the subscriber installs

---

[1] Remember that 3GPP System is the more general term which includes UMTS, cf. Chapter 4, Section 4.6.

navigation software on the terminal that can operate without network support. Some operators in the US offer location-based services based on a combination of GPS and, e.g. Google Maps, i.e. by collaborating with companies outside the operator community. In Japan, location-based services are integrated into mobile web-portals such as i-mode.

20.1.1.1 Regional Differences

The analysis above of today's mobile services is only true on average. In fact, very pronounced regional differences exist. The South-Korean and particularly the Japanese market are much more developed. A large variety of mobile data services—emailing, web surfing, downloading of games or music, and mobile commerce in general—is widely used. The most successful data service is the Japanese i-mode, an operator-owned mobile web-portal based on an adaptation of the **Hypertext Markup Language** (HTML) standard. Operations began in 1999; in June 2007, i-mode had 52 million users in Japan. Competing operators have launched similar services. i-mode and its relatives offer sports results, weather forecast, games, financial services, ticket booking and location-based services. An important feature is integrated payment functionality: the user does not need to enter credit card details on a miniscule keypad; rather the amount is directly debited to his account with the mobile operator. This payment functionality can also be used by third parties offering their service in the i-mode portal. Currently, the functionality is being expanded and i-mode enabled mobile phones can generally be used as a payment card or ID card. Interestingly, while also available outside Japan, i-mode and similar services are not nearly as successful abroad.

Another example of a successful mobile service both in Japan and South Korea is mobile TV. In November 2007, there were 38 million subscribers in these two countries. Mobile TV can be delivered in a variety of ways. Normally the uplink control-channel is via 3G and the downlink delivery via a one-way dedicated broadcast network based on standards such as **Digital Radio Transmission** (DRT) or **Digital Video Broadcasting** (DVB).

20.1.1.2 Summary of Today's Services and Technology Trends

We may conclude that there exists today a variety of mobile network, and that the packet-switched mobile networks are—in most parts of the world—used mainly to provide connectivity. The original vision for 3G Networks—high-quality anywhere anytime connectivity, support for data transport, Internet and Intranet access, as well as a variety of services—is only slowly becoming a reality, and possibly in different forms than those anticipated originally.

20.1.2 Today's Business Models

It may be observed that by 2007, the business model for 3GPP Systems had evolved as compared to the 1990s when it was first developed: The original business model was intended to offer an integrated end-to-end solution—high-quality services based on carefully controlled connectivity. As discussed in more detail below, this business model is not as successful as planned, partly due to services and connectivity offered by others, based on competing models. The interesting question which we will discuss is how operators may develop their business model further, thereby avoiding the threat of becoming "bit pipes", i.e. sellers of mere connectivity: 3GPP Systems are not designed for this purpose. There are simpler—and cheaper—ways of providing wireless connectivity.

We will discuss the evolution of business models for connectivity provisioning and services separately, in the following two subsections.

20.1.2.1 Connectivity Provisioning

The business landscape for provisioning wireless broadband connectivity has changed since the 1990s. In particular, diversity, availability and competition has increased dramatically in recent years:

- Commercial WLANs are commonplace. They are normally operated as individual hot-spots owned by an operator; access is bought on a per-time basis. Mobile telecommunication operators often own both a 3G Network and WLAN Hotspots as independent entities. With I-WLAN (not deployed yet) and GAN, 3GPP standardized a joint operation that allows for the sharing of subscribers and movement between the network types, cf. Chapter 18.
- WiMAX has wider coverage than WLAN and is often marketed as wireless broadband access, i.e. a wireless alternative to DSL or cable. A more recent version of the WiMAX standard, however, also allows for truly mobile network access, positioning the technology as an alternative to 3.5G Networks; in Chapter 22 we will hear more about Mobile WiMAX.

One should also keep in mind the providers of fixed broadband connectivity:

- The traditional fixed-line telecommunication operators have extended their business beyond telephony and use their existing copper-wire infrastructure to offer broadband Internet access in the home via a technique called **Digital Subscriber Line** (DSL).
- Cable operators, whose original business was the unidirectional broadcasting of TV, have also moved into providing broadband Internet access over their existing cable infrastructure by connecting cable modems. This technology is especially popular in North America and Oceania.

Both fixed-line operators and cable operators are also currently positioning themselves to move into the mobile market on the basis of a concept called **Fixed-Mobile Convergence** (FMC) about which we will hear more in Chapter 20, Section 20.2 and in Chapter 22, Section 22.3.

Of course, this also works the other way round. We have already encountered in Chapter 18.4 how femtocells and GAN allow for an extension of the mobile operator's business into the realm of fixed-line operators.

At the same time, more different business models are emerging. For example, the "operator" FON does not even own a network or hotspots. Instead, its subscribers own the infrastructure: they can volunteer to install a special network box, a WLAN-Access-Point-plus-router, and attach it to their fixed broadband connection at home, e.g. DSL or cable. This box allows for sharing the wireless access with other subscribers in a secure fashion. Subscribers contributing to the infrastructure are provided with free access to all FON hotspots. All other subscribers pay for their access. Interestingly, a number of traditional mobile telecommunication operators partner with FON.

20.1.2.2 Service Provisioning

Services—telephony, multimedia services, location-based services, etc.—are not the exclusive domain of mobile telecommunication operators. Third parties also create services very successfully which they offer via the Internet instead of the IMS. They are independent of the underlying network, and more often than not the basic service is free of charge. A prominent example is the (fixed) telephony service offered by Skype. As such services can undermine the business model of mobile telecommunication operators their usage from 3G Networks is often restricted: mobile telecommunication operators traditionally pursue a so-called **walled-garden** approach.

However, greater openness and flexibility is starting to develop:

- Sometimes services available in the Internet can be combined fruitfully with an operator service. As mentioned above, some operators in the US offer location-based services based on GPS and Internet-based mapping services.
- Another strategy is pursued by the operators offering the i-mode portal: third parties can offer their services via i-mode; however the operator participates in the revenue.
- Alternatively, operators can enter into an exclusive business relation with third-party service providers such as Apple and Amazon for services such as iPhone Web Applications and the Amazon Kindle eBook. Note how the operator's role in this case is indeed providing a "bit-pipe"; however the operators normally participate in the service's revenue.
- A number of operators have announced an open interface to their IMS, allowing third party developers to create new services.

An interesting and potentially disruptive move goes even beyond an open interface to the IMS or a particular mobile phone model such as the iPhone: Google collaborates with other companies, particularly several major mobile telecommunication operators and manufacturers, to develop an open-source software platform for mobile phones called Android.[2] This platform is Linux-based and will combine operating system, applications, user interface and a Software Development Kit that allows anybody to program new applications, and have these applications interwork. In other words, the operators supporting this project are providing easy access for third parties to develop and offer services, presumably without IMS, and without coordination with operator or handset manufacturer—for example for Google to provide location-based services. While operators employing Android on their mobile phones may still restrict the additional software which subscribers may install, they are certainly abandoning the walled-garden approach. The potential advantages from their point of view are a larger number of applications and a faster development and deployment of services. Possibly in a reaction to Android, in June 2008 Nokia announced that it will buy Symbian, a widespread operating system for mobile phones, and make it open source.

Another business strategy for Telecommunication Operators is investing in the ongoing development of **Quadruple Play**, delivery of high-speed Internet Access, TV and telephony to the user's fixed and mobile devices, with which mobile Telecommunication Operators challenge the business of both fixed line and digital broadcast and cable (i.e. TV) operators.

---

[2] At this point in time (mid-2008), some parts of the platform are in fact still proprietary, although Google have announced that this will change in the future.

Conversely, Quadruple Play also opens the mobile market to fixed-line operators and cable operators.

20.1.2.3 Summary of Today's Business Models

We have seen that the traditional integrated business model of mobile telecommunication operators is being challenged. New players offer connectivity and services separately and allow the user to abandon the walled garden. It is true that the combination still lacks the comfort and quality of an integrated solution—for example mobility support and QoS are often not offered. However, this piece-meal solution has at least two important advantages: it often costs less, and the users have freedom in choosing and combining services, just as they are used from their Personal Computers.

How do mobile telecommunication operators react to this threat?

- One reaction is the continuous search for successful operator-owned services that are simple to use. This strategy has proved to be promising in Japan and South Korea.
- Or operators may sell exclusive connectivity to third parties and participate in the revenue. Apple's iPhone can be operated in that way.
- Another possibility is a careful opening up to and integration of potentially competing technologies, e.g. with I-WLAN, GAN, or the use of Google Maps to provide location-based services. The open IMS interface (cf. Part I Epilogue) and the Android project are other examples of this approach.
- Yet another possibility is to move into new markets—other player's markets, e.g. with femtocells, GAN and Quadruple Play.

20.2 Short-term Evolution Towards 4G

After discussing the status quo in the previous section, the topic of this section is the ongoing evolution of 3G Networks towards 4G Networks. The analysis is based on current standardization activities: both standards that have already been finalized and that are poised to enter the market and standards that are still in the making but which can be expected to be implemented within the next five years. As before, we discuss services, technology trends and business models.

20.2.1 Short-term Service and Technology Trends

A view of what lies ahead was expressed in [Saunders 2007]: the years up to 2011 are "best characterized as simply being more of everything—more operators, more frequency bands, more networks, more air interface technologies, more devices and especially greater volumes of data being carried wirelessly". We may add "more services" and "more bandwidth". This quantitative change is certainly a distinctive feature. However, concurrently standardization organizations are working on qualitative changes that go beyond "more". We will hear about both below.

20.2.1.1 User Services

Of course, it is anybody's guess as to which user services will be successful in the near future. Presumably the goal of 3G Networks will—ultimately—become a reality in one form or another: it is likely that videoconferencing, mobile TV, video streaming, location-based information, navigation and mapping services, etc. will become commonplace in many parts of the world.

20.2.1.2 Radio Interface and Bandwidth

It may be assumed that the near future will bring us "more" bandwidth; the corresponding specifications have already been finalized, or are about to be finalized. For Rel-8, 3GPP has just specified the **evolved UTRAN** (E-UTRAN) [3GPP TR 25.913], the analogous development in 3GPP2 is called **Ultra Mobile Broadband** (UMB) [3GPP2 C.S0084]. E-UTRAN and UMB include completely new specifications of the radio interfaces which will yield theoretical bandwidths between 50 and 100 Mb/s. In other words, 3GPP and 3GPP2 draw level, approximately, with the bandwidth offered by Mobile WiMAX, see also Figure 2.4. E-UTRAN, UMB and WiMAX are based on OFDMA plus MIMO, cf. Chapter 5, Sections 5.2.3 and 5.2.5.

20.2.1.3 Access Network

The Access Network is an area of much innovation. We will cover two topics: the collaboration of Access Networks of different technologies, and new, more flexible topologies for Access Networks.

20.2.1.3.1 Convergence of Heterogeneous Access Networks

More radio interface technologies and also more networks will be deployed. At first sight, this seems to point towards an exacerbation of today's situation: in order to be always connected users need more hardware, more multi-homing, a multitude of subscriptions and, in particular, technical expertise. Looking more closely, however, it becomes apparent that the situation might indeed improve. The general tendency is towards a *collaboration* of heterogeneous technologies and networks, something known as **convergence of heterogeneous Access Networks**. The above-mentioned **Fixed-Mobile Convergence** (FMC) is one facet of this.

The ultimate vision is a high-level architecture for 4G Networks as seen by the ITU [ITU M.1645] and as shown in Figure 20.1: The *Access Networks* of the different technologies are connected to a common IP-based *Core Network*. These Access Networks include *mobile* Access Networks of many denominations as well as *fixed* Access Networks such as DSL and cable, and pure digital broadcast networks, i.e. digital TV.

We need to investigate the details of the vision depicted in Figure 20.1, because the terms **Access Network** and **Core Network** have different meanings in different technologies.

From a WLAN perspective, the Access Network is an ESS with Access Points, Access Router and possibly an AAA Server; by contrast, the Core Network is a connectivity backbone, see Chapter 4, Section 4.7.

But what is an Access Network from a 3GPP perspective? The 3G architecture in fact does not define the term (cf. Chapter 4, Section 4.2). Should we assume that the Access Network is the RAN? This would imply that the IP-based Core Network must provide certain functions, e.g. subscriber management. Or is the Access Network an IP-CAN, i.e. RAN plus PS Domain? In this case, the IP-based Core Network is a connectivity backbone just as for WLAN. In Chapter 21 we will see that 3GPP actually settled for an intermediate solution. Furthermore, in Chapter 22 we will see that, generally, the different standardiza-

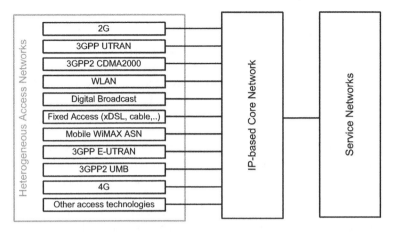

Figure 20.1 High-level architecture of 4G Networks

tion bodies are working out different, sometimes also incompatible, solutions to this general vision.

Beyond this architectural view, what does it mean that the heterogeneous Access Networks converge? In the 4G context, different degrees of convergence are discussed, in particular:

• Common Service Network

In its simplest form, convergence means access to a common network supporting services, which for the purpose of this book we call **Service Network**. In other words, no matter what the access technology, the user can access the same services.

As we already saw in Part I (cf. Chapters 10 and 15), 3GPP designed their service delivery support infrastructure, the IMS, to be access independent. A variety of standardization organizations are working intensely on specifying how their "Access Networks" can collaborate with the IMS. For example, the WiMAX Forum [WFArch] specifies how to access the IMS from WiMAX, and 3GPP2 have adapted the IMS specifications to be compatible with cdma2000. The fixed world also supports access to the IMS in the context of Fixed-Mobile Convergence, for example ETSI, the ITU, the Japanese ARIB as well as the consortium of cable operators, CableLabs. The IMS is therefore an example of a common Service Network

• Single Identity and Subscription

The subscriber has a single subscription and the MS has a single identity—a URI or a telephone number—in all Access Networks. The subscriber can be reached in all Access Networks and can roam between them if his MS supports the corresponding technology. He receives a single bill.

From an operator's perspective, this scenario raises an interesting question: with which operator does the user have the subscription? Subscriptions usually generate more value to the operator than providing roaming to somebody else's users. Technically, this question translates into who owns the HSS/AAA Server? Where—in the Core Network or in (which?) Access Network—is this server located?

3GPP in the **Evolved Packet Core** (EPC) [3GPP 22.278], 3GPP2 as **Converged Access Network** (CAN) [3GPP2 X.S0054] are already working on detailed solutions for Core Networks which feature an HSS/AAA Server to which heterogeneous ANs can be attached. The WiMAX Forum complements this approach neatly by specifying a Mobile WiMAX

Access Network which can be attached to any Core Network and which it expects to feature a AAA Server or equivalent [WF Arch].

The concept of a single subscription for heterogeneous ANs is also on the books of the standardization organizations for fixed networks, e.g. ETSI with **Next Generation Network** (ETSI NGN) and CableLabs with their **PacketCable 2.0** technology. They typically envisage an AAA Server respectively HSS to be located both in the Core Network and the Service Network.

- Automatic selection of Access Network
Here, the user no longer needs to select the most appropriate Access Network manually. Instead, a multi-homed MS chooses what is most appropriate automatically, in collaboration with the network. For example, it would pick DECT or Bluetooth while at home, UMTS when moving and WLAN when in reach of a hotspot. Obviously, the algorithm for selecting what is the "most appropriate" Access Network is of utmost interest: who decides according to which criteria? Which configuration options will be available to the user? Criteria for selecting the Access Network could be cost, QoS necessary to the desired application, current network load and how long the access is expected to be available—there is no use in choosing WLAN when the user is only driving through a hotspot.

Automatic selection of Access Networks in a network-assisted, i.e. operator-controlled way will be part of the next release of the 3GPP System. While the other standardization organizations can be expected to include this functionality as well, the respective specifications are not yet at an advanced stage. A summary of the problems and issues is provided in [RFC 5113].

- Handover and seamless handover between Access Networks
In the most sophisticated form of convergence, handover—or even seamless handover, is possible between heterogeneous Access Networks. This includes the basic mobility functionality, and, furthermore, support for transferring security credentials—i.e. a single authentication for all Access Networks—and QoS reservation to the new access.

Seamless handover is currently being addressed by 3GPP EPC and 3GPP2 CAN. seamless mobility between circuit-switched and packet-based Access Networks is even being specified. Full details are, however, still missing. Other standardization organizations are also planning to work on the issue.

20.2.1.3.2 Multi-hop Relay Topology

The Access Networks which we have encountered so-far have a fixed, tree-like structure, illustrated in Figure 20.2a, consisting of antennae with Base Stations or Access Points and a gateway node which, in 3G Networks, is also responsible for controlling the downlink Base Stations.

In many situations a more flexible arrangement would be desirable that allows one to easily—permanently or temporarily—extend or modify coverage and capacity, without installing additional wires. The areas of application are extension of coverage to remote areas, in-building coverage, e.g. in tunnels, or temporary coverage of, e.g. conventions or sports events. A solution to this problem are **multi-hop relay topologies**, where the Access Point is attached to the Controller/Gateway via one or more additional wireless hops, the **relays**, see Figure 20.2b.

The IEEE is working on 802.16j, an amendment to the WiMAX standard to allow for such multi-hop relaying. It is expected to be approved and published soon.

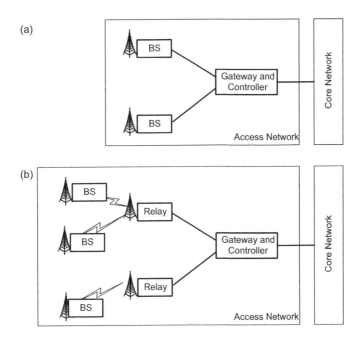

Figure 20.2   Access Network topologies: (a) Single-hop and (b) multi-hop relay

20.2.1.4 Mobile Stations and Networks of Mobile Stations

In the mobile world, the subscriber traditionally attaches a single mobile device, the Mobile Station, to the operator's network. As we see below, the capabilities of these single devices evolve. More recently, 3GPP specified how a subscriber can attach a network of devices behind a NAT, with only one device featuring a UICC (cf. Chapter 19, Section 19.5). Note that in the world of fixed networks it is in fact quite normal for a subscriber to attach her home network of laptops and PCs via a router to the DSL line or cable. We will describe below how this trend is likely to develop further.

20.2.1.4.1 Single Devices
The Mobile Station, particularly the "mobile phone", is likely to become a phone-sized full-scale computer in its look and feel and also in the way it is used and the application that it provides. The Apple iPhone, released 2007, provides a first taste of this: In addition to telephony, address book and calender, applications such as email and web browsing are available in a form especially suitable to a small mobile device.

One can also expect Mobile Stations to feature increasingly the openness and flexibility associated with computers: today's 3GPP UEs are usually of the type Internet-enabled-phone-plus-organizer-plus-camera, integrating applications, operating system and the chip with the radio communication capability. Technically speaking, TE and MT cannot be separated (cf. Chapter 9, Section 9.1). The Android platform promises to open up the system so that users can develop their own applications and generally are enabled to reprogram the system. WiMAX is going even further. The upcoming "WiMAX radio chips" can be integrated into any kind of Mobile Station—including 3GPP UEs—analogous to WLAN capability. In other words, WiMAX providers do not strive to sell Mobile Stations or particular applications. True to their heritage, they just sell connectivity.

Another computer-like feature of Mobile Stations will be **always-on**, i.e. IP-layer connectivity whenever the Mobile Station is switched on. We remember that with a 3GPP System this is currently not the case because an IP address is associated with an active PDP Context (cf. Chapter 11, Section 11.6). However, this will change as part of the ongoing EPC update (cf. Section 20.2.1.3).

20.2.1.4.2 Networks of Mobile Stations

We may observe an additional technology trend that is developing somewhat in parallel to the ongoing evolution of Telecommunication Networks. This trend relates to the emergence of new mobile network types. These new network types are able to attach, or will soon be able to attach to the mobile telecommunication infrastructure. It is likely that support of these networks will be a requirement on 4G Networks.

20.2.1.4.2.1 Personal Area Network and Personal Networks

Users increasingly own a multitude of mobile electronic devices that are able to network. For example, the digital camera can be connected to the laptop in order to upload photographs. MP3 players can connect to the Internet in order to download music. Nowadays, these personal electronic devices are networked mostly intermittently. It is, however, expected that **Personal Area Network** (PAN), more or less permanent networks formed of the electronic devices of a user, will soon become commonplace.

The devices in a PAN are in close physical proximity, and the networking is often wireless, e.g. based on Bluetooth. Devices in the PAN can communicate among themselves and, furthermore, if one device, e.g. the laptop or the mobile phone, has connectivity to other networks, e.g. the Internet or a 3GPP System, all other devices in the PAN can use this connectivity as well. Figure 20.3 illustrates a PAN.

3GPP acknowledges the trend towards PANs. We have already seen that with Rel-7, the IMS is enabled to deal with a UE that is in fact a network of devices behind a NAT

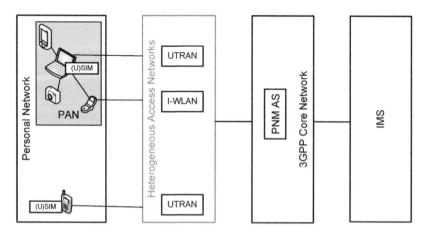

Figure 20.3   PAN (grey rectangle) and Personal Network

(cf. Chapter 19, Section 19.5). Concurrently, 3GPP is working on outright support of PANs and proposes an expansion of the concept to **Personal Networks** (PNs) [TS 22.259], see also Figure 20.3. A PN is composed of the networked electronic devices of a single user or a group of users, e.g. a family. Unlike in a PAN, however, the devices forming a PN do not need to be in physical proximity. For example, the mobile phone may travel to work with the user, while the rest of the PN stays at home. In this case, the PN can remain networked over a VPN via the 3GPP System.

In their investigation of PANs and PNs, 3GPP is, of course, especially interested in the support which the network can offer to the user. This support takes the form of a **Personal Network Management** (PNM) Server where the user can register the PN devices and their capabilities. This allows all PN devices to access other Access Networks, based on a single, locally available USIM. For example, the mobile phone in the middle of Figure 20.3 can access the I-WLAN, based on a USIM in the laptop, although the laptop itself is not connected to the I-WLAN but to a UTRAN. Furthermore, the PNM Server automates establishment of the VPN between a physically separated PN, and can direct incoming sessions to particular devices.

20.2.1.4.2.2 Moving Network

A **Moving Network** is a group of network elements—Mobile Stations as well as Access Points, routers etc.—on the move. Connectivity to the outside, i.e. the Internet or a mobile Telecommunication Network, is provided through one or more links via (a) dedicated Access Point(s). In other words, Mobile Stations do not individually connect to the outside. Instead, they route along a star-topology via the Access Point, as illustrated in Figure 20.4. The advantage of this approach is that mobility management can be performed in a bundled fashion for the entire Moving Network.

A Moving Network can, for example, be installed on a train, to faciliate Internet and Intranet access for passengers with laptops. The laptops attach to, e.g. a WLAN Access Point installed on the train. This in-train Access Point in turn could route traffic to an in-train 3GPP Mobile Termination (MT, cf. Chapter 9, Section 9.1.2) which connects to Node Bs along the train tracks, and performs handover when necessary. All traffic from the laptops is routed over this link.

From the perspective of the Node B, the entire Moving Network is a single UE. The train company, i.e. the UE owner, is responsible for any traffic originating from this UE. The individual laptops in turn have no clue that they are mobile, and they do not need to support mobility. Moving Networks in trains have been deployed based on proprietary solutions. The IETF has developed a protocol for this problem called **Network Mobility** (NEMO) [RFC 3963]; however it is not yet clear how this protocol would collaborate with, e.g. the 3GPP System.

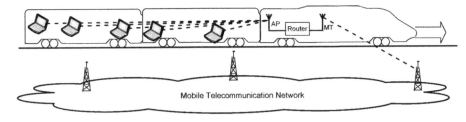

Figure 20.4 Moving Network; dotted lines denote radio links

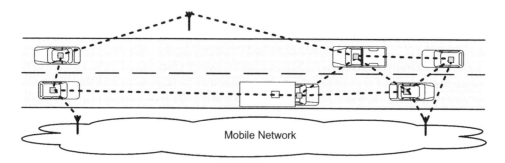

Figure 20.5   Vehicular Ad-hoc Network; dotted lines denote radio links

Another example of a Moving Network is in an aircraft that supports mobile telephony or web surfing. Of course, when in reach of terrestial Base Stations, airborne passengers can, in principle, use their MSs as usual. However, as we all know, this is forbidden because of concerns that the MS's radio signal may interfere with the aircraft's own wireless communication and control. Recent solutions for GSM and GPRS therefore offer a mobile Base Station which can be installed in the aircraft itself and which connects via a satellite link to the terrestrial mobile Telecommunication Network. The in-plane mobile Base Station emits an interference signal that effectively blocks reception of signals from other, terrestrial, Base Stations. The passengers' MSs thus connect to the local mobile Base Station. As a result, the MS's radio signals are not as strong and interfering, as if they were trying to communicate with a more distant terrestrial Base Station.

20.2.1.4.2.3 Vehicular Ad-hoc Network

A **Vehicular Ad-hoc Network** (VANET) is formed autonomously on-the-fly by vehicles, e.g. cars on the motorway, see Figure 20.5. Each Network Element (car) can be an end point, i.e. a receiver of information, and at the same time maintains a routing table in order to relay information passed by others. Via the roadside infrastructure, each car can be connected to the Internet. Since the network topology changes constantly, the routing tables must be continuously updated. Compared to the static star-topology of a mobile network, a VANET with its dynamic mesh-topology is more complicated.

A VANET allows data exchange between the vehicles and between the vehicles and roadside infrastructure, regarding, for example, traffic jams, road conditions, etc. This information could be fed automatically into the car's control system. There are also plans to use VANETs in a more broad-band fashion for user services, e.g. to facilitate inter-car games, or to broadcast the audio played in each car so other travellers can also benefit from it (some may doubt the benefit).

The IEEE standardizes VANET communication in their IEEE 1609 family of standards together with 802.11p as **Wireless Access in Vehicular Environments** (WAVE). 802.11p is scheduled to be released in 2009. In the US, highway tests of the system are already being planned. The topic of routing protocols for general (i.e. not just vehicular) ad-hoc networks is covered in the IETF Mobile Ad-hoc Networks (MANET) Working Group [IETF MANET].

20.2.1.4.2.4 Wireless Sensor Networks and RFIDs

A **Wireless Sensor Network** (WSN) is another form of ad-hoc, autonomously forming network, see Figure 20.6. WSNs can have a star or a mesh topology. The network nodes in WSNs are sensors that measure, e.g. temperature, pressure, etc. Sensors are stand-alone devices of very

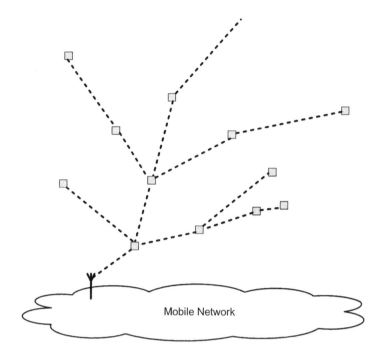

Figure 20.6   Wireless Sensor Network; dotted lines denote radio links

limited processing power and memory. They have their own power supply and should operate for years without manual intervention. Therefore all protocols must be designed for low resource consumption. Often, the sensors can only be end-points and nodes of more capability must act as relays. Also, unlike VANETs, WSNs usually only transmit limited information, i.e. the sensor readings. They do not need to transport large amounts of data. Therefore, the network need not support seamless mobility. WSNs are, for example, employed in public transport systems, in order to transmit the estimated arrival time of the transport vehicle to passenger stations.

IEEE has standardized a PHY and MAC layer for WSNs (which, incidentally, they regard as low data-rate PANs), in 802.15.4 [IEEE 802.15.4]. A protocol for joining and leaving an 802.15.4 network, for neighbour discovery, establishing routes and securely routing frames was specified by the ZigBee Alliance in [ZigBee 2006]. The IETF has developed an alternative that also allows one to run IPv6 over 802.15.4 networks [RFC 4944].

A special form of "sensor" employed in WSNs is **Radio Frequency Identification** (RFID) tags. RFIDs can be integrated into ID cards or be attached to products or containers. They store identification information that they pass wirelessly to RFID Readers. Hence, RFIDs can be used for contact-less identification and tracking, which of course raises a number of challenging security issues.

RFIDs come in two forms. They can be **passive RFIDs** without their own power supply. When prompted by the RFID Reader, they use the energy of the incoming radio signal to power up and transmit their response. Simple WSNs with passive RFIDs are, for example, access control systems and systems for tracking books in libraries or generally in inventories. In the longer term, passive RFIDs are expected to replace the bar codes that are used today to identify consumer products.

Active RFIDs have their own power supply. They have a longer reach and are equipped with a larger memory. WSNs with active RFIDs are, for example, being trialled in hospitals: patients receive a wrist-band with an RFID tag that stores all the relevant patient information. Statically installed RFID readers allow for the tracking of the patient's whereabouts if need be. Mobile readers carried by hospital staff allow for the secure identification of each patient and his needs.

20.2.1.5 Service Creation

How are user services going to be created in the future? One possible scenario is a continuation of today's popular model: "operators create services for Telecommunication Networks; everybody creates services for the Internet". However, as described in [Etoh 2005] and [Saunders 2007], and considering the recent moves of operators (cf. Section 20.1.2.3), in particular the open IMS interface, the Android platform and the collaborations with Apple's iPhone, Amazon's Kindle, etc., it seems likely that operators will continue opening up to services created—and hosted—beyond their realm, even to services created by individual users.

20.2.1.6 Summary of Short-term Services and Technology Trends

In the short term we expect first of all quantitative changes: more physical networks, more operators, more technologies, more network types, more bandwidth and more services. Regarding services, we expect those planned originally for 3G Networks to finally become widely accepted: multimedia services, location-based services, etc.

Furthermore, we expect another would-be feature of 3G to become reality: anytime anywhere communication. Based on convergence of heterogeneous Access Networks, including Fixed-Mobile Convergence, the user will no longer have to worry about which access technology to use. This also parallels an earlier development in Computer Networks: today's users deal with their web browser. They normally do not care about the Physical Layer.

Finally, we expect a tendency towards more openness in service creation, Mobile Station configuration and technology choice. This topic will be taken up in the next section on business models.

20.2.2 Short-term Business Models

We have already observed that today's business models of mobile Telecommunication operators have evolved from the original end-to-end service idea and are becoming more diverse (see Section 20.1.2.3). The technology described in the previous section points in the same direction.

New business models will include providing connectivity based on heterogeneous access technologies; on the other hand, they will include an opening of the networks to heterogeneous services from a variety of sources. We can therefore observe how the original mono-technological "vertical silo-type" business model (cf. Chapter 1, Section 1.2.1) becomes more flexible, accepting more technologies and collaboration with more players.

At the same time, operators will continue to avoid becoming bit-pipes. It is likely that the actual subscriptions, i.e. the knowledge of user data, will increasingly be an asset and a key differentiating factor: The operator, by virtue of having a relationship of trust with subscribers, is able to identify, authenticate and bill its subscribers. As **identity providers** they can provide this capability as a service to third parties, e.g. to operators of other networks and service providers. A classical, already existing identity provisioning service is roaming: the Home PLMN authenticates and bills the subscriber on behalf of the Visited PLMN. Along the same lines, we saw how the in the i-mode web portal leverages its relationship of trust with the subscriber (cf. Section 20.1.1.1): the operator owning the portal ultimately sells billing and authentication services. 3GPP already works on a further expansion and abstraction of the concept. The **Generic Authentication Architecture** [3GPP 33.220] allows a third party Application Server to bootstrap a Security Association with the UE on the basis of the relationship of trust of the 3GPP operator with the subscriber. Thus, for 4G one of the crucial questions is the location of the server containing the subscriber data, i.e. the HSS or AAA Server. In Chapters 21 and 22 we will see how many standardization organizations define such a server in their own Network.

An additional trend appears to be emerging which is also likely to broaden the traditional operator's business model: new, previously unexpected, players enter the operator market, namely consumer brands, retail chains and the like [Saunders 2007]. Presumably, these "retail-chain operators" will sell subscriptions to their traditional customers. However, beyond the occasional hot-spot, they will not invest heavily in actual infrastructure. Instead, they will have Roaming Agreements (cf. Chapter 11, Section 11.1) with an allied, infrastructure-owning operator. This agreement allows subscribers of the retail-chain operator to roam into the PLMN of the allied operator and—developing this thought further—maybe even into the PLMNs of all operators with which the allied operator, in turn, has a Roaming Agreement. These alliances may be regarded as another expansion of roaming, on the basis of **proxied Roaming Agreements**.

Figure 20.7 illustrates this situation. The retail-chain operator has a Roaming Agreement with an allied operator X. This Agreement allows the subscribers of the retail-chain operator to

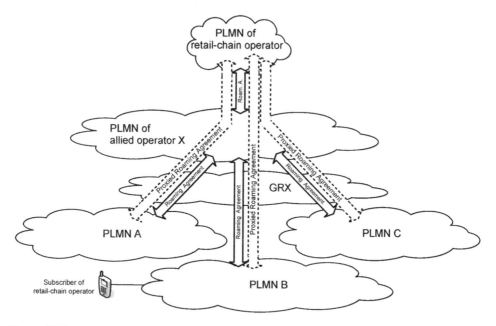

Figure 20.7   Roaming Agreement of a retail-chain operator with an allied PLMN operator X, and proxied Roaming Agreements with operators of PLMNs A, B and C

also roam into PLMNs of operators A, B and C with which operator X has a Roaming Agreement (for further information on the role of the GRX (see Chapter 11, Section 11.8)). From the perspective of operators A, B and C, operator X is responsible for the roaming subscribers of the retail-chain operator with whom they do not have a business relationship. In other words, operators A, B and C settle their bill with operator X and operator X settles the accumulated bill with the retail-chain operator.

20.3 IMT-Advanced

In this last section of our 4G overview chapter, we will look at **IMT-Advanced**, the upcoming official definition of 4G Networks by the ITU. Since 2002, the ITU has been publishing vision and framework papers on **Systems Beyond IMT-2000**—later renamed as IMT-Advanced, both regarding radio interface and Access Network [ITU M.1645] and network aspects in general [ITU Q.1702, ITU Q.1703]. In March 2008, an open call was published for candidate technologies. In July 2008, the ITU was planning to release the actual requirements which an IMT-Advanced technology must satisfy. Suitable candidates will be recognized as belonging to the IMT-Advanced family. Deployment is expected to commence in 2010.

Since the requirements are not yet available, we base our IMT-Advanced analysis on the capabilities published in the aforementioned vision papers by the ITU. We will show that the short-term technology trends described in the previous section already point towards IMT-Advanced and hence 4G.

20.3.1 IMT-Advanced Services and Technologies

IMT-Advanced systems will of course include all of the capabilities of IMT-2000 systems (cf. Chapter 2, Section 2.3.2) which we do not reiterate in this section. Furthermore, IMT-Advanced will be inter-operable with IMT-2000.

Below, we will go through the envisaged IMT-Advanced capabilities in the same sequence as we discussed near-term capabilities in Section 20.2.1.

20.3.1.1 Radio Interface and Bandwidth

A IMT-Advanced Network it is expected to support data rates of up to approximately 100 Mbit/s for high mobility such as mobile access, and up to approximately 1 Gbit/s for low mobility such as nomadic and local wireless access.

A crucial question is which spectrum the new technologies will occupy. The ITU-R has already allocated the spectrum for IMT—both IMT-2000 and IMT-Advanced, however there are worries that this may not suffice. In order to reduce spectrum needs, the ITU vision therefore expects 4G to incorporate technology that increases spectrum usage efficiency and allows for spectrum sharing, i.e. non-exclusive assignment of spectrum. Spectrum sharing capability would also reduce the need for centrally-planned spectrum allocation.

A technology for increasing spectrum usage efficency is beamforming antennae, e.g. MIMO, which is already incorporated in E-UTRAN, UMB and WiMAX.

An enabling technology for spectrum sharing is **Software Defined Radio** (SDR). Whereas traditional radio receivers and transmitters are mostly defined in terms of hardware, SDR performs the signalling processing in software as much as possible. This means that SDR devices can, ideally, tune to any frequency, and work with any modulation scheme, even across different technologies. Of course, an SDR device can also be re-programmed. One can even envisage on-the-fly adaptation of mobile devices: when the mobile device comes into a region with an unknown radio technology it downloads the appropriate software before handing over.

SDR can be used for spectrum sharing by incorporating it into so-called spectrum-sensing **Cognitive Radio**. The idea is for a device to sense the spectrum and then to move to a frequency that is currently available, i.e. performing a kind of cross-spectrum medium access control. Cognitive Radio thus allows for, e.g. utilizing spectrum licensed to someone else as long as it does not interfere with the primary user. A WiMAX addition, 802.16h [IEEE 802.16h] that builds on Cognitive Radio is currently being developed.

20.3.1.1.1 Current Evolution of Radio Interface and Bandwidth

As we saw in Section 20.2.1.2, a number of standardization organizations are working towards satisfying the bandwidth requirements by IMT-Advanced, i.e. 3GPP with an evolved UTRAN (E-UTRAN), 3GPP2 with UMB as well as WiMAX. While they have not yet achieved it, future versions of these technologies are considered to be contenders for the IMT-Advanced family. For example, in May 2008 3GPP announced the plan to develop **LTE Advanced** [3GPP 36.913] and submit it as IMT-Advanced candidate technology. Along the same lines, WiMAX is working on 802.16m, and 3GPP2 is also working on technology that will satisfy IMT-Advanced requirements.

20.3.1.2 Access Networks

20.3.1.2.1 Convergence of Heterogeneous Access Networks

IMT-Advanced systems support convergence, i.e. heterogeneous Access Networks, both wireless and wireline. The basic architecture with diverse Access Networks and a common, access-independent, IP-based Core Network was illustrated in Figure 20.1. The convergence envisaged should encompass the entire spectrum listed in Section 20.2.1.3: common Service Network, single identity and subscription—including roaming between heterogeneous networks, automatic selection of Access Network and seamless handover between Access Networks.

20.3.1.2.1.1 Current Evolution of Convergence of Heterogeneous Access Networks

As we have seen, convergence of heterogeneous Access Networks is already being worked on in 3GPP and other standardization organizations.

20.3.1.2.2 Multi-hop Relay Topologies

The other technology trend regarding Access Networks described in Section 20.2, multi-hop relay topologies, is not mentioned explicitly in current IMT-Advanced documents. It is not clear whether it will be part of the final requirements.

20.3.1.3 Mobile Terminals[3]

20.3.1.3.1 Single Mobile Terminals

Mobile Terminals are expected to become more diverse. Some Mobile Terminals will have somewhat limited capabilities, in an extreme case they are sensors which do not even have a user interface and which are capable only of machine-to-machine communication. Other

---

[3] Mobile Terminal is the ITU-specific term for what we generically call Mobile Station, see Appendix A Terminology.

Mobile Terminals are machines in general, and yet others are fully capable phone-sized mobile computers, or specialized devices, e.g. electronic books. Mobile Terminals are always-on and support a large range of heterogeneous access technologies, they can network among themselves and some are multi-homed and may be SDR-capable.

20.3.1.3.1.1 Current Evolution of Mobile Terminals
We have seen in Sections 20.2.1.4 and 20.2.1.5 that there is indeed a tendency towards more diverse Mobile Terminals, always-on, multi-homing, networking and more capable terminals, which mirrors the expectations of the ITU.

In Section 20.2.1.4 we saw also that in the short term Mobile Terminals may become more open, flexible and reprogrammable, just as computers are. For IMT-Advanced, the ITU postulates modular structures and open interfaces. However, from the documents published so far it is unclear whether modularity and openess refers not only to the network level but also to individual components such as Mobile Terminals.

20.3.1.3.2 Networks of Mobile Terminals
The ITU also expects network types to become more diverse and, moreover, IMT-Advanced networks to inter-operate with these networks. Support for mobile networks, PANs and ad-hoc networks are mentioned specifically: for example, users are expected to own more than one Mobile Terminal and to network them in an ad-hoc manner so as to result in PANs. Mobile Terminals belonging to different users can also form ad-hoc networks and communicate directly, i.e. without routing via an operator's network. All Mobile Terminals are able to access the backbone networks, possibly in a multi-hop fashion via other Mobile Terminals.

20.3.1.3.3 Current Evolution of New Types of Networks
We saw in Section 20.2.1.5 that standards are emerging for creating and running new network types, e.g. the WAVE standard for VANETs, the ZigBee standard for Wireless Sensor Networks, or the routing protocols for mobile and ad-hoc networks by the IETF. However, work is only just starting on connecting networks of these new types to mobile telecommunication Networks, e.g. with network-based management of Personal Networks in 3GPP.

20.3.1.4 Service Creation

An important capability of IMT-Advanced systems will be rapid service creation—i.e. service creation in real-time or at least within weeks. Not only operators but also many different players, e.g. users and independent third party service providers, will be able to create services and cooperate to this end. Service creation will be based on an open service platform with standardized interfaces. This platform must provide access to building blocks vital for service creation, e.g. security and charging solutions—just as, e.g i-mode provides a charging solution which can be used by third party service providers (cf. Section 20.1.1.1). Otherwise, each service creator must re-invent its own security and charging solution.

20.3.1.4.1 Current Evolution of Service Creation
The envisaged openness and modularity of the service platform goes beyond what the IMS standard offers today. The IMS offers a session control service to selected third parties, with

hooks to security and charging, as well as advanced service support, e.g. a Presence Service or a Push Service, which can be used to create new services (cf. Chapter 10, Section 10.1). On the other hand, the current and near-term de-facto development outside standardization, e.g. an open interface for developing IMS services (cf. Sections 20.1.2.2 and 20.2.1.4), seems to point in the direction envisaged by the ITU.

20.3.1.5 Other Technical Features

IMT-Advanced systems exhibit a number of other interesting technical features only two of which we will mention here:
- An IMT-Advanced network is an IP network. Circuit-switched technology is not mentioned in the existing documents, which means that it does not need to be supported.
- User-plane and control-plane functions will be separated by an open and standardized interface in order to allow for an independent evolution of the two planes. The division into control-plane and user-plane may be embodied by a separation of network elements into control nodes and user-plane nodes. However, it is also possible to locate both types of functions on the same Network Element, with an open intra-Network Element interface.

Looking back at UMTS, user-plane and control-plane are handled by different network elements in the CS Domain (cf. Chapter 7, Section 7.1 and Figure 7.2). However, in the PS Domain, a clean split of user-plane and control-plane was not performed: GGSN and SGSN each handle both planes, and an open interface between the planes is defined for neither of the two network elements. The architectural evolution of the PS Domain, EPC, exhibits a partial split, cf. Chapter 21. In Chapter 22 we will see that most other network architectures relevant for 4G do not provide for a clear division, either.

Interestingly, Computer Networks exhibit a clear division into user-plane and control-plane: a generic Network Element, e.g. a router has a user-plane, and control-plane functions can be added modularly. The interface between user-plane and control-plane functions is open and standardized.

20.3.1.6 IMT-Advanced Architecture

Finally, in Figure 20.8, we present the likely IMT-Advanced architecture (already sketched in Figure 20.1) in comparison to the architecture of IMT-2000 (Figure 4.1). We note the following differences:
- Support of the attachment of **networks of Mobile Terminals** rather than of single Mobile Terminals.
- Support of heterogeneous ANs rather than technology-specific RANs.
- The CN is IP-based only.
- A Service Network[4] is added as an independent entity to the architecture. IMT-2000 did not explicitly feature a Service Network.

---

[4] Note that "Service Network" is not official IMT-Advanced terminology.

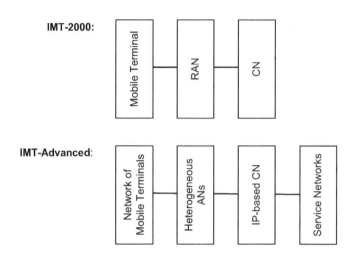

Figure 20.8   Comparison of the architectures of IMT-2000 and IMT-Advanced

20.3.2 *Summary of IMT-Advanced*

The ITU is currently specifying the requirements for IMT-Advanced systems, i.e. 4G. Based on the vision and framework documents published so far, it is likely that IMT-Advanced systems will have the following capabilities:

- 100 Mbit/s for high mobility and up to 1 Gbit/s for low mobility.
- Convergence of heterogeneous Access Networks.
- Rapid service creation by operator, third party service provider and users, based on an open service platform.
- Support of diverse Mobile Terminals, ranging from sensors to phone-sized computers.
- Mobile Terminals are always-on.
- Attachment of Moving Networks and networks of Mobile Terminals instead of single Mobile Terminals.
- IP-based network.
- Separation of user-plane and control-plane functions by an open, standardized interface.

Relating these capabilities to the ongoing evolution of mobile Telecommunication Networks described in Section 20.2, we may conclude that most of these capabilities are being worked on. It is likely that 3GPP, 3GPP2 and Mobile WiMAX will apply for 4G status when the ITU opens the call.

20.4 Discussion

What makes a mobile Telecommunication Network a 4G Network? Mobile Telecommunication Networks have traditionally been classified according their air interface technology and their air interface bandwidth. For example, a 2G Network must have a digital air interface, whereas in 1G the air interface is analogue. A 3G technology must offer up to 2Mb/s. UMTS plus HSPA is termed a 3.5 G technology because its bandwidth exceeds 2 Mb/s (cf. Chapter 19).

4G however, is different. Unlike previous generations, it is not defined predominantly by the radio technology. As [Saunders 2007] put it: "air interfaces become as interesting as the Ethernet PHY in 2007". The key point of 4G is the integration of heterogeneous technologies.

Another interesting point for discussion is the *convergence of Access Networks*. In Chapter 1 we observed that mobile communication technologies are said to converge and asked what this means exactly. The erstwhile, straightforward hypothesis was that it means evolution towards technical identity (cf. Chapter 1, Section 1.2.5). Convergence of ANs, however, does not imply eventual technical identity of ANs. Applying a somewhat pragmatic twist, it means indiscernability and indeed transparency of ANs from a layperson's perspective.

20.5 Summary

This chapter described the ongoing evolution towards 4G regarding services, technology and business models. It concluded with the capabilities which the ITU is expected to require from 4G Networks.

We first looked at current 3G Networks. The 3G vision was offering anytime anywhere connectivity, multimedia services as well as location-based services. Currently, however, anytime anywhere communication is hampered by the boundaries of technology. Regarding services, in most parts of the world, 3G Networks today are mostly used for circuit-switched telephony and packet-based connectivity to Intranets and the Internet. In Japan mobile web-portals are very popular, and in both Japan and South Korea mobile TV is an increasingly successful service.

We then investigated the foreseeable short-term evolution, i.e. ongoing standardization activities and expected business developments. First of all, what are noticeable are quantitative changes: more bandwidth, more services and more heterogeneity. Multimedia services and location-based services are expected to become widely used. At the same time, qualitative changes are emerging, such as always-on Mobile Stations. The most characteristic change, however, is convergence and integration of heterogeneous Access Networks, including fixed access and digital broadcast networks. As a result, the erstwhile vision of 3G should become a reality. Additional changes include greater openess in service creation and Mobile Station configuration, and interworking with new network types such as PANs, Moving Networks and VANETs, albeit that these developments are still mostly proprietory.

Finally, we looked at the expected ITU definition of IMT-Advanced systems as 4G. Presumably, 4G will build on 3G and include the changes that are already becoming apparent in the short-term evolution, although in a more perfected fashion. Interestingly, while previous generations were distinguished by their air-interface technology, 4G is likely to be distinguished by its integration of heterogeneous technologies.

Regarding the operator's business models, we observed how the traditional "vertical silo" is becoming challenged. It is likely to become more open, accepting heterogeneous technologies and heterogeneous services from a variety of sources. It is also likely that the relationship of trust with subscribers will become a key asset and that provisioning of identity will play an important role in future business models.

21

Evolution Towards 4G: 3GPP

UMTS R99 is a 3G Network. With the introduction of HSDPA in Rel-6, the 3GPP System became a 3.5G Network because the radio interface bandwidth was substantially more than that required for 3G. With Rel-8, which is the subject of this chapter, 3GPP makes the transition to a system that can be evolved to 4G. Rel-8 differs substantially from the previous releases. The result is called the **Evolved Packet System** (EPS).

The EPS is illustrated in Figure 21.1. Compared to earlier releases, the radio interface is completely new, as is the architecture of the RAN. These updates are known as **Evolved UTRA** (E-UTRA) and **Evolved UTRAN** (E-UTRAN), respectively. They result from a project known as **Long Term Evolution** (LTE). The main goals of the work in LTE were a higher data rate and a lower packet delay. LTE thus develops the 3GPP System towards satisfying the IMT-Advanced requirements on the radio interface, cf. Chapter 20, Section 20.3.1.1. Furthermore, the complexity of the Radio Access Network has been reduced in order for it to become competitive as regards other technologies such as WiMAX.

For EPS, the packet-switched Core Network has also been redesigned. It is called the **Evolved Packet Core** (EPC) and is the result of a project called **System Architecture Evolution** (SAE). The main goal of the EPC work is the convergence of heterogeneous Access Networks (ANs), which is another IMT-Advanced requirement; cf. Chapter 20, Section 20.3.1.2 This convergence is supported by an increased usage of IETF Protocols such as Mobile IP and its relatives. A side effect of the accommodation of diverse Access Networks and protocol options is indeed an increase of CN complexity. Figure 21.1. illustrates the difference between RAN and AN.

EPS integrates earlier 3GPP RANs, in that UTRAN and GERAN plus SGSN are supported as ANs. A smooth network upgrade from pre-Rel-8 to Rel-8 is, however, not possible. An evolution of the CS Domain is not a subject of the EPS work, beyond making sure that theCS Domain and EPS can collaborate.

The work on Rel-8 EPS is expected to be finalized by the end of 2008.

This chapter introduces EPS by way of the structure employed in Part I to introduce the 3GPP System pre-Rel-8: We first take an overall look at the architecture and then discuss the radio interface, i.e. the E-UTRA, and the main architectural components, i.e. EPC and E-UTRAN, in more detail. We will then discuss functionality. The EPS functionality is explained, followed by

UMTS Networks and Beyond   Cornelia Kappler
© 2009 John Wiley & Sons, Ltd

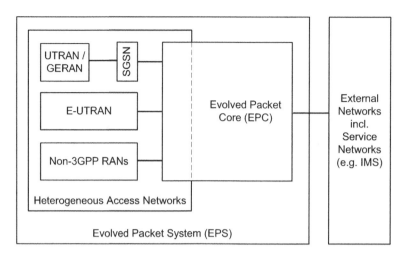

Figure 21.1   The Evolved Packet System (EPS) as evolution of the 3GPP System

specific control functions: mobility, security, QoS, charging and policy control. We conclude with a brief discussion of the motivation for the EPS changes.

A note on how to read this chapter: the description of EPS presented here is not "stand-alone". Instead, we focus on the differences from earlier releases of the 3GPP System and therefore we will reference previous chapters frequently. It might be helpful to turn back to these chapters and to refresh one's memory, if need be.

The service requirements for the EPS are documented in [3GPP 22.278]. The E-UTRA(N) can be studied in [3GPP 36.300], and the details for the EPC are in [3GPP 23.401, 3GPP 23.402].

Terminology discussed in Chapter 21:	
3GPP Access Network	3GPP AN
Bearer Establishment	
Controlling Gateway	
Dedicated Bearer	
Default Bearer	
Downlink Shared Channel	DL-SCH
Dual-Stack Mobile IPv6	DSMIPv6
Evolved Node B	eNB
Evolved Packet Core	EPC
Evolved PDG	ePDG
Evolved Packet System	EPS
Evolved UTRA	E-UTRA
Evolved UTRAN	E-UTRAN
Generic Routing Encapsulation	GRE
Global mobility	
Guaranteed Bitrate	GBR
Heterogeneous Access Network	Heterogeneous AN
Home Routed	

Link-layer mobility	
Local Breakout	
Localized mobility	
Local Mobility Anchor	LMA
Long Term Evolution	LTE
LTE Advanced	
Mobile Access Gateway	MAG
Mobile Node Home Address	MN-HoA
Mobility Management Entity	MME
Network Attach	
Non-3GPP Access Network	non-3GPP AN
Packet Data Network Gateway	PGW0
Proxy Mobile IPv4/IPv6	PMIPv4/v6
QoS Class	
Serving Gateway	SGW
System Architecture Evolution	SAE
Tracking Area	
Trusted Non-3GPP AN	
Untrusted non-3GPP AN	
Uplink Shared Channel	UL-SCH

21.1 3GPP Rel-8—Architecture and Protocols

The EPS architecture is shown in Figure 21.2. We may discern heterogeneous Access Networks (ANs) which are connected to an IP-based Core Network (CN). Heterogeneous ANs include, of course, 3GPP ANs. Heterogeneous ANs, however, cover the entire range depicted in Figure 20.1, e.g. 3GPP2 ANs, WiMAX AN and fixed ANs. Note that a 3GPP AN consists of a 3GPP RAN—such as UTRAN, GERAN or E-UTRAN—plus a **Controlling Gateway**[1] such as the SGSN or its evolution. The CN contains the AAA functionality including the subscriber database. From the CN, a gateway provides access to other networks, in particular Service Networks such as the IMS. This architecture represents a specific case of the IMT-Advanced architecture (cf. Figures 20.1 and 20.8).

The architecture displays an additional feature that is not mentioned by IMT-Advanced: Policy and Charging Control is applied to AN, CN and Service Network in a coherent fashion by a network element called PCRF. In other words, the Policy and Charging Control (PCC) mechanism developed in Rel-7 is extended in Rel-8, and now exerts centralized control on the entire network.

21.2 E-UTRA

The E-UTRA supports 326.4 Mbps downlink and 86.4 Mbps uplink. It is optimized for low-speed mobility; however, it also supports mobility at speeds up to 350 or 500 km/h. Note the shift in emphasis: the original UMTS UTRA was designed for low-speeds and high-speeds equally. The E-UTRA uses the same spectrum as UTRA.

---

[1] Book-specific terminology.

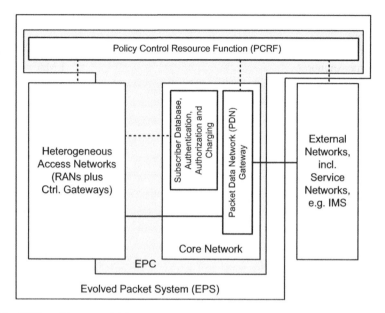

Figure 21.2 EPS Architecture with heterogeneous Access Networks. Dashed lines denote pure control interfaces, solid lines data and control interfaces

The E-UTRA as defined currently does not yet satisfy the IMT-Advanced requirement of 1 Gb/s in low-speed, and 100 Mb/s in high-speed environments. An update of the E-UTRA is, however, already underway under the title **LTE-Advanced** [3GPP 36.913].

The following technical aspects of the E-UTRA build on the more detailed discussion in Chapter 5.

The modulation schemes (cf. Chapter 5, Section 5.1) are QPSK, 16-QAM and 64-QAM, i.e. six bits are coded in each phase-amplitude shift.

Radio resources on the *downlink* are divided between users using OFDMA (cf. Chapter 5, Section 5.2.5.1). Furthermore, MIMO (cf. Chapter 5, Section 5.2.3) is used for spatial multiplexing and transmit diversity, with two transmission antennae in the E-UTRA, and two reception antennae in the UE.

Radio resources on the *uplink* are divided between users using SC-FDMA (cf. Chapter 5, Section 5.2.5.2) combined with MIMO. In the uplink, however, the UE has only one transmission antenna, i.e. MIMO is used for spatially multiplexing several UEs on the same frequency, in order to increase the cell capacity.

As an aside, neither OFDMA nor SC-FDMA suffers from the near-far effect; different senders send on different frequencies and can be distinguished in this way. Without near-far effect, macrodiversity, i.e. attaching the UE to more than one cell at the same time, is not necessary. Without macrodiversity, there is no soft handover. In order to simplify matters, 3GPP therefore decided that the E-UTRA supports only hard handover. Notwithstanding, this hard handover is seamless, as we will see in Section 21.4.2.1.

Of course, much more could be said about the E-UTRA technology. However, since the focus of this book is the network aspects we abandon the topic at this level of detail and suggest that the interested reader study other books, e.g. [Holma, 2007].

21.3 EPC—Architecture and Protocols

In this section we look at the EPC architecture in detail. We introduce the main network elements and the most important protocols. We approach the topic in a two-stage fashion by first giving a high-level overview and then taking a closer look at the detail including the protocols employed.

21.3.1 High-level View of the EPC Architecture and Protocols

The high-level EPC architecture shown in Figure 21.3 distinguishes 3GPP ANs and non-3GPP ANs:

- 3GPP ANs are, of course, E-UTRAN, UTRAN and GERAN. They rely on the HSS as a subscriber database and connect to external networks, e.g. Service Networks such as the IMS, via a gateway, the **Packet Data Network Gateway** (PGW).
- Non-3GPP ANs are Access Networks of any other technology, e.g. WLAN or WiMAX. They utilize an AAA Server. HSS and AAA Server coordinate over an interface. Non-3GPP ANs connect to external networks via the PGW, just as 3GPP ANs.
  Non-3GPP ANs come in two qualities, trusted and untrusted.
  - o An untrusted non-3GPP AN is quite similar to a generalized I-WLAN. While the untrusted non-3GPP AN and the 3GPP Network indeed have a relationship of trust involving a secure tunnel connecting them (so to some extent "untrusted" is a misnomer), the UE additionally builds a secured tunnel through the untrusted non-3GPP AN to the gateway into the 3GPP Network, the ePDG.
  - o A trusted non-3GPP AN generally belongs to the same operator who owns the 3GPP Network, or to a "friendly" operator. In this case, UE and 3GPP Network do not need the secure tunnel through a trusted non-3GPP AN.

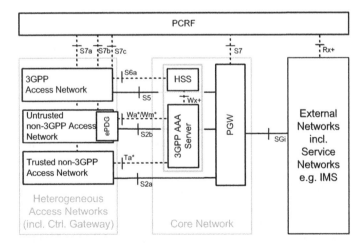

Figure 21.3 High-level EPC Architecture. Dashed lines denote pure control interfaces, solid lines data and control interfaces. Note that interface names may still be subject to change

It is instructive to compare the EPC architecture to the I-WLAN architecture shown in Figure 18.1a:

An I-WLAN, as non-3GPP AN, utilizes an AAA Server located in the 3GPP System CN and interfacing with the HLR, whereas UTRAN or GERAN, as 3GPP ANs, utilize an HLR directly. Both I-WLAN and UTRAN/GERAN connect to external networks via a gateway, namely the PDG in case of I-WLAN, and the GGSN in case of UTRAN/GERAN.

EPC changes this architecture in three major ways:

- it applies to any kind of AN, not just to WLAN.
- it consolidates the gateway to external networks so that 3GPP AN and non-3GPP ANs both use the PGW.
- It introduces a control interface from the policy infrastructure to the AN.

A note about EPC protocols: a major change as compared to previous releases is the introduction of IETF Protocols to complement and to some extent replace the traditional 3GPP-specific protocol stack, in particular for mobility control. Indeed, several alternative IETF mobility protocols are standardized by 3GPP so that on some interfaces up to three alternative protocol stacks exist! While this allows for good interoperability with other ANs—EPC is able to handle almost any of their mobility solutions—it also increases complexity because the number of possible protocol combinations within EPC is quite large.

Note that the fact that the standard specifies all of these protocol options does not mean that a given 3GPP Network will support all of them. While it is too early to make predictions, it is in any case conceivable that some operators will select one option, while other operators select another option, leading to interoperability issues and, in any event, to the necessity of selecting the mobility mechanism—all of which is also treated by the standard. It is also conceivable that a future Rel-9 will decrease the number of protocol options once operational experience has been gained.

21.3.2 Detailed EPC Architecture and Protocols

The architecture in Figure 21.4 illustrates the internal structure of EPC including the ANs in more detail, leaving out, however, some of the finer points. We cover the three cases—3GPP AN, untrusted non-3GPP AN, trusted non-3GPP AN—in separate subsections.

21.3.2.1 3GPP Access Network Architecture

The internal structure of a 3GPP AN distinguishes between the new E-UTRAN and the legacy UTRAN/GERAN. We will focus firstly on the E-UTRAN.

We should start by observing a structural pattern: a 3GPP System data packet always traverses the same basic set of network elements, illustrated in Figure 21.5a: starting from the UE, it passes the RAN, a Controlling Gateway, and finally continues via a gateway to the external networks. In traditional UMTS, i.e. UMTS up to Rel-7, this corresponds to the sequence UTRAN–SGSN–GGSN, cf. Figure 21.5b.

For 3GPP ANs in EPC, the same pattern is followed except that the Controlling Gateway is split into a control-plane and user-plane element, the **Mobility Management Entity** (MME) and **Serving Gateway** (SGW), respectively. The user-plane sequence thus becomes E-UTRAN–SGW–PGW, cf. Figure 21.5c.

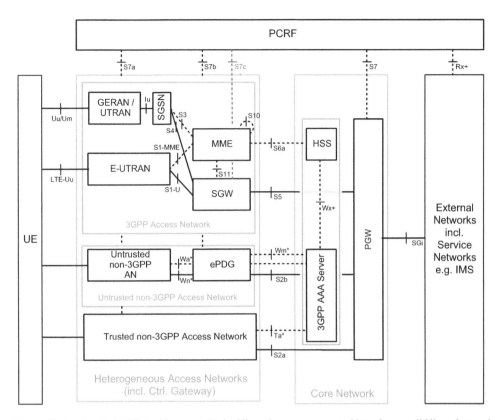

Figure 21.4   Detailed EPC Architecture. Dashed lines denote pure control interfaces, solid lines data and control interfaces. Note that the names of the interface may still be subject to change

Many of the other details of the E-UTRAN–EPC architecture are similar to the traditional 3GPP System: the MME pulls subscriber data from the HSS, just as the SGSN pulls subscriber data from the HLR (see Figure 6.1). The EPC policy architecture is also to a large extent adopted from Rel-7 PCC (see Figure 17.6): in EPC, the PCRF controls a PGW, in 3GPP System Rel-7, the PCRF controls its equivalent, the GGSN. Besides, the PCRF may interface with an AF in an external Service Network. One detail of the policy architecture, however, goes beyond Rel-7 PCC: In EPC, the PCRF may also control a PEP in the SGW. We will discuss the details in the following sections.

"Legacy" 3GPP RANs, UTRAN and GERAN also connect to the EPC, via the SGSN, cf. Figure 21.4. The UTRAN/GERAN–SGSN combination is treated almost like a E-UTRAN, i.e. it has a control interface with the MME and a user-plane interface with the SGW.

21.3.2.1.1 MME

The MME is the main control-plane node in the EPC. Generally, it takes on the same control functions as the SGSN: when the UE powers up and connects to a 3GPP AN, it attaches itself to a specific MME. When the UE moves, the MME may change. The MME authenticates and authorizes the UE based on information pulled from the HSS. It selects the proper PDN(s) and establishes one or more bearers from E-UTRAN via SGW to the PDN(s) with QoS as

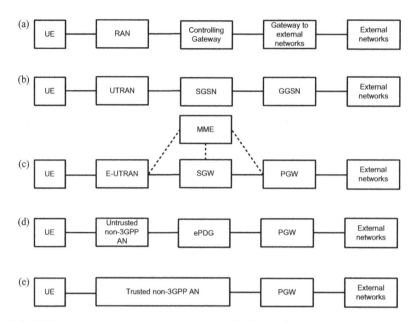

Figure 21.5 (a) Typical sequence of user-plane network elements in 3GPP Systems, (b) UMTS up to Rel-7, (c) EPC and 3GPP AN (including MME, a control-plane element) (d) EPC and untrusted non-3GPP AN, (e) EPC and trusted non-3GPP AN

appropriate. The MME is also involved in mobility management and collects Charging information.

21.3.2.1.2 SGW

The SGW is the user-plane gateway from AN to CN and is controlled by the MME. Whenever the UE is attached to a 3GPP AN, it is associated with a specific SGW. When the UE moves, the SGW may change.

The SGW also contains a PEP policed by the PCRF, which however is only used in combination with a non-traditional, non-GTP protocol option, which we will discuss in the next subsection.

21.3.2.2 3GPP Access Network Protocols

We have seen above that architecturally, 3GPP System Rel-7 and EPC are quite similar. When it comes to protocols, however, an important change was introduced: bearer set-up and bearer control between SGW and PGW may be performed by the traditional GTP, or— alternatively—by an IETF Protocol. We thus have two protocol stacks connecting a 3GPP AN with the CN, illustrated in Figures 21.6 and 21.7.

Figure 21.6 illustrates the traditional GTP option. In the control-plane, GTP-C is the protocol of choice. Between E-UTRAN and MME, a new SS7-family protocol, S1-AP, replaces RANAP. The user-plane protocol stack is identical to the pre-Rel-8 3GPP System when we ignore the recasting of gateways.

Figure 21.7 presents the new "IETF Protocol option". In the control-plane, it differs from the GTP option by replacing GTP-C between SGW and PGW with a new IETF mobility protocol,

(a) **S5 / S8a Control Plane, GTP option**

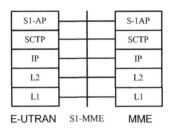

MME	S11	SGW
SGW	S5/S8a	PGW
MME	S10	MME

E-UTRAN   S1-MME   MME

(b) **S5 / S8a Control Plane, GTP option**

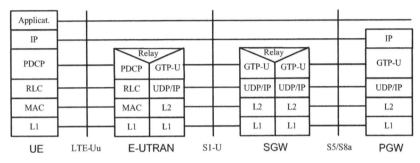

UE   LTE-Uu   E-UTRAN   S1-U   SGW   S5/S8a   PGW

Figure 21.6   GTP option of protocol stack 3GPP AN–CN (a) control-plane (b) user-plane

Proxy Mobile IPv6 (PMIPv6) [RFC 5213] which will be described in more detail in Section 21.3.3.1. Of course, GTP-C provides more than just mobility control. It also provides QoS control, charging control and some security functions. In the IETF Protocol option, these latter functions are taken care of by the Diameter protocol via the new policy control interface S2c between SGW and PCRF.

In the user-plane, GTP-U—which is basically a tunnelling protocol—is replaced by an IETF-defined tunnelling protocol, e.g. **Generic Routing Encapsulation** [RFC 2784] between SGW and PGW.

It is noteworthy that the IETF Protocol option is not end-to-end. Indeed, PMIPv6 is only applied between SGW and PGW. In the E-UTRAN, 3GPP decided to stick to their own protocol solution as interoperability was not an issue there.

Another interesting feature of the protocol stacks in Figures 21.6 and 21.7, and indeed of Rel-8 protocols stacks generally, is that both IPv4 and IPv6 are allowed by the standard, increasing the number of possible protocol combinations even further.

21.3.2.3 Untrusted non-3GPP Access Network Architecture

Untrusted non-3GPP ANs connect to the CN following the typical 3GPP structure: untrusted non-3GPP AN—**evolved Packet Data Gateway**—(ePDG)–PGW (cf. Figure 21.5d). The ePDG is a development of the PDG from I-WLAN (cf. Chapter 18, Section 18.1.2.1).

(a)   **S5 / S8b Control Plane, PMIPv6 option**

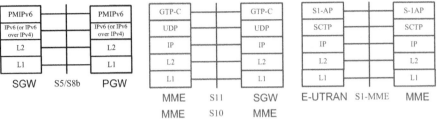

(b)   **S5 / S8b Control Plane, PMIPv6 option**

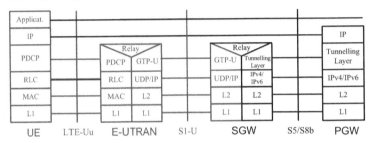

Figure 21.7   PMIPv6 option of protocol stack 3GPP AN–CN (a) control-plane (b) user-plane

21.3.2.3.1 ePDG

The ePDG has features of both WAG and PDG that are used in I-WLAN architectures (cf. Chapter 18, Section 18.1.2) It is the Controlling Gateway to the CN, and establishes a secure tunnel with the UE. Going beyond PDG functionality, the ePDG is a PEP for the PCRF.

21.3.2.4 Untrusted non-3GPP Access Network Protocols

A protocol stack for untrusted non-3GPP ANs is illustrated in Figure 21.8. The mobility protocol is again PMIPv6. The resulting protocol stack resembles that of the PMIP-option for 3GPP ANs, with the ePDG taking on the role of the MME/SGW.

Indeed, an alternative protocol stack is also possible based on another new IETF mobility protocol, **Dual-Stack Mobile IPv6** (DSMIPv6) [ID DSMIPv6]. DSMIPv6 is a new IETF mobility protocol that is still at the Internet-Draft stage. It builds on Mobile IPv6 and allows a MS with an IPv6 Home Address to use Mobile IPv6 transparently while being attached to an IPv4 Network. The DSMIPv6 protocol option is not shown in Figure 21.8 and is not treated in detail in this book because it does not add any substantial new insight.

21.3.2.5 Trusted non-3GPP Access Network Architecture

From an architectural perspective, trusted non-3GPP ANs differ from untrusted ones in that there is no ePDG (cf. Figures 21.4 and 21.5e): It is assumed that a secure tunnel through the AN is not necessary. Furthermore, the PEP that is governed by the PCRF is located directly in the trusted non-3GPP AN.

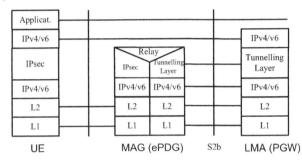

Figure 21.8 Protocol stack of untrusted non-3GPP AN–CN (a) control-plane (b) user-plane. Mobile Access Gateway (MAG) and Local Mobility Agent are PMIP-specific entities, cf. Section 21.4.1.1

21.3.2.6 Trusted non-3GPP Access Network protocols

Trusted non-3GPP ANs provide three options for the protocol stack: just as in untrusted non-3GPP ANs, mobility can be supported with PMIPv6 or DSMIPv6, or with MIPv4 (cf. Chapter 12, Section 12.3.2). In the case of DSMIPv6 and MIPv4, the UE is also involved in mobility control.

As we see in Figure 21.9,[2] it is assumed that the proxy performing network-based mobility control on behalf the UE, or a MIPv4 Foreign Agent (cf. Chapter 12, Section 12.4.1.2) respectively, are located in the non-3GPP AN. The PGW is a mobility anchor which, in case of MIPv4, acts as Home Agent (HA).

21.3.2.7 PGW

The PGW parallels the GGSN in earlier 3GPP System releases. The PGW is chosen depending on the destination network of a session. It assigns the IP address to the UE. The PGW is a PEP governed by the PCRF and performs deep packet inspection in order to block, filter, police and QoS-mark packets.

Unlike the GGSN, the PGW has an important role in mobility control. We have already seen that it terminates the IETF-option of mobility protocols, PMIPv6, DSMIPv6 and MIPv4.

---

[2] The details of DSMIPv6 are not discussed here.

(a) S2a Control Plane, PMIP option   **(b) S2a Control Plane, MIPv4 option**

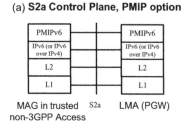 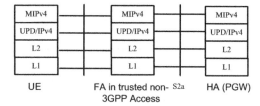

MAG in trusted   S2a   LMA (PGW)   UE   FA in trusted non- S2a   HA (PGW)
non-3GPP Access   3GPP Access

(c) S2a, User Plane

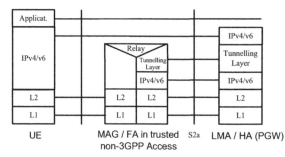

UE   MAG / FA in trusted   S2a   LMA / HA (PGW)
                        non-3GPP Access

Figure 21.9   Protocol stack of trusted non-3GPP AN–CN (a) control-plane PMIPv6 option (b) MIPv4
option (c) user-plane

21.3.3 E-UTRAN: Architecture and Protocols

21.3.3.1 E-UTRAN Architecture

An important goal of the E-UTRAN architecture is a reduction of the cost and complexity of the
equipment. This is achieved by reducing the number of options, and by removing one layer of the
hierarchy in the E-UTRAN architecture, as illustrated in Figure 21.10: compared to the UTRAN
architecture (cf. Figure 8.1), the main control node, the RNC, has been dismissed. Its functions—
radio resource control, QoS control and mobility control—have been integrated into the **evolved
Node B** (eNB). All eNBs are connected by an IP Network and can communicate with each other
using a protocol from the SS7 family-over-IP (see Figures 21.6 and 21.7).

21.3.3.2 Protocols and Channels

The protocol stack between UE and E-UTRAN is very similar to that between UE and UTRAN,
with the RRC protocol providing the main control functions. The channel structure, however,
has received a major update. The reader will recall that channels (cf. Figure 8.3) are located
between layers and describe how data is transported. Logical Channels build on Transport
Channels which in turn build on Physical Channels.

In order to decrease complexity, the number of Transport Channels was reduced; in particular
the concept of dedicated Transport Channels was abandoned. In E-UTRAN, all Transport

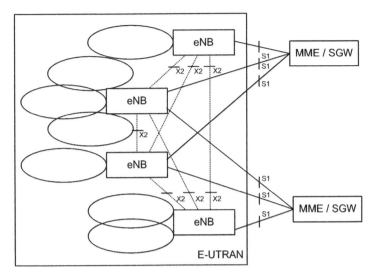

Figure 21.10   E-UTRAN architecture

Channels are shared. Remember that in GPRS the new concept of Downlink Shared Channel (DSCH) was introduced, allowing for the multiplexing of user-plane data to several UEs. In E-UTRAN, the DSCH is slightly adapted and becomes the DL-SCH. Furthermore, it is complemented by a new **Uplink Shared Channel** (UL-SCH). Note that dedicated Logical Channels continue to exist. They are, however, mapped onto shared Transport Channels.

21.4 3GPP Rel-8 Functionality

21.4.1 Basic Functionality

This section explains how basic communication functionality is realized: the user switches on the UE and starts a communication session. In pre-Rel-8 3GPP Systems, the following steps are involved (cf. Chapter 11):
- UE preparation—i.e. finding a suitable access point in a suitable 3GPP System PLMN,
- establishing radio connectivity,
- GPRS Attach, i.e. network attachment,
- PDP Context Establishment, i.e. establishing IP connectivity.

Overall, this sequence is still valid in Rel-8. It is, however, expanded slightly:
- UE preparation includes finding the access points of heterogeneous technologies and selecting both the most suitable technology and the most suitable Access Point of this technology. When the UE first powers up, the network obviously cannot assist in this task. The UE must scan for what is available and typically needs user input or a user policy telling it what technology to try first.
- The procedure for establishing radio connectivity is performed depending on the AN technology;

- GPRS Attach is extended to provide always-on IP connectivity. It becomes **Network Attach**;
- PDP Context Establishment is also extended and becomes **Bearer Establishment**.

We will discuss Network Attach and Bearer Establishment in greater detail below because these are the steps where 3GPP had to define a substantially new functionality. We start, however, with a more detailed description of the PMIPv6 protocol which is used optionally—as an alternative to GTP-C, DSMIPv6 or MIPv4—in the Network Attachment process and for localized mobility control. More information on PMIP and its relation to EPS is also available from [Emmelmann, 2007].

21.4.1.1 Proxy Mobile IP

In Chapter 12, Section 12.4 on IETF mobility protocols, we classified Mobile IP as a **global mobility** protocol that modifies global routing; it allows a MS to maintain reachability when its globally routable IP address changes due to mobility. Often, however, a MS only moves locally. It makes sense to account for local movement with locally confined modifications. This way, signalling overhead is reduced and handover delay is decreased. Indeed, the 802.11 WLAN standard offers **link-layer mobility**, invisible to the IP layer, as long as the MS stays within one ESS (cf. Chapter 12, Section 12.3.1). On the IP layer, Hierarchical Mobile IP (HMIP, Chapter 12, Section 12.4.2.2) provides for both, global and localized route changes.

Since the development of HMIP, however, the IETF has recognized that a completely separate protocol for **localized mobility** on the IP layer is a better and more modular solution [RFC 4830], because it allows for the independent evolution of global and localized mobility protocols. In fact, recently, new protocols have been developed that could be used as global mobility protocols, as an alternative to Mobile IP; for example the above-mentioned HIP [RFC 4423] (cf. Chapter 12, Section 12.4.1) or **Mobile IKE** (MOBIKE) [RFC 4555], a mobility and multi-homing extension of IKE. At the same time, Proxy Mobile IP is being specified as a localized mobility protocol that operates independently of the global mobility solution and can collaborate with any of them. Indeed, from the perspective of the Home Network, PMIP local mobility is invisible. PMIP is designed to complement link-layer mobility and to support localized mobility within the same "Access Network". The IETF is currently working on PMIP solutions for both IPv6 and IPv4. Since PMIPv6 is the protocol of interest to 3GPP this is what we will concentrate upon.

It is noteworthy that the definition of Access Network in PMIPv6 is slightly different from the definition of Access Network used in IMT-Advanced and 3GPP: while the latter is homogeneous regarding its radio technology, a PMIPv6 Access Network can include multiple radio technologies. In any case, however, an Access Network must belong to a single operator. Thus, what is termed "heterogeneous Access Networks" for the 3GPP System Rel-8 in Figure 21.3 would be called "Access Network" in a PMIPv6 context.

Besides providing a localized mobility solution, PMIPv6 has another striking feature. The reader will recall that all mobility protocols developed by the IETF so far locate mobility control in the MS, while 3GPP favours network-controlled mobility as provided by GTP (cf. Chapter 12, Section 12.5). PMIPv6 is the first IETF Protocol to offer network-controlled mobility—the UE only provides information on signal quality, location, etc. and the network makes the decision to handover. The advantage of this design is that it allows the deployment of

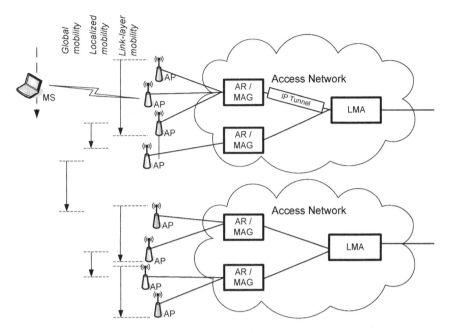

Figure 21.11   PMIPv6 architecture

PMIPv6 without an update of UE software. Furthermore, it makes PMIPv6 very interesting to 3GPP who decided to adopt the protocol when it was still at an early design stage.

Figure 21.11 illustrates the PMIPv6 architecture. It depicts link-layer mobility between Access Points attached to the same Access Router, localized mobility supported by PMIPv6 between Access Points attached to different Access Routers but belonging to the same Access Network and global mobility between different Access Networks.

PMIPv6 is modelled after Mobile IPv6 but with two important changes. The Home Agent equivalent, called **Local Mobility Anchor** (LMA), is located close to the ingress of the Access Network. Furthermore, a proxy, the **Mobile Access Gateway** (MAG), signals on behalf of the MS. The MAG is typically co-located with the Access Router (cf. Figure 4.7) and governs several Access Points. For the record, in EPS, the MAG is located in the Controlling Gateway, and the LMA is located in the PGW.

When the MS first attaches to an Access Network, the MAG which controls the corresponding Access Point sends a Binding Update to a local LMA. The LMA establishes a bi-directional tunnel to the MAG, using, e.g. IpSec or Generic Routing Encapsulation (GRE) [RFC 2784] and assigns a local IPv6 network prefix to the MS. The MS uses this prefix to generate a full IPv6 address under which it can be reached. This address is—somewhat confusingly—called the **Mobile Node Home Address** (MN-HoA).

When the MS moves into the realm of a new MAG, this new MAG sends a Binding Update to the LMA and the tunnel between MAG and LMA is adjusted. The MN-HoA of the MS, however, does not change as long as it moves within the same PMIPv6 Access Network, and the MS can be oblivious of its own movement.

The MS is expected to employ additionally a global mobility protocol. If the global mobility protocol is MIPv6, the MS would have an actual Home Address assigned to it by the Home

Agent in its Home Network. The collaboration of PMIPv6 and MIPv6 quite naturally proceeds as follows: the MS would communicate its MN-HoA as Care-of-Address to its Home Agent. When a packet destined for the MS reaches the Home Agent, it sends it to the Care-of-Address. The packet is caught by the LMA, who in turn tunnels it to the MAG. The MAG keeps track of the MS's Access Point using a link-layer mobility protocol and delivers the packet to the MS.

21.4.1.2 Network Attach

When entering the Network Attach phase, the UE has already established a Radio Bearer. The UE now registers with the network just as with GPRS Attach. Going beyond GPRS Attach, it also establishes always-on IP connectivity. In 3GPP language, one says that it establishes a **Default Bearer** by obtaining an IP address immediately and connectivity with best-effort QoS. Any packet may travel on the Default Bearer, its Traffic Flow Template (TFT, cf. Chapter 11, Section 11.6.3) is unspecific. The Default Bearer is maintained until the UE detaches. Always-on is a substantial enhancement compared to the previous releases of the 3GPP System and satisfies an IMT-Advanced requirement, cf. Chapter 20, Section 20.3.1.3.1.

The details of the attachment procedure depend on the type of AN and on the protocol option. Note that, it is not yet clear as to how it is decided which protocol option is to be used—a choice which also depends on UE capability, because MIPv4 can only be chosen if the UE supports it—nor as to how the network and UE communicate their choice.

In the next subsection we will cover the basic procedure. In subsequent subsections, we will highlight the interesting access-specific details.

21.4.1.2.1 Basic Message Flow for Network Attach

The Network Attach procedure is depicted in Figure 21.12. It builds on GPRS Attach (cf. Chapter 11, Section 11.5 and Figure 11.7) and the equivalent procedure for I-WLANs (cf. Chapter 18, Section 18.1.3).

Network Attach starts with access-specific steps in which the UE contacts the Controlling Gateway of its AN—e.g. the MME or ePDG, cf. Figure 21.5—to authenticate and authorize with the AAA/HSS. In the case of a 3GPP AN, the UE also communicates location information such as Cell Global Identifier and **Tracking Area Identifier** (Tracking Area is the Rel-8 equivalent of the Routing Area) and the HSS informs the MME about Subscriber Data—non-3GPP ANs must work without this information. Additional access-specific configuration procedures may follow.

There now follows the establishment of the Default Bearer, in particular the network side between controlling Gateway and PGW. The steps involved are by-and-large access independent. Also, on a high-level, there is little dependence upon the protocol option. Before delving into the details, however, we should consider what the "Default Bearer" involves: the PGW must assign an IP address to the UE, and a tunnel must be established between the Controlling Gateway and the PGW so that when the UE later moves into the realm of a new Controlling Gateway, the tunnel—which is anchored in the PGW—moves to the new Controlling Gateway.

With the GTP-option, Default Bearer establishment is straightforward: the MME/SGW signals to the PGW to set-up a PDP Context with minimum QoS.

With the IETF Protocol option, the messaging between Controlling Gateway and PGW to set-up the Default Bearer is performed by PMIPv6 or MIPv4. With PMIPv6, the SGW is a

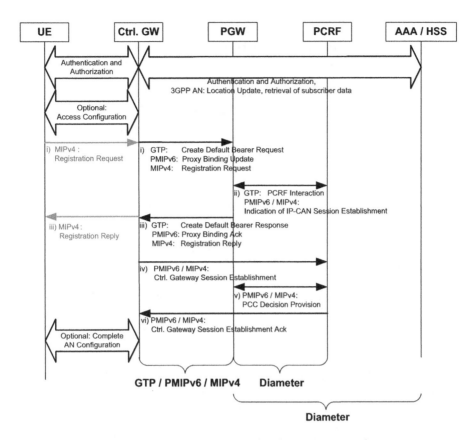

Figure 21.12 Basic message flow for Network Attach

mobility signaling proxy, i.e. the MAG, for the UE, and the PGW is the local Home Agent, i.e. the LMA. In order to establish the Default Bearer, the SGW/MAG sends a Binding Update to the LMA/PGW. The LMA/PGW assigns a Care of Address to the UE and sets up a bi-directional GRE tunnel to the MAG/SGW.

With MIPv4, the Default Bearer is established as follows: the UE sends a Registration Request (the MIPv4-equivalent of the MIPv6 Binding Update) to its Home Agent, which is collocated with the PGW.

We now can look at the message flow in more detail.

(i) The Controlling Gateway contacts the PGW in order to establish the Default Bearer. With MIPv4, which supports UE-controlled mobility, the Controlling Gateway is triggered by the UE. Note that when GTP is used, the Controlling Gateway can include a variety of information such as the UE's MSISDN or the Access Network type, which it cannot transmit with the IETF mobility protocols.
(ii) The PGW informs the PCRF of the new session and may pull policies that shall be applied, e.g. charging policies.
(iii) The PGW acknowledges establishment of the Default Bearer to the Controlling Gateway (respectively the UE with the MIPv4 option) and assigns an IP address to the UE.

(iv–vi) The message exchange with the PCRF which now follows applies only to the IETF Protocol option. It is necessary because the IETF mobility protocols in step (i) do not allow for the sending of the same information as GTP from the Controlling Gateway to PGW. In order to compensate, the Controlling Gateway establishes a control session with the PCRF that lasts as long as the Controlling Gateway is in charge of the UE. The Controlling Gateway and PCRF use this session in order to exchange information, and the PCRF can push policies to the Controlling Gateway. After establishing the session with the Controlling Gateway, the PCRF may use the new information to update the policy which it sent to the PGW. The details of which information is provided by the Controlling Gateway to the PCRF, however, have not yet been elaborated at the time of writing.

Depending on the AN type, the Network Attach process is concluded by a completion of the AN configuration, which includes the establishment of the Default Bearer between the UE and Controlling Gateway.

21.4.1.2.2 Network Attach for 3GPP Access Networks
As we have seen, the Network Attach procedure for 3GPP ANs is very similar to the GPRS Attach process (Chapter 11, Section 11.5) plus a PDP Context set-up (Chapter 11, Section 11.6). Note, however, that the Controlling Gateway, the former SGSN, is split into a control-plane node, the MME, and a user-plane node, the SGW. Authentication and authorization is performed by the MME. The MME then triggers the SGW to perform steps (i) to (vi).

An interesting point about the Default Bearer: when the UE starts the Network Attach process via a 3GPP AN, it has established a RRC Connection just as with pre-Rel-8. Also as with pre-Rel-8, when the UE idles (technically it goes into the corresponding idle Mobility Management state), the RRC Connection is torn down in order to save radio resources. The Default Bearer, however, stays established. In other words, when the UE needs to send or receive packets, the RRC Connection for the Default Bearer is re-established.

21.4.1.2.3 Network Attach for Untrusted non-3GPP Access Networks
The authentication and authorization procedure for untrusted non-3GPP ANs builds on the equivalent procedure for I-WLANs (cf. Chapter 18, Section 18.1.3). The UE receives a local (inner) IP address from the untrusted non-3GPP AN and authenticates with the AAA Server in the EPC. It uses IKE to establish a Security Association with the ePDG in order to establish an IPsec tunnel between UE and PDG and to authenticate each other.

21.4.1.2.4 Network Attach for Trusted non-3GPP Access Networks
The authentication and authorization procedure for trusted non-3GPP ANs is outside the scope of 3GPP.

21.4.1.3 Dedicated Bearer Establishment

After performing Network Attach, the UE has best-effort connectivity, i.e. a Default Bearer. Of course, this may not always be sufficient. Sometimes better QoS is necessary, and sometimes a different APN (i.e. PGW), connecting to a different network, needs to be used. Therefore, additional **Dedicated Bearers** with better QoS can be established.

The reader will recall that for pre-Rel-8 3GPP Systems, a similar concept exists, namely secondary PDP Contexts. While PDP Context set-up is usually started by the UE, in Rel-8 the establishment of a Dedicated Bearer is expected to be triggered by the PCRF in the PGW. In other words, a Dedicated Bearer is assumed to relate to a service request in a Service Network such as the IMS, which contains an Application Function (cf. Chapter 17, Section 17.3.2) interfacing with the PCRF.

21.4.1.3.1 Dedicated Bearer Establishment for 3GPP ANs

The establishment of a Dedicated Bearer starts with the PCRF triggering the PGW. The subsequent procedure depends on the protocol option.

With GTP, the PGW sends a Create Bearer Request to the SGW, containing QoS parameters as appropriate.

With the PMIPv6 option, a different approach is necessary—PMIPv6 simply does not offer an appropriate message: originally, for the Default Bearer, the SGW sent a Proxy Binding Update to the PGW in order to establish a mobility-insensitive bi-directional tunnel. When it comes to establishing a Dedicated Bearer the UE thus already has a tunnel. From a PMIPv6 perspective, this is all that can be done for the UE. A second tunnel is not foreseen. Furthermore, PMIPv6 does not anyway carry QoS parameters. As a consequence, the Dedicated Bearer is carried in the same tunnel as the Default Bearer. In fact, for establishing the Dedicated Bearer, no PMIPv6 signalling is performed at all. Rather, the PCRF informs not only the PGW but also the SGW that a Dedicated Bearer be set up—both towards the UE and towards the PGW, and that QoS be assigned to the PMIP tunnel as appropriate.

21.4.1.3.2 Dedicated Bearer Establishment for non-3GPP ANs

For non-3GPP ANs, the same method is applied as for 3GPP ANs with the PMIPv6 option: the PCRF informs the respective Controlling Gateway that the Dedicated Bearer be set up. What the Controlling Gateway makes of this information—in particular whether and how it reserves resources towards the UE—is outside the realm of 3GPP and is therefore not specified further.

21.4.1.4 Detaching

The detach process can be triggered by UE or network:
- With GTP, a Delete Bearer Request is sent from the MME via SGW to the PGW.
- With PMIPv6, a deregistration message is sent from the Controlling Gateway to the PGW.
- With MIPv4, a deregistration message is sent from UE to PGW or, when the network initiates the detach, the Controlling Gateway revokes the registration of the UE at the PGW.

The Controlling Gateway sooner or later (depending on when it expects to see the UE again) deletes UE state and informs the AAA/HSS. The PGW tears down all bearers of the UE. PGW and—in case of the IETF-option also the Controlling Gateway—inform the PCRF of the event. Finally, the radio resources are torn down.

21.4.1.5 Roaming

What does it mean for the UE to roam when it is attached to a non-3GPP AN? Put simply, the UE is roaming when the "next" CN to which it is attached is not part of its HPLMN. Putting it more elaborately and more precisely, the UE is roaming when the AN to which it is attached does not have a relationship of trust with its HPLMN and cannot access the AAA Server or HSS in the HPLMN directly (remember that a non-3GPP AN, even an "untrusted" one, always has a relationship of trust with a 3GPP AN (cf. Section 21.3.1)). See also the discussion on roaming with I-WLANs in Chapter 18, Section 18.1.2.

Where are network elements located in the Rel-8 architecture when the UE is roaming? Their locations are derived straightforwardly from the other roaming architectures which we have encountered: pre-Rel-8 3GPP System (Figure 11.10), PCC (Figure 17.7) and I-WLAN (Figure 18.2b):

- The HSS and AAA Server are always located in the HPLMN because they store the subscriber's data. In agreement with the AAA roaming architecture (cf. Chapter 13, Section 13.5.1.3), a Proxy AAA Server is added in the VPLMN which directs the UE's requests to the AAA Server in the HPLMN.
- The AN, including the Controlling Gateway, is always located in the VPLMN.
- The PGW, can be located in the VPLMN (**Local Breakout**) or the HPLMN (**Home Routed**)—analogous options already existed pre-Rel-8, cf. Chapter 11, Section 11.8. The two options are illustrated in Figures 21.13a and 21.13b, respectively.

  Note how in the Home Routed scenario, the LMA—the "Local Mobility Anchor"—finds itself at quite a distance from UE and MAG in the HPLMN; while the IETF specified PMIPv6 as local mobility protocol (cf. Section 21.4.1.1), the specification obviously also allows for its usage as global mobility protocol.
- The PCRF is split into a V-PCRF in the VPLMN and an H-PCRF in the HPLMN. The H-PCRF controls the PGW in the Home Routed scenario. The V-PCRF proxies policy decisions from the H-PCRF and can add its own policies on top. It controls the PEP in the AN, and in case of Local Breakout, also in the PGW.

With a more detailed elaboration of the roaming scenarios a number of difficulties emerge due to the large number of possible combinations. We will not cover these problems here and will only illustrate by way of example what happens when the HPLMN and VPLMN support different protocol options, e.g. the HPLMN only supports GTP and the VPLMN only supports PMIPv6. When the UE attempts to perform a Network Attach procedure in a Home Routed scenario, the Controlling Gateway (in the VPLMN) sends a Proxy Binding Update to the PGW (in the HPLMN), while the PGW, of course, expects to receive a GTP Create Default Bearer Request message. Obviously this cannot work. We need a dual PMIP/GTP enabled SGW or PGW, otherwise roaming is not possible.

21.4.2 Mobility

Rel-8 mobility support depends on whether the UE moves within 3GPP ANs or whether a non-3GPP AN is involved.

Within 3GPP ANs, the full 3GPP System machinery providing advanced mobility support can be used. Indeed, the pre-Rel-8 procedures for Routing Area Update (now named Tracking Area Update) and handover have only changed moderately.

Mobility between a 3GPP AN and a non-3GPP AN is less straightforward because location information such as Cell Identifier or Tracking Area Identifier is not available in non-3GPP ANs and because features such as context transfer are not supported.

In the next subsections, we first cover intra-3GPP AN mobility and then move on to mobility involving non-3GPP ANs.

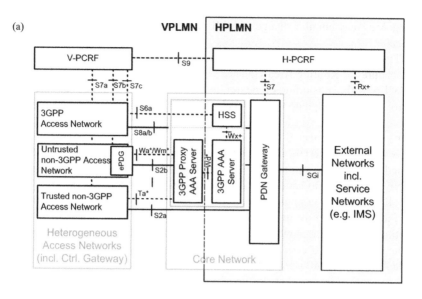

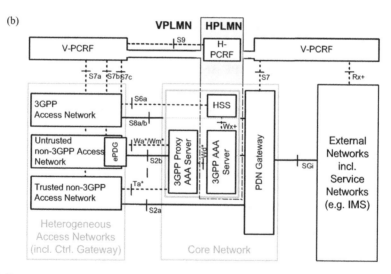

Figure 21.13 Location of network elements in different roaming scenarios, (a) Local Breakout, (b) Home Routed. Not shown is the GRX or IPX connecting the PLMNs

21.4.2.1 Mobility within 3GPP Access Networks

Analogous to Pre-Rel-8 3GPP System (cf. Chapter 12, Section 12.2), EPS defines mobility management states, in particular an idle state and a connected state.

When the UE is in idle state, i.e. it is not communicating, it tracks its location with the accuracy of a Tracking Area and updates the MME when the Tracking Area changes. A change of Tracking Area may entail a change of MME and SGW, and hence a rerouting of the UE's Default Bearer.

With GTP, the procedure for Tracking Area Update is a straightforward adaptation of the former procedure for Routing Area Update.

With PMIPv6, the rerouting of bearer is understood as a handover: the new SGW sends a Proxy Binding Update to the PGW and then establishes a control session with the PCRF.

When the UE is in connected state, i.e. it is actually sending and receiving data, seamless handover is supported: the UE measures the reception quality in the current and neighbouring cells and reports the result to its serving eNB. On the basis of these measurements, the serving eNB may decide to handover the UE to a new cell. It contacts the eNB in charge of the new cell in order to configure radio resources, transfers context information such as details on the UE's radio link, and then asks the UE to attach to the new cell. Meanwhile the SGW continues to delivers downlink packets to the old eNB, which buffers and forwards them to the new eNB. To conclude the transaction, the old eNB requests the MME (the control-plane node) to advise the SGW (the user-plane node) to reserve resources and henceforth route the UE's packets to the new eNB. The result is a seamless handover with context transfer.

Compared to handover in pre-Rel-8 3GPP Systems, we may note how soft handover is no longer supported, and how the roles of Node B and RNC are combined in the eNB.

21.4.2.2 Mobility Involving non-3GPP Access Networks

Regarding mobility involving non-3GPP ANs, 3GPP is of course only concerned with mobility between non-3GPP ANs and 3GPP ANs. Mobility within a single non-3GPP AN, or between different non-3GPP ANs is not a 3GPP issue. Note also that the concept of Mobility Management States does not apply. Therefore, we do not distinguish mobility in idle state and actual handover in connected state. There is only handover.

For mobility between non-3GPP ANs and 3GPP ANs, we cover three topics: discovery and selection of the new access, handover with make-before-break and so-called "optimized" handover supporting also context transfer and forwarding of user-plane packets to the new AN.

21.4.2.2.1 Access Network Discovery and Selection

In the context of convergence of heterogeneous ANs, the automatic discovery and selection of a new AN, without user interaction, is an IMT-Advanced requirement (cf. Chapter 20, Section 20.3.1.2.1). The method that is used within 3GPP ANs, which is based on a coordinated operation of the network, does not work in this case: within 3GPP ANs, information on neighbouring cells and Tracking Areas is readily available to the UE on known broadcast channels. Furthermore, network elements are acquainted with their neighbours: when the UE moves into another cell, the serving eNB knows which of its colleagues is responsible for this cell. Likewise, when the UE moves into a new Tracking Area, the MME knows the MME responsible for this Tracking Area.

Of course, the UE could discover new accesses by scanning for radio signals from all of the technologies which it supports. However, this would be rather wasteful. Indeed, network-based functionality that provides information to the UE on where to find neighbouring non-3GPP ANs would be helpful. The actual handover decision could be manual or automatic based on user and operator preferences and policies.

3GPP has already studied the problem and has identified potential requirements [3GPP 22.912]. The actual specification will, however, only be part of Rel-9.

21.4.2.2.2 Handover

After discovering the new AN, the UE performs steps similar to a Network Attach procedure, i.e. it triggers the configuration of the access resources, authenticates and authorizes. During the authentication procedure, the Controlling Gateway, however, learns from the AAA Server/HSS that the UE is already attached, and in particular learns the IP address of the PGW. This enables the Controlling Gateway to contact the PGW—the UE's mobility anchor—and to ask it to reroute the UE's bearers—using, e.g. a Proxy Binding Update when PMIPv6 is the protocol of choice. The reply from the PGW also contains the UE's IP address, which is thus maintained. Both PGW and Controlling Gateway (unless the Controlling Gateway is a MME and GTP is used) contact the PCRF to adapt their policies to the new access. Finally, the PGW or PCRF—the details have not yet been determined—trigger the release of the bearer in the old AN.

The handover just described is therefore seamless to the extent that it supports make-before-break.

21.4.2.2.3 "Optimized" Handover

3GPP would like to do better than just make-before-break. On its wish-list are context transfer, e.g. of authentication credentials, in order to save radio-interface resources and to decrease handover time, as well as buffering and forwarding of user-plane packets from old AN to new AN in order to reduce packet loss. Of course, greater operator control of the process would also be welcome...

Optimized handover requires support in the non-3GPP AN, therefore it must be specified individually for each non-3GPP access technology, in collaboration with the standardization body responsible for the technology. At the time of writing, optimized handover to and from cdma2000 ANs is being worked on. A specification for optimized handover to and from WiMAX ANs is planned.

The basic idea for optimized handover is the establishment of a connection between MME/SGW and the Controlling Gateway in the non-3GPP AN. This connection allows the transfer of context information and the forwarding of user-plane packets. Unlike in "non-optimized" handover, in this case the network makes the handover decision, it controls and times the process.

21.4.3 Security

Since Rel-8 alters the 3GPP System considerably, 3GPP analysed the new security threats. The focus is on threats resulting from non-3GPP ANs, and in particular mobility between 3GPP ANs and non-3GPP ANs, as well as threats resulting from the simplified, lower cost E-UTRAN which will presumably lead to deployment in more vulnerable locations, e.g. public indoor sites as opposed to roof-tops, and to less trusted transmission links, e.g. regular Ethernet cables, connecting the eNBs. The details of the enhanced Rel-8 security are still being worked on. EPS security is documented in [3GPP 33.401] and [3GPP 33.402].

21.4.3.1 Secret Keys

Generally, it is assumed that an EPS subscriber has a UICC with an identifier and a Master Key. Identifier and Master Key are used for authentication with the CN, and for deriving temporary keys for UE–CN communication. EPS secret keys can have a length of 256 bit, i.e. compared to the pre-Rel-8 3GPP System; the possible key length was doubled—while 128-bit keys are still considered secure, even from a perspective of ten to twenty years, the additional overhead to support 256 bits is seen as small compared to an even longer-term future security. 256-bit keys are, however, not mandated.

When the AN is an E-UTRAN, temporary secret key are derived from the Master Key. Unlike with the pre-Rel-8 3GPP System, where a single temporary key (IK) for integrity protection and a single temporary key (CK) for encryption were employed, in EPS two sets of temporary keys can be created: one set for securing the link between UE and eNB and another for securing the signalling between UE and MME. In this way the a security breach in the potentially more vulnerable E-UTRAN does not endanger the security of the UE–MME relationship.

When the AN is a non-3GPP AN, the temporary key can also be derived from the 3GPP System Master Key. Alternatively, AN-specific credentials may be used. In any case, the exact nature of the derived keying material is access-specific and outside the scope of 3GPP—for example, the 3GPP System keying material includes the Authentication Vector (cf. Chapter 13, Section 13.3.2), whereas other ANs have other requirements. This also means that a handover between different ANs necessitates re-authentication in order to derive the access-specific keying material. Such a re-authentication takes time, and therefore for optimized handover both early pre-authentication with ANs which are handover candidates as well as a transfer of security information as context information are investigated.

21.4.3.2 Authentication and Authorization

In the case of a 3GPP AN, mutual authentication and authorization is performed between UE and HSS, via the MME. At this point in time it has not yet been decided whether the procedure will be based on UMTS-AKA (cf. Chapter 13, Section 13.3.2), or on EAP-AKA (cf. Chapter 18, Section 18.1.5) which was already introduced in the I-WLAN specification, and which will also be used for non-3GPP ANs. While UMTS-AKA is tried-and-proven, EAP-AKA would have the benefit of streamlining the authentication procedure in 3GPP ANs and non-3GPP ANs. To identifiy itself, the UE presents the IMSI.

In the case of a non-3GPP AN, mutual authentication and authorization is performed between UE and (Proxy) AAA Server, based on the EAP-AKA method. To identify itself, the UE presents a Network Access Identifier in the format user@realm, where "user" contains the IMSI of the subscription.

21.4.3.3 Encryption and Integrity Protection

As a general rule in EPS, signalling messages are integrity-protected, and both signalling messages and user-plane traffic are encrypted on the air interface and when crossing operator boundaries. There are, however, several additional points to consider.

In the case of a 3GPP AN, signalling messages on the air interface are protected on the RRC layer, just as in pre-Rel-8. This protection is performed with the first of the above-mentioned key sets. Additional security may, however, be introduced in order to protect the CN from a possibly more exposed E-UTRAN: in this case, the E-UTRAN is handled formally as a different network. It has a static relationship of trust involving a secret key with the MME/SGW, and communication is via a Security Gateway (cf. Chapter 13, Section 13.3.4.2). Furthermore, the signalling messages between UE and MME are protected using the second key set.

In the case of a non-3GPP AN, 3GPP cannot mandate whether and how lower-layer signalling traffic between UE and non-3GPP ANs is secured. Presumably, a non-3GPP AN qualifies as "trusted" only if adequate integrity protection and encryption is applied to the signalling. Otherwise, in the case of untrusted non-3GPP ANs, the entire traffic between UE and ePDG is protected in an IPsec tunnel, so that a security problem in the AN does not infect the UE–CN relationship.

As a final point, when the mobility protocol is not GTP, mobility signalling messages may be protected using a variety of protocol-specific mechanisms. This implies that the entities exchanging the signalling messages must have a relationship of trust and share secret keys. For example, when the mobility signalling is between the UE and PGW, as in the case of MIPv4, the UE and PGW must derive an additional temporary key from the Master Key. The details of these procedures have not yet been defined, nor how this additional protection relates to that which already exists such as the IPsec tunnel between the UE and ePDG in the case of an untrusted AN.

21.4.4 QoS

21.4.4.1 QoS Parameterization

The QoS parameterization concept of 3GPP Systems was simplified in Rel-8. Formerly, four Traffic Classes were defined, each with its own value range for each QoS parameter such as Guaranteed Bit Rate, Maximum Bit Rate, delay, packet loss, etc., see Chapter 14, Section 14.3.1 and Table 14.1. Thus, when a PDP Context is set up, the Traffic Class and exact values for about a dozen parameters need to be negotiated. With hindsight, this solution is considered to be unnecessarily complex. Freedom of choice in parameter values is not really needed.

Instead, for Rel-8, the concept of **QoS Classes** was introduced. A QoS Class prescribes a fixed value for parameters such as delay and packet loss so that they do not need to be negotiated. The standard distinguishes **Guaranteed Bitrate** (GBR) QoS Classes and non-GBR QoS Classes. A GBR QoS Class is associated with a guaranteed and a maximum bit rate which can be chosen freely. A non-GBR QoS Class does not guarantee a bandwidth. Obviously, the Default Bearer is associated with a non-GBR QoS Class.

21.4.4.2 QoS Signalling

Rel-8 changes the QoS-signalling causality-chain as compared to earlier releases. In pre-Rel-8, the UE could request a particular QoS be assigned to its bearers (see Chapter 14, Section 14.3.2.1); when the bearer relates to an IMS service, the GGSN double-checks with the PCRF whether the QoS signalled in the PS Domain is the same as that requested in the IMS (cf. Chapter 17, Section 17.3).

As we saw in Section 21.4.1.3, in Rel-8 only the PCRF can request QoS for a bearer—a Dedicated Bearer—using the Diameter protocol.

With the GTP option in 3GPP ANs, the PCRF signals the QoS information as PCC rules to the PGW, from where it is distributed further to SGW with GTP. With the PMIPv6 option in 3GPP ANs, the PCRF signals the QoS information, i.e. PCC rules, to both PGW and SGW.

This latter approach could also be used to signal QoS information to non-3GPP ANs. This topic has, however, not yet been elaborated upon at the time of writing.

21.4.5 Charging

The PCC-based charging architecture defined in Rel-7 is re-used in Rel-8, protocols and parameterization needs only to be adapted slightly. PGW and Controlling Gateways are CTFs, i.e. they receive Charging Rules from the PCRF and meter Chargeable Events. Further detail is currently being worked on.

21.4.6 Policy Control

The PCC architecture defined in Rel-7 is re-used, with an additional interface to the Controlling Gateway. No further changes are foreseen at this time.

21.5 Discussion

Let us step back to take a wider view. With EPS, 3GPP makes a serious move towards satisfying the expected IMT-Advanced requirements, with a substantial increase of air interface bandwidth, convergence of heterogeneous Access Networks, always-on and an IP Network.

We may observe how 3GPP realizes the convergence of heterogeneous ANs: the network is defined so that the value-generating network elements (cf. Chapter 20, Section 20.2.2)—control of subscriber data (HSS), control of subscriber movement (PGW) and the gateway to external networks (PGW)—are located in the 3GPP-defined EPC. By the same token, there is only limited interest in defining how the E-UTRAN could become an alternative AN to other Core Networks. In Chapter 22 we will see to what extent other standardization bodies are doing the same.

There is something else which is noteworthy about EPS. Simplification and lowering costs are important aims for E-UTRAN design, while at the same time former goals are abandoned, such as optimal support of mobility at very high speeds. This shift of emphasis is inspired presumably by the success of the competition, especially WiMAX. WiMAX equipment is typically less costly, but also—so far—offers less sophisticated functionality. We thus observe another, somewhat different, case of convergence: convergence of standards.

21.6 Summary

In this chapter we presented the short-term evolution of the 3GPP System towards IMT-Advanced, as specified in Rel-8. This evolution is called EPS or LTE/SAE and supports higher data rates, convergence of Access Networks and always-on based on an IP Network. EPS brings about a major redesign of air interface technology, overall packet-switched architecture and protocol stacks.

The air interface, E-UTRA, supports 100 Mb/s downlink and 50 Mb/s uplink. It is optimized for low-speed mobility. E-UTRA adopts the modulation scheme 64-QAM, which is used in addition to 16-QAM and QPSK. Radio resources are divided between users on the downlink with OFDMA, and on the uplink with SC-FDMA. Additionally, MIMO is used.

The architecture of the evolved UTRAN, E-UTRAN, is simplified as compared to that of the UTRAN, it consists solely of eNBs. The eNBs integrate the functionality of the former Node B and RNC. They are connected by an IP Network. The overall EPS architecture features a CN connecting ANs of different technologies, both 3GPP ANs and non-3GPP ANs. The CN harbours the HSS/AAA, and the PDN, which is a LMA/Home Agent to support mobility as well as the gateway to external networks. ANs consist of the RAN, e.g. E-UTRAN, plus a Controlling Gateway, e.g. the MME/SGW. Policy control regarding QoS authorization and charging is performed by a PCRF on PEPs located in the Controlling Gateways and the PGW.

IETF mobility protocols—PMIPv6, MIPv4 and DSMIPv6—are introduced which, together with Diameter for QoS signalling and tunnelling protocols on the user-plane, provide an alternative to GTP in the CN, and allow handover between 3GPP ANs and non-3GPP ANs.

Convergence of ANs includes access to heterogeneous ANs based on a single 3GPP subscription, network-supported selection of handover-candidate ANs and mobility supporting make-before-break between 3GPP ANs and non-3GPP ANs. 3GPP collaborates with other standardization bodies such as 3GPP2 in order to optimize the handover further, based on technology-specific means.

Always-on is realized by establishing a best-effort Default Bearer automatically when the UE attaches to the CN. Additional Dedicated Bearers with guaranteed bit-rate can be established later.

Because of the considerable changes, 3GPP performed an analysis of new security threats to the 3GPP System, threats which result from non-3GPP ANs, mobility between 3GPP ANs and non-3GPP ANs and from the simplified E-UTRAN which is expected to be deployed in more vulnerable locations. As a consequence, link-layer security between UE and AN, the security of mobility protocols and the security of the UE-CN relationship are separated strictly by employing different keys.

QoS handling is simplified by introducing the concept of QoS Classes. With the choice of QoS Class, the value of most QoS parameters is pre-determined and does not need to be negotiated.

3GPP System Rel-8 is expected to be released by the end of 2008.

22

Evolution Towards 4G: Non-3GPP Technologies

In this chapter we will expand our horizons and look at the Communication Networks being defined by standardization bodies other than 3GPP, in particular the technologies which are expected to play a role in 4G. Interestingly, these are technologies which come both from the mobile and from the fixed world. In particular, we will cover cdma2000 by 3GPP2, Mobile WiMAX, ETSI NGN and PacketCable 2.0.

For each technology, we will provide an overview of functionality, radio interface (if applicable), architecture and protocols. In order to help the reader contextualize, the overview is given with reference to the 3GPP System and to the IMT-Advanced architecture.

Terminology discussed in Chapter 22:	
3rd Generation Partnership Project 2	3GPP2
Access Gateway	AGW
Access Service Network	ASN
Access Terminal	AT
ASN Gateway	
CableLabs	
Cable Modem	
Cable Modem Termination System	CMTS
Code Division Multiple Access 2000	cdma2000
Code Division Multiple Access One	cdmaOne
Connectivity Service Network	CSN
Converged Access Network	CAN
Core IMS	
Data Over Cable Service Interface Specification	DOCSIS
Evolved Packet Data Interworking Function	ePDIF
Fixed Mobile Convergence	FMC

UMTS Networks and Beyond   Cornelia Kappler
© 2009 John Wiley & Sons, Ltd

IEEE 802.16	
IEEE 802.21	
Media Independent Handover	MIH
Mobile WiMAX	
Network Access Provider	
Network Attachment Subsystem	NASS
Next Generation Networks	NGN
PacketCable 2.0	
PacketCable Application Manager	
Policy Server	
PMIPv4 Proxy Mobility Agent	
Resource and Admission Control Subsystem	RACS
Service Layer	
Ultra Mobile Broadband	UMB
WiMAX ASN Profile	
Wireless Broadband	WiBro

22.1 cdma2000

The **cdma2000** standard is being developed by 3GPP2 [3GPP2]. It includes the 3G technologies **cdma2000-1xRTT** and **cdma2000-1xEV-DO**. Historically, cdma2000 evolved from the 2G technology cdmaOne, see Figure 22.1. Currently, 3GPP2 is developing cdma2000 towards meeting the IMT-Advanced requirements. This development is called **Ultra Mobile Broadband** (UMB). It includes a new radio interface with higher data rates and lower delay, always-on and support of converged Access Networks under the name of **Converged Access Network** (CAN).

In fact, cdma2000 by and large offers the same services and satisfies the same requirements as a 3GPP system. However, the technologies are incompatible; it is only with the support of "convergence of Access Networks", i.e. with 3GPP EPS and 3GPP2 UMB, that interworking becomes possible.

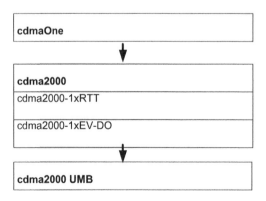

Figure 22.1   cdma2000 Evolution

Despite this incompatibility, however, the technologies have many commonalities—after all they basically solve the same problem. As we will see below, the architectures developed by 3GPP and 3GPP2 are similar, and a number of building blocks developed by 3GPP—e.g. IMS and PCC—have been picked up and are being adapted by 3GPP2.

At the same time, there are pronounced differences, for example in protocol usage. In cdma2000, IETF Protocols are employed quite commonly, whereas 3GPP tends to prefer customized solutions. For example, 3GPP specified MAP and GTP whereas 3GPP2 uses RADIUS and Mobile IP.

In the following, we will provide a brief review of cdma2000-1xRTT and cdma2000-1xEV-DO, and then present UMB, including interworking with IMS and the integration of a 3GPP E-UTRAN into 3GPP2 UMB.

22.1.1 cdma2000-1xRTT and cdma2000-1xEV-DO

cdma2000 is a family of standards. It includes the 3G technologies cdma2000-1xRTT and cdma2000-1xEV-DO. The architecture of cdma2000-1xRTT resembles that of UMTS in that it features a circuit-switched domain and a packet-switched domain, both connected to the same RAN. By contrast, cdma2000-1xEV-DO—DO stands for "Data Only"—is exclusively packet-switched, i.e. it leaves out the circuit-switched domain.

In the packet-switched domain, cdma2000 employs IETF Protocols wherever it is deemed to be feasible. An example of an area where nothing feasible was found at the time of development is seamless handover. While MIPv4 is used for global mobility support, 3GPP2-specific protocols were developed for local mobility and context transfer.

22.1.2 UMB

Ultra Mobile Broadband (UMB) is the most recent cdma2000 standard. Its network aspects had not yet been finalized at the time of writing. The overall idea is documented in [3GPP2 X.S0054].

22.1.2.1 Radio Interface

UMB has a theoretical bandwidth of 288 Mb/s downlink and 75 Mb/s uplink. The radio interface is based on OFDMA for the downlink and on OFDMA plus CDMA for the uplink. MIMO and beam forming (cf. Chapter 5, Section 5.2.3) are supported.

22.1.2.2 Architecture and Protocols

* Network structure
  A simplified UMB architecture is shown in Figure 22.2.[1] The similarity to the EPS archi- tecture is evident: the MS, here called **Access Terminal** (AT), is attached to heterogeneous

---

[1] We restrict our discussion to the "single-AGW" architecture. There is an alternative called "split AGW" with the AGW split between 3GPP2 AN and CN.

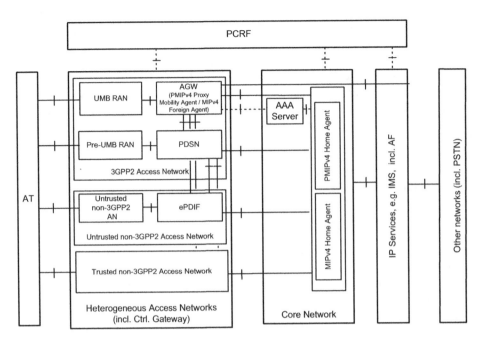

Figure 22.2 Simplified 3GPP2 UMB architecture

ANs, which in turn are attached to a CN, which in turn is connected to Service Networks such as the IMS.

- Terminal

  The terminal is assumed to be a single network element.

- Access Network

  In the Access Networks, the 3GPP2 applies the same classification as 3GPP: they distinguish 3GPP2 ANs, trusted non-3GPP2 ANs and untrusted non-3GPP2 ANs. ANs contain a RAN and a Controlling Gateway.

  In the case of the UMB RAN, the corresponding Controlling Gateway is called **Access Gateway** (AGW). Mobility control is currently performed on the basis of PMIPv4 or MIPv4, thus the AGW contains a **PMIPv4 Proxy Mobility Agent**—the equivalent of the PMIPv6 MAG (cf. Chapter 21, Section 21.4.1.1), or a MIPv4 Foreign Agent (cf. Chapter 12, Section 12.4.1.2). For the record, although the mobility protocols are IPv4 based, the mobility of IPv6-based hosts is also supported by a protocol translation mechanism using the DSMIP protocol [ID DSMIPv6]; we encountered a similar solution for 3GPP EPS, cf. Chapter 21, Section 21.3.2.4.

  The trusted non-3GPP2 AN includes a Controlling Gateway, its specification is outside of the scope of 3GPP2.

  For the untrusted non-3GPP2 AN, the Controlling Gateway is called the **evolved Packet Data Interworking Function** (ePDIF), which obviously corresponds to the ePDG of 3GPP.

  Note that control interfaces are provided between the Controlling Gateways of each AN type. These interfaces have not yet been specified in detail; they will support seamless handover between ANs, e.g. transfer of context information and user-plane packets.

- Core Network
  The UMB Core Network architecture also resembles the EPS Core Network architecture in that it contains an AAA Server and a gateway to external networks. The gatway is home to a mobility control function, a MIPv4 Home Agent or a PMIPv4 Home Agent, the latter being the equivalent of a PMIPv6 LMA.
- Service Network
  The Service Network can be the IMS or another IP-based network, e.g. the Internet.
- Policy infrastructure
  Policy and Charging Control is applied to AN, CN and the Application Function in the Service Network from the PCRF. PCEFs in the AN are located in both the UMB RAN and the AGW. In the CN they are located in the PMIP/MIP Home Agents.
- Protocols
  UMB supports both IPv4 and IPv6. 3GPP2, true to its tradition, employs IETF Protocols wherever possible, i.e. outside the UMB RAN; note however that for some interfaces the protocols have not yet been determined. Mobility is supported by MIPv4 and PMIPv4; QoS with DiffServ; and AAA functionality with EAP and RADIUS/Diameter. Since they have been using IETF Protocols all along, 3GPP2 can sidestep many of the deliberations which 3GPP is facing currently about interworking issues with the different protocol families used for 3GPP ANs and non-3GPP ANs: we saw in Chapter 21 that GTP can be used in a 3GPP AN, whereas MIPv4 can be used in a non-3GPP AN. Likewise, 3GPP has not yet decided whether authentication in a 3GPP AN will be based on UMTS-AKA as in pre-Rel-8 UMTS, or on EAP-AKA as in non-3GPP ANs.

22.1.2.3 Interworking with Other Technologies

The details of the interworking of UMB with non-3GPP2 ANs have not yet been worked out. The E-UTRAN will be integrated as one possible non-3GPP2 AN. For the E-UTRAN, requirements such as seamless handover support have already been specified.

22.2 Mobile WiMAX

WiMAX is an IEEE standard. Originally, it was developed as an air interface for *nomadic* Internet access in 802.16-2004 [802.16-2004]. The most important practical difference between 802.16-2004 and WLAN is the coverage radius of a single base station, which is in the range of tens of kilometres rather than one hundred metres. At about the same time, in Korea, the wireless broadband standard **WiBro** was being standardized, which also supported *mobile* access. Finally, the IEEE decided to update the WiMAX standard, including mobile access, on the basis of WiBro in 802.16e [802.16e]. What is more, 802.16e was used as a basis for specifying an entire mobile network [WF Arch] including support for security, QoS, charging and roaming. This work was performed by the WiMAX Forum. The resulting technology is called **Mobile WiMAX**; for an overview see the white paper by the WiMAX Forum [WF Overview 2006].

In the autumn of 2007, the ITU recognized Mobile WiMAX as member of the IMT-2000 family, i.e. Mobile WiMAX is a 3G Network, alongside UMTS and cdma2000. For Mobile WiMAX operators, this has the pleasant consequence that they can deploy their network in the frequency bands set aside for IMT-2000: WiMAX is specified so that it can work in a variety of frequency bands, both licensed and unlicensed. For commercial deployment, licensed bands are often preferred in order to prevent interference. Licensed bands are, however, normally assigned to specific technologies. By being recognized as an IMT-2000 family member, Mobile WiMAX sidestepped elegantly the lengthy frequency-assignment processes.

22.2.1 Radio Interface

Mobile WiMAX has a theoretical bandwidth of 63 Mb/s downlink and 28 Mb/s uplink. The radio interface is based on OFDMA and employs MIMO. It supports link-layer mobility up to vehicular speeds. The handover decision may be taken by the MS or by the network. The link-layer handover can be of two kinds: non-seamless hard handover with break-before-make—or, alternatively, seamless soft handover on the basis of macrodiversity. Seamless link-layer handover also includes context transfer, e.g. transfer of authentication state, between Base Stations.

22.2.2 Architecture and Protocols

- Network structure
  The WiMAX Forum defined the Mobile architecture depicted in Figure 22.3. A MS—which may include a User Identity Module such as a UICC—is attached to an Access Network, here called **Access Service Network** (ASN), which in turn is attached to a Core Network, here called **Core Service Network** (CSN), which in turn is connected to other networks, e.g. the Internet, the IMS, other telecommunication networks or other CSNs—e.g. for roaming.
  WiMAX is an IP Network—as its IEEE parentage suggests, a CS Domain is not included.
- Terminal
  The terminal is assumed to be a single network element.
- Access Network
  The functionality of the ASN is as one would expect: It is responsible for radio resource control, admission control for QoS, paging (there is an idle state), link-layer mobility,

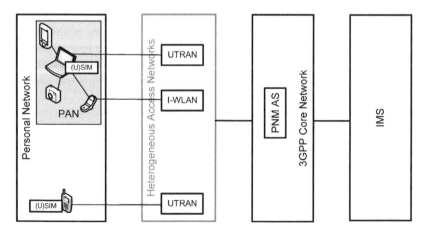

Figure 22.3 Mobile WiMAX architecture. Shown in grey are possible physical network elements

network discovery and selection, metering and data transport. The admission control may be based on external policy decisions, i.e. the ASN includes a PEP.

- Core Network
The functionality of the CSN is also straightforward: it includes authentication and authorization, IP address configuration, network-layer mobility control, collection of Metering Data and of course data transport.
- Service Network
Mobile WiMAX is a technology for providing connectivity to IP Networks, including of course the IMS.
- Policy infrastructure
The CSN includes a Policy Function which controls the admission decisions of the ASN, based, for example, on input from an Application Function (AF) in the IMS. This policy infrastructure is modelled after the PCC of 3GPP; however the details are not yet clear as the full specification has not yet been published.
- Protocols
Outside the ASN, Mobile WiMAX is based entirely on IETF Protocols. It supports IPv4 and IPv6. Network-layer Mobility is supported by MIPv4, MIPv6 or Proxy MIPv4. QoS can be provisioned with DiffServ. The policies for admission control are communicated from the Policy Function to the ASN by Diameter or RADIUS. The AAA Server also employs RADIUS or Diameter. Even online charging is included, based on an AAA Server and a Diameter extension [RFC 4006] or a RADIUS extension (still at the individual Internet Draft stage [ID RADIUS ext]).

Looking at the architecture more closely, it reveals interesting differences with respect to the 3GPP System and cdma2000 architectures which we have encountered so far:

- The WiMAX standard prescribes the distribution of functionality down only to the level of MS, ASN, CSN and the reference points between these functional groups. The further distribution of functionality into physical network elements is an implementation choice—more detailed specification is not deemed necessary—after all, the internal architecture of an ISP's Core Network is also not prescribed.

  For example, the ASN could be a single device, or it could be subdivided further into a Base Station and an ASN Gateway. In the case of such a subdivision, the distribution of the overall ASN functionality into Base Station and ASN Gateway is not standardized.

  Of course, this flexibility has a cost, namely potential interoperability problems, e.g. between ASN Gateways and Base Stations produced by different manufacturers. This problem is alleviated by defining a number of **WiMAX ASN Profiles**. A WiMAX ASN Profile determines a particular functionality configuration in the ASN, for example radio resource control is located in the Base Station in one profile and in the ASN Gateway in another profile.
- Only MS and ASN contain WiMAX-specific components. The CSN relies explicitly on IETF specifications such as an AAA Server and a Home Agent—without any additional embroidery. In principle, the CSN network elements can be commercial-off-the-shelf components.
- The ASN is envisaged as an independent entity that can be operated by a **Network Access Provider**, i.e. an operator which does not own a Core Network.

22.2.3 Interworking with Other Technologies

Mobile WiMAX differs from 3GPP Systems and cdma2000 in its approach to interworking: the focus in 3GPP and 3GPP2 is on defining a Core Network that allows for the attachment of heterogeneous ANs. How an E-UTRAN or a UMB RAN can be attached to the Core Networks standardized by others is of secondary importance. The focus in Mobile WiMAX, by contrast, is on how a Mobile WiMAX network can be attached to a generic Core Network. The root of this difference lies of course in the business models. 3GPP and 3GPP2 operators, as telecommunication operators, see their margin in the provision of high quality connectivity, user services and—in the future—identity services. The operators of Mobile WiMAX networks, however, see themselves as connectivity providers. They do not address the service or telephony market.

Concretely, the WiMAX Forum currently specifies WiMAX interworking with Pre-Rel-8 3GPP Systems by adapting the IWLAN solution. In parallel, 3GPP and the WiMAX Forum are discussing how a WiMAX ASN could be attached to the Rel-8 EPC. One of the issues raised is the usage of a draft IEEE standard, IEEE 802.21 [IEEE 802.21] on **Media Independent Handover** (MIH), which the WiMAX community would like to employ, while 3GPP already have their own solution to the problem.

The aim of the work on MIH is specifying infrastructure in the network which supports the preparation of handover between heterogeneous ANs. The MIH infrastructure provides information on available ANs, their properties and how they can be accessed; it can perform resource availability checks and pass messages between MS and new Access Points such as handover commands. The actual handover is expected to be performed by other mechanisms such as Mobile IP. An overview of MIH and its usage is available in [Lampropoulos 2008].

22.2.4 Mobile WiMAX and IMT-Advanced

Is the Mobile WiMAX architecture an IMT-Advanced architecture? In fact, the most distinguishing feature, support of heterogeneous ANs, is missing in Figure 22.3. While the Home Agent in the WiMAX CSN can, in principle, enable inter-technology handover to non-WiMAX ANs, the current Release 1 of Mobile WiMAX does not yet support it. Therefore, Mobile WiMAX Release 1 has indeed an IMT-2000 architecture (cf. Figure 4.1). Future releases of Mobile WiMAX are, however, expected to provide for seamless inter-technology handover and thus to be IMT-Advanced conformant.

Regarding bandwidth, the IEEE is working currently on 802.16m, an amendment to the WiMAX standard which will satisfy IMT-Advanced requirements on the radio interface.

22.3 Next Generation Networks

Just as the ITU defines the next generation of mobile Telecommunication Networks, the ITU also works on the replacement of PSTN and ISDN. This work bears the name **Next Generation Networks** (NGN). A NGN is IP-based, and supports VoIP as well as multimedia services in general; regarding control functionality, QoS and of course security are supported. Even mobility and roaming are on the agenda. NGN also allows for access from heterogeneous ANs [ITU Y.2001]. In fact—apart from a high-bandwidth air interface—the requirements for a NGN are very similar to the requirements for IMT-Advanced networks. This phenomenon is also known as Fixed-Mobile Convergence (FMC), cf. Chapter 20.

Standardization organizations active in the area of fixed networks are currently developing their specifications to become NGN-conformant. Below, we will look at NGN as an evolution of fixed telecommunication networks as defined by the European ETSI, and at NGN as an evolution of the TV-delivering cable network as defined by CableLabs.

22.3.1 ETSI NGN

ETSI has been developing their NGN concept in parallel and in coordination with the ITU. They have also been very interested in the IMS from the start and early on began defining policy coordination between the IMS and their own Access Network. The 3GPP work on PCC came later so that currently the policy infrastructures are not fully aligned. However, in the summer of 2007, ETSI and 3GPP decided to collaborate on the development of a common IMS so that a streamlining of specifications is to be expected.

22.3.1.1 Architecture and Protocols

Figure 22.4 depicts a simplified ETSI NGN architecture [ETSI 180 001, ETSI 282 001]. We discern, from left to right, terminals, connected to IP-CANs, connected to a Service Network, connected to other networks. The entire system is governed by a policy infrastructure. Compared to the architectures we have seen so far in this chapter, and to the 3GPP EPS architecture, we notice, however, a number of differences:

- Network structure
  The overall network structure of an ETSI NGN differs slightly from the IMT-Advanced structure we have been dealing with so far: the subdivision into Access Network and Core Network is not pronounced. Instead, ETSI adopts the concept of IP-CANs which 3GPP introduced together with the IMS: the IP-CAN provides connectivity and hides mobility from the Service Networks, in particular the IMS.
- Terminal
  The user equipment can be a single terminal such as a desktop computer, a multimedia terminal, a set-top box or a nomadic terminal. It can also be a residential gateway connecting several additional terminals and generally an entire customer network.
- Access Network and Core Network (IMT-Advanced terminology)
  The fixed broadband IP-CAN defined by ETSI exhibits a separation of control plane and transport plane. The control-plane consists of **the Network Attachment Subsystem** (NASS). The NASS's responsibilites include authentication and authorization on the basis of a subscriber data base that is also part of the NASS, allocation of the IP addresses and generally configuration of the access. The NASS currently supports nomadic UEs. Full mobility support is planned for later releases.
- Service Network
  The focus of the network architecture is on the **Service Layer**. Its main element is the IMS. ETSI defines a "**Core IMS**" that is a subset of the 3GPP IMS: the Core IMS includes only session control functionality, i.e. CSCFs, and a number of gateways. Application Servers, Multimedia Resource Function, etc. are not part of the Core IMS.
  - The Service Layer also maintains its own a HSS-like subscriber data base called the **User Profile Server Function**. Note how the Service Layer does not re-use the subscriber data

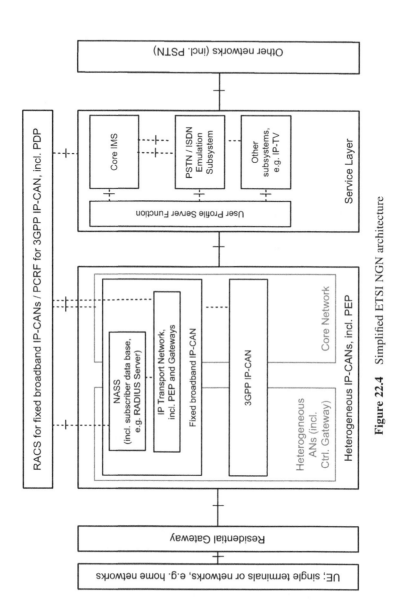

Figure 22.4 Simplified ETSI NGN architecture

base of the IP-CAN, on the assumption that the operator of the IP-CAN is different from the operator of the Service Layer. By contrast, 3GPP and 3GPP2 do not make explicit a separation of subscriber data for the AN and Service Network; both may be managed by a single HSS/AAA Server.

o Inside the Service Layer, a number of subsystems offer the actual services, e.g. IP-TV. These subsystems use the Core IMS for session control. An important subsystem is the PSTN/ISDN Emulation Service which supports legacy PSTN and ISDN terminals and thus allows for a gradual migration to NGN.

o Policy infrastructure

Admission control and QoS provisioning is policy controlled by an infrastructure called the **Resource and Admission Control Subsystem** (RACS). RACS is similar to PCC in that it controls the gateway from the IP-CAN to external networks based on input from an AS. As opposed to PCC, however, RACS does not deal with charging rules. Also, its internal architecture is different from PCC. RACS interfaces with the control function of the IP-CAN, the NASS, in order to exchange subscriber information. Note that RACS, unlike PCC, does not perform policy control on the NASS.

RACS also includes functionality for supporting the traversal of Network Address Translators (NATs). In Chapter 19, Section 19.5 we discussed the problems which a NAT can cause.

• Protocols

ETSI NGN supports both IPv4 and IPv6. Generally, IETF Protocols are used such as SIP, Diameter, RADIUS and EAP. The MEGACO protocol, also known as H.248 [ITU H.248], plays an important role. It is used for communication between control-plane network elements and user-plane network elements, e.g. between NASS and the IP transport network.

22.3.1.2 Interworking with Other Technologies

The ETSI NGN supports access to the services network from heterogeneous IP-CANs. When the IP-CAN is a 3GPP System, it is assumed that PCC is used instead of RACS. The details have yet to be provided.

22.3.2 PacketCable

The Cable operators' standardization organization is called CableLabs. Originally, cable usage was related to plain TV. Later, the standard was updated to cover Internet access. This Internet access via cable is facilitated by a **Cable Modem** installed at the user site, which communicates with a **Cable Modem Termination System** (CMTS) on the operator's premises, see Figure 22.5. The corresponding infrastructure on the operator's side is called the **Data-Over-Cable Service Interface Specification Cable Network** (DOCSIS Cable Network).

Later again, CableLabs defined a Service Network that supports digital telephony over the DOCSIS Cable Network, and more recently, multimedia services were added on the basis of the IMS. Terminal mobility and access via heterogeneous ANs will also be possible, although the details have not yet been worked out. The resulting overall system is called **PacketCable 2.0**[PKT FW 2.0 2007].

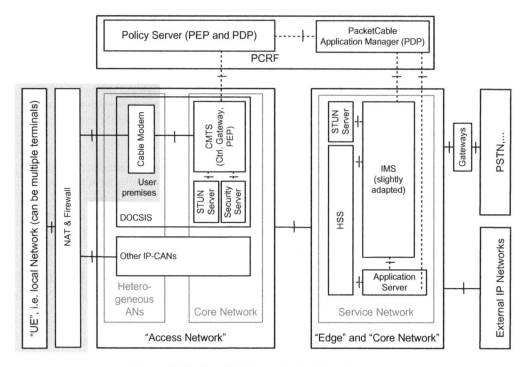

Figure 22.5 Simplified PacketCable 2.0 architecture

22.3.2.1 Architecture and Protocols

Figure 22.5 depicts a simplified PacketCable 2.0 architecture, presented in the usual structure. From left to right, are the terminal(s), an "Access Network", a "Core Network" and other networks. The system is controlled by a policy infrastructure. On a more detailed level, however, considerable differences exist as compared to mobile systems such as EPS, UMB and WiMAX, while similarities exist to ETSI NGN:

- Network structure
  The overall network structure of PacketCable 2.0 is similar to that of ETSI NGN, and is slightly different from the IMT-Advanced structure, and also in terminology: the DOCSIS "Access Network" includes what in the mobile world—including IMT-Advanced—is regarded as an Access Network plus Core Network, i.e. it is an IP-CAN. The PacketCable 2.0 "Core Network", in turn, is what for IMT-Advanced we called a Service Network. For ease of reference we employ the PacketCable terminology within quotation marks.
- Terminal
  As with ETSI NGN, the subscriber can connect a single terminal or—as the more typical set-up—a local network.
  o In a free adaptation of 3GPP terminology, ETSI calls **individual** terminals "User Equipment", although they do not necessarily include a removable UICC which a 3GPP UE always does—traditionally, for authentication with the DOCSIS AN, the MAC address of the Cable Modem and a security certificate installed in the Cable Modem,

which is already provided by the manufacturer, are used. To add to the confusion, 3GPP calls UE the *set* of all user terminals connecting on the basis of a single UICC (cf. Chapter 9, Section 9.1 and Chapter 19, Section 19.5).

o The user's local network is expected to feature a firewall and NAT.
o The local network interfaces with the operator's cable network via a Cable Modem. While located at the user's premises, the Cable Modem is controlled by the operator and is therefore considered to be part of the DOCSIS AN.

- Access Network and Core Network (IMT-Advanced terminology)
  The main network element in the DOCSIS "Access Network" is the **Cable Modem Termination System** (CMTS). It performs authentication and authorization on the basis of subscriber data in a Security server, and generally configuration of the access.

  o The DOCSIS "Access Network" interfaces with a DHCP server for providing IP addresses to UEs.
  o Care is taken to solve the problems caused by a NAT. The DOCSIS "Access Network" includes a STUN Server (cf. Chapter 19, Section 19.5) [ID STUN] that allows the UE to determine the external addresses which can be used for its user-plane media flows.

- Service Network
  The focus of the PacketCable network is providing multimedia services. They are located in what PacketCable calls the "Edge" and "Core Network", and what in this book we call the Service Network. Its main ingredient consists of a slight adaptation of IMS Rel-7: for example, support for the PacketCable-specific policy infrastructure was added. Subscriber data for services are managed in an HSS. Note that subscriber data for the DOCSIS AN is managed in an independent security server located in the AN just as with ETSI NGN.
  The Service Network also contains a STUN Server – normally co-located with the P-CSCF—which takes care of the NAT and firewall issues related to IMS signalling.

- Policy infrastructure
  Admission control and QoS provisioning is controlled by a policy infrastructure. This policy infrastructure is functionally equivalent to the PCRF, however it clearly structures and separates control of the Access Network and Service Network: The Access Network, in particular the CMTS, is controlled by a **Policy Server**. The Services Network, in particular P-CSCF and Application Server, are controlled by the **PacketCable Application Manager**. The Policy Server and PacketCable Application Manager, of course, need to coordinate: among themselves, they also have a policy control relationship, with the PacketCable Application Manager as overarching PDP and the Policy Server as PEP.

22.3.2.2 Protocols

PacketCable 2.0 supports both IPv4 and IPv6. For the Service Network, IETF Protocols are used such as SIP and Diameter. For communication with and within the DOCSIS "Access Network", a number of PacketCable specific protocols are used in combination with IETF Protocols such as RADIUS; likewise for interfacing with PSTNs.

22.3.2.3 Interworking with Other Technologies

Support of heterogeneous ANs is envisaged, however the details have not yet been worked out.

22.4 Discussion

This chapter has shown how technology coming from very different backgrounds—mobile Telecommunication Networks, fixed telecommunication networks and wireless access to Computer Networks—is evolving towards IMT Advanced. Note that this is even true for ETSI NGN and PacketCable 2.0 which set out to qualify as ITU NGN. It is just that the ITU's vision of NGN and IMT Advanced happens to be quite similar. It differs in that ITU NGN does not require advanced mobility features such as a high-bandwidth air interface and support of Moving Networks.

As a result, we note greater convergence:

- Functionality

  On a high level, the technologies discussed in this chapter aim to provide the same functionality—although the technical solutions have not yet been worked out in all cases. The difference lies in the detail. For example, mobility is always supported, but, e.g. for PacketCable 2.0, it is not as much the focus as it is for UMB. We may therefore ask whether mobility is seamless? Up to which speed does it work? Is it only possible within the same technology or also between heterogeneous Access Networks?

- Architecture

  The high-level architecture—terminal or network of terminals, Access Network, Core Network and Service Network—is the same in all technologies. This is partly because the problem lends itself to this solution, and partly because IMT-Advanced systems are in fact supposed to have this structure. The difference between technologies lies again in the detail. What is the substructure of this high-level architecture? How is functionality distributed in functional elements? Is a substructure elaborated at all (for example, in WiMAX it is not)? We have also frequently encountered inconsistencies in the terminology where the same term, e.g. Access Network or Core Network, is interpreted differently by different technologies. These inconsistencies, apart from leading to confusion, make the architectures look more similar than they really are.

  However, we may observe that a number of architectural elements are—with minor variations—re-employed in all of the technologies, for example PCC and the IMS.

- Protocols

  All of the technologies use IETF Protocols. While in the RAN and on the radio interface, however, technology-specific protocols are often preferred, in the CN and the Service Network, IETF Protocols are employed almost exclusively. 3GPP is even developing an alternative solution to GTP.

We note a certain concensus among the different standardization bodies as to which IETF Protocols should be used: PMIP immediately became a favourite; SIP, RADIUS, Diameter, DiffServ and MIP are also very popular. However, there are some differences that will limit interoperability, for example 3GPP standardizes PMIPv6, while 3GPP2 and WiMAX tend towards PMIPv4.

One could wonder about this general consensus for using IETF Protocols, which typically do not satisfy the design principles for Telecommunication Networks (cf. Chapter 1, Section 1.2). Have a look, however, at which protocols are favoured: for example, Diameter can be used to exercise telecommunication-style centralized control, and the development of PMIP is strongly influenced by the telecommunication operators who need network-controlled mobility.

We may therefore conclude that the technology convergence inspired by NGN, IMT Advanced and 4G goes beyond the convergence already claimed for 3G (cf. Epilogue Part I).

22.5 Summary

In this chapter we have discussed Communication Networks other than the 3GPP System which are expected to play a role in 4G. These Communication Networks have very heterogeneous origins: a mobile Telecommunication Network background such as UMB, standardized by 3GPP2; a Computer Networks background such as Mobile WiMAX, standardized by the IEEE and the WiMAX Forum; a fixed Telecommunication Network background such as ETSI NGN; and a cable TV background, such as PacketCable 2.0. These technologies do not yet qualify for IMT Advanced, however they are certainly moving in that direction.

The technologies which have been discussed offer comparable functionality, namely, compared to 3G functionality, a higher bandwidth, convergence of heterogeneous ANs, access to Service Networks and always-on. They pursue the same basic architectural approach with a subdivision of (network of) MS(s), AN, CN and Service Network such as the IMS, and policy control. They employ more or less compatible sets of IETF Protocols, in particular outside the AN. Inside the AN and towards the MS, technology-specific protocols are often used.

However, the focus of each technology differs, so that the resulting systems are of course still very diverse:

- 3GPP2's UMB is a mobile Telecommunication Network. Just as for 3GPP, the goal is providing high-bandwidth, high-quality connectivity and services for—traditionally—single mobile devices. The emphasis in UMB is therefore on the convergence of Access Networks. Detailed work goes into the definition of 3GPP2 ANs, trusted ANs and untrusted ANs, The integration of the IMS is a major topic.
- Mobile WiMAX is a mobile Computer Network. The goal is to provide high-bandwidth connectivity to—traditionally—single mobile devices. Mobile WiMAX mandates a high-level architecture in which only radio interface and AN are WiMAX-specific. The CN and the Service Network are generically built from IETF-defined network elements whose exact choice is up to the operator. The focus is on serving as heterogeneous AN for others rather than on integrating other's heterogeneous ANs.
- ETSI NGN is a fixed Telecommunication Network. The goal is to provide high-bandwidth, high-quality services to a subscriber's network of personal devices. As a new feature, these devices may be nomadic, or even mobile, and access the operator's services via heterogeneous ANs. The ETSI NGN focus is thus currently on services, i.e. the IMS, the basic possibility of access via a heterogeneous AN, and on complications such as NAT and firewalls arising from the fact that the subscriber owns a network rather than a single device.
- PacketCable 2.0 has similar goals to ETSI NGN, and a similar focus.

23

Beyond 4G?

A great deal of research has been invested in the preparation of what will eventually become 4G. Concurrently, research has been ongoing into the future of the Internet in general. Meanwhile, the ITU is defining 4G with the IMT-Advanced concept. Some of the concepts that were studied originally for 4G are too advanced to be included into IMT-Advanced requirements, or are not yet sufficiently mature, or are too controversial, or are not deemed necessary.

In this chapter we will present a selection of concepts, features and ideas which are not part of the current version of the IMT-Advanced framework. Bear in mind that this framework dates back to 2004. Additional features may be included in the final requirements, which are expected to be released some time in 2008. Or they may become part of 5G, or of something different, or they may never become reality at all.

We will now progress through three topics, in an ascending order of futurism.

Terminology discussed in Chapter 23:
All-IP Network   AIPN
Ambient Intelligence
Clean-Slate Approach
Internet of Things
Knowledge Plane
Network Composition
Protocol heap
Self-managing network
Ubiquitous Computing
Wildly successful protocol

23.1 Self-managing Networks

Since their conception, the complexity of Communication Networks has increased. As we have witnessed in the course of this book, today's Communication Networks cover a greater variety

---

of uses, deliver more services, integrate a greater diversity of technology, serve more customers and feature a greater variety of terminals and networks of terminals. However, the user ought not to be aware of this complexity. The operators of the Communication Networks would also like to conceal this complexity from their staff, and minimize the need for manual configuration and maintenance in order to reduce operational costs. Another, very important, aspect is that minimized human interaction simplifies the deployment and administration of a network in regions where highly specialized personnel tends to be scarce, e.g. in rural, remote areas. Therefore, **self-managing networks** which run more-or-less autonomously are a topic for energetic research and discussion.

Generally speaking, a self-managing system monitors its environment as well as itself and adapts its configuration, state and functions accordingly, see [Kephart 2003]. In particular, it exhibits the following capabilities:

- **Self-Configuration**
  The ability to detect dynamically and automatically a change in environmental conditions and to configure itself according to high-level policies.
- **Self-Healing**
  The ability to discover, diagnose and correct problems and malfunctions automatically.
- **Self-Optimization**
  The ability to monitor its own performance and efficiency, to evaluate it and adjust it according to pre-set criteria.
- **Self-Protection**
  The ability to identify proactively security threats and system failures, to issue early warnings and to defend itself.

We may observe that a self-managing system features an intelligent closed control loop, as depicted in Figure 23.1, with sensors that detect environmental- and self-conditions, an

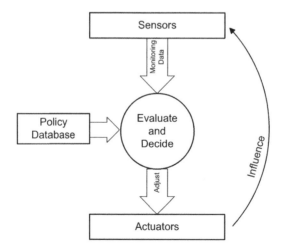

Figure 23.1   Schematic design of a self-managing system

evaluation and decision engine which is fed with high-level policies and actuators which are able to adjust system parameters state and, ultimately, sensor readings.

The key terms in the last paragraph are "*intelligent* closed control loop" and "*high-level policies*". Today's Communication Networks feature many closed control loops — for example the power control in UE, RNC and Node B (cf. Chapter 8, Section 8.1) — and there is also no lack of policy-based decisions (cf. Chapter 17). A truly self-managing network, however, derives its decisions from high-level policies that specify what is desired, rather than from a detailed description of how the desired state is accomplished.

A fully-fledged self-managing network, by definition, does not need any human help beyond someone who decides on the high-level policies, someone to physically install it and someone to press the start button. However, partial solutions are often possible, and are indeed more likely. A promising starting point for the introduction of self-management, in particular self-configuration, is the plug-and-play installation of new network elements or new functionality. An example is the autonomous bootstrapping of Base Station, as discussed in [Zimmermann 2005] which could simplify the installation — possibly temporary — of multi-hop relay Access Networks, cf. Chapter 20, Section 20.2.1.3.2. Indeed, the IEEE is discussing whether to include self-configuring Base Stations in the next WiMAX amendment, 802.16m (cf. Chapter 20, Section 20.3.1.1.1). Another example of self-configuration, though much simpler, is the envisaged MIPv6 bootstrapping capability (cf. Chapter 12, Section 12.4.1 and [RFC 4640]) which enables the MS to acquire dynamically both its own Home Address and the address of its Home Agent. At the other end of the spectrum, in Chapter 20, Section 20.2.1.4.2, we have already encountered autonomously forming Vehicular Area Networks and Wireless Sensor Networks. Further examples are provided in the next subsection where we study 3GPP activities in the area of self-management.

23.1.1 Self-management in a 3GPP System

For LTE-Advanced, 3GPP generally plans support for self-configuring network elements, in particular eNBs. Amongst others, the following use cases are being considered [3GPP 36.902], [3GPP 32.821]:

- Self-optimization of cell capacity and cell coverage where the network monitors itself and adjusts its control parameters according to current system load and system status: when load is low, selected eNBs fall asleep; when load is high, additional eNBs activate. When a eNB disappears–because of falling asleep or because of a failure, or when an additional eNB appears, the neighbouring eNBs adjust themselves so as to cover the affected area.
- Self-optimization of handover parameters, e.g. in order to decrease handover probability immediately after RRC Connection set-up, or in order to perform load-balancing between cells while at the same time minimizing the number of handovers.
- Auto-configuration of Home (e)NBs, i.e. miniature (e)NBs deployed in a subscriber's home (cf. Chapter 18, Section 18.4 on femtocells), upon being connected with the PLMN.

In parallel, 3GPP has been evaluating the feasibility of **Network Composition**, i.e. a dynamic, self-managed interworking of networks [3GPP 22.980]. The concept of Network Composition is in fact the result of one of the 4G research projects funded by the European Union (see Chapter 2, Section 2.3) called "Ambient Networks" [Ambient Networks]. Examples of Network Composition [Kappler 2005] are:

- the dynamic attachment of a non-3GPP AN, e.g. an IWLAN to the 3GPP Core Network. Currently, the interworking between a 3GPP Network and, e.g. an IWLAN is static in the sense that the IWLAN is expected to be immobile. One could easily imagine the business case of a nomadic IWLAN, however, whose operator offers WLAN coverage at mass-events. For such an operator it is desirable to perform the IWLAN attachment in a plug-and-play manner.

- on-demand establishment or modification of Service Level Agreements between PLMNs (cf. Chapter 14, Section 14.3.4.1).

Network Composition is, however, not yet on the list of features to be included in an upcoming 3GPP Release.

23.1.2 Discussion

The alert reader will, of course, have noticed how the grand vision of an autonomous system acting on the basis of high-level policies alone becomes somewhat more modest when applied in the real world. Indeed, "self-management" often just refers to the ability of a network element to adjust "meta information", i.e. its own control parameters such as address or online-status. The algorithm governing this change is based on detailed instructions rather than on high-level policies.

In fact, a true self-managing network exhibits certain pitfalls; how can one control a network that is self-controlled?—to paint an extreme scenario, the administrator may attempt to shut down an ill-behaved network, and the network—robust as it is—views this as an external attack and defends itself successfully. On a less sci-fi note, today's much simpler electronic systems are known to exhibit absurd behaviour when left alone without human monitoring and control: the author's favourite real-life example is the case of the thermostat-controlled air conditioning of an office building whose room temperature always resulted in a pleasant 20 °C. Everybody was happy, nobody bothered to look, until at some point a detailed energy monitoring system was installed. It revealed that in fact the cooling and the heating systems were running concurrently, the cooling system cooling, the heating system heating, resulting in an even overall temperature, but also in enormous energy waste and energy bills. The point is that it is hard to foresee all of the complicated ways in which a system can interact and malfunction, and that it is not feasible to human-control proper performance by results alone. Consequently, finding ways to control the undesired instability and unpredictability of self-managed networks is currently an area of active research.

23.2 Ubiquitous Computing, the Internet of Things and Ambient Intelligence

With the original mainframe computers, the relation of humans to machine was many-to-one; later, with Personal Computers, this relation became one-to-one; and now we witness a slow inversion of the original relation towards one-to-many: computers are becoming embedded in our environment, in all kinds of machines, devices and indeed in everything—the average car is full of embedded computers, the "smart home" with remotely controlled appliances which was envisaged for UMTS (cf. Chapter 20, Section 20.1.1) is becoming a realistic option—albeit on the basis of different technologies, see for example [digitalstrom], and the "intelligent fridge" which warns its owner that the milk supply is low became legendary even before it entered the market. We interact frequently with the numerous computers in our environment. However, while today there is still a surprisingly high number of infelicitous user interfaces, the goal is that the ubiquitous computers shall blend into the environment and their user interfaces shall be so unobtrusive that we will not even notice that they are there. Of course, these ubiquitous, embedded computers are networked amongst themselves—and indeed they are self-managing. Mark Weiser coined the term **Ubiquitous Computing** for this vision and he stated "The most profound technologies are those that disappear. They weave themselves into the fabric of everyday life until they are indistinguishable from it" [Weiser 1991].

When we now extrapolate this concept to all objects surrounding us, and think of very simple computers embedded in these objects, we arrive at the vision of the **Internet of Things** as described in an ITU Report [ITU Report 2005]. In this vision, the computer embedded in each object allows one to identify and locate it remotely and to read out simple parameters of relevance such as the energy consumption of machines, the temperature of containers carrying perishable goods, or the quantitiy of goods still available for sale. A simple version of the Internet of Things can be realized today on the basis of RFID Tags (cf. Chapter 20, Section 20.2.1.4.2.4).

We can go one step further, and add **Ambient Intelligence** explicitly to Ubiquitous Computing. In this vision, the embedded computers react to the users in a personalized way, considering always their current situation and context. In 2001, the European Commission initiated a description of Ambient Intelligence scenarios which at the time were thought to be a prediction for 2010.

To give an example, in one of these scenarios a business woman travels to a foreign country carrying only a single electronic device, a watch-sized personal assistant, on her wrist. This device had obtained her visa autonomously and she is able to stroll through immigration without stopping because her electronic assistant deals with any ID checks as she goes along. She proceeds to the parking lot, where a car—reserved of course by the same electronic assistant—is waiting for her and starts at the push of a button. While she drives, her daughter phones and the call is directed automatically onto the car's audio system. And so it goes on.

Today, in 2008, we still have numerous difficulties with regard to realizing this scenario. To name just a few:

- The ubiquitous network is self-managed to a degree far beyond what is possible today.
- The scenario raises numerous security issues which are difficult to resolve: apparently the business woman can be tracked wherever she goes and her electronic agent negotiates on her behalf, leaving traces of her preferences with little external control. Managing consistently which information shall be available to whom is indeed very challenging. As an aside, immigration into a foreign country based on autonomously acting electronic devices is somewhat more unrealistic today than it was back in 2001.
- Redirecting telecommunication sessions (e.g. the daughter's telephone call) automatically to an audio system available to the user on short notice, without explicit user-initiated configuration of, e.g. a SIP Proxy (cf. Chapter 15, Section 15.2.2), is not yet state-of-the-art. It requires additional networking and exchange of information regarding the user's location and the local environment between the personal electronic assistant and the user's

telecommunication agent (e.g. SIP Proxy). The entire procedure must be standardized in order to function in all countries and between the devices of different vendors.

Ubiquitous Computing, the Internet of Things and Ambient Intelligence are at this point beyond the realm of the standardization of 3GPP Systems. However, the 3GPP community is discussing how to support and integrate these ideas in a feasibility study on a so-called **All-IP Network** (AIPN) [3GPP 22.978].

23.3 Clean Slate Approach

Last but not least we must describe an interesting new focus in Internet research called the **Clean Slate Approach** for designing a Future Internet. It has the potential to affect Communication Networks in general. Work began in the 1990s, e.g. in the American **New Arch** project [Clark 2003a]. Meanwhile, the American National Science Foundation with **Future Internet Design** (FIND) and—some time later—the European Union with **Future Internet Research and Experimentation** (FIRE) set up programmes to fund corresponding research projects and large-scale experiments.

The motivation is as follows: the Internet was designed about 30 years ago, and as we all know it was intended for linking a handful of research groups, i.e. for a purpose and on a scale far more restricted than that for which it is being used today. According to the memorable classification by [Thaler 2007], the IP protocol is thus a **wildly successful protocol**, i.e "one that exceeds its original goals either in terms of purpose (it is used in scenarios that extend beyond the initial design) or in terms of scale (it is deployed on a scale much greater than originally envisaged) or in terms of both; that is, the protocol has overgrown its bounds and has ventured out into the wild".

Not surprisingly, the original Internet technology does not address today's requirements, e.g. mobility support, high security, QoS, separation of identifiers for identity and location, etc. As we have witnessed in the course of this book, numerous additions—protocols, architectures, etc.—have been developed in order to solve these problems. On the one hand, these solutions often have a hard time being adopted in the first place: with such an enormous deployment base, the Internet and Computer Networks have considerable inertia; and, unlike Telecommunication Networks, they do not experience regular coordinated technology updates when a new generation is introduced (cf. Chapter 1, Section 1.2.2). On the other hand, many solutions have indeed been adopted. We have witnessed how the resulting network is of considerable complexity—some say of incoherent patchiness—and still does not fully satisfy all requirements. Instead, the interaction of new features tends to create new problems.

According to the Clean Slate Approach, the problems of the current Internet have their root in fundamental architectural issues; these issues must be resolved from the ground up. The idea is to think radically, and to design and build a new network, based on a new architecture, satisfying the new requirements in a coherent manner. Backwards compatibility with the current Internet is not required.

The FIND and FIRE programmes have only just commenced, so we cannot yet report on the outcome. The NewArch project resulted in many inspiring new concepts, see [Clark 2003a] for a summary. They include questioning the layer structure and suggesting **protocol heaps** in order to increase flexibility and decrease interaction problems [Braden 2002]; an architectural model for decoupling of the identifier for identity and location [Clark 2003b] which is more general and flexible than HIP [RFC 4423, RFC 5205] (cf. Chapter 1); and the idea of instilling the network with a high-level cognition of its purpose, the **Knowledge Plane**, which amongst other things supports self-management [Clark 2003c].

23.3.1 Discussion

The idea of a Future Internet technology with all fundamental architectural issues resolved, tailored to solving problems that arose because not only the IP protocol but also the entire Internet is wildly successful, is certainly appealing. At the same time, some obvious doubts come to mind.

- It is not realistic to expect that the enormous base of "legacy technology" will be replaced. IPv6 is another clean slate design, albeit one able to coexist with the legacy IPv4. A global roll-out of IPv6 would be a comparatively simple affair, but even this has so far not happened.
- The Future Internet will be ready for deployment maybe ten years from now. Doubtless, many new requirements will have come into existence by then, raising new issues that may make the "Future Internet" at this point somewhat outdated.

In an interesting recent paper [Dovrolis 2008], a parallel is drawn with biology. It is argued that an evolutionary approach to resolving the Internet's fundamental design issues is also possible, just as the biological evolution has so far always solved life's design issues: "the evolutionary process is able to adapt and survive in radically changing environments and to produce surprising innovation, without ever needing a clean-slate restart". It goes on: "One may argue that extinction events represent examples of clean-slate restarts. On the other hand, it is safe to say that there are no signs of an upcoming Internet extinction". Furthermore, evolutionary solutions tend to be more robust and overall are more feasible than the revolutionary Clean Slate Approach.

Indeed, the motivation for Clean Slate research is often not so much replacing the current Internet but instead obtaining a fresh view and novel ideas which can be absorbed by the evolution of the Internet and, of course, of Communication Networks in general.

23.4 Summary

In this chapter we presented three areas of research on Communication Networks whose results are currently not included in the IMT-Advanced framework. However, some features are already likely to be part of 3GPP's LTE-Advanced as well as of other prospective 4G Networks. It is up to the reader to analyse to what extent future generations of networks will take up these ideas.

- Self-managing Networks
  A self-managing network monitors its environment as well as itself, and adapts its configuration, state and functions intelligently on the basis of high-level policies. It runs to a large extent autonomously, with minimal human interaction. Self-managing networks are thought to help both user and administrators to cope with the ever-increasing complexity of Communication Networks.

We will probably not see fully self-managing networks in the near future. More modest adaptations are currently being integrated into WiMAX and 3GPP's LTE Advanced, where "self-management" typically refers to the ability of a network element to adjust its own control parameters.

- Ubiquitous Computing, the Internet of Things and Ambient Intelligence
  Computers are decreasing in size and increasing in power. They are being embedded in more and more everyday devices; obviously they are nomadic or even mobile and network among themselves. The concept of Ubiquitous Computing refers to these computers being integrated into the environment to such an extent as to become unnoticeable.

  The Internet of Things describes a particular case of Ubiquitous Computing, where the embedded computers help to identify and locate objects, and read out simple object-specific or environmental parameters. RFIDs are a promising approach for the Internet of Things.

  When the ubiquitous computers become context-aware and personalized, we arrive at Ambient Intelligence.
- The Clean Slate Approach
  Expanding a classification scheme which was devised originally for protocols, the Internet is "wildly successful". It is being used in scenarios that extend the original design. Upon examination, it has been shown that for supporting today's requirements, the Internet has in fact a number of fundamental architectural shortcomings. The Clean Slate Approach is intended to design a "better Internet" with these issues resolved.

Appendix A

Terminology

This Annex summarizes the definitions of the most important technical terms introduced in this book. For other terms please consult the index and look up the corresponding chapter.

Note that a number of the terms introduced in this book are used in a different, often more loose, meaning elsewhere! The reason for this is that we are travelling at the intersection of Telecommunication Networks and Computer Networks. Some terms have a Telecommunication Networks origin; some have a Computer Networks origin. Sometimes a term coined by one of the communities was taken up and used by the other community—however, with a slightly different meaning. This terminology section therefore intends to clarify in which sense a particular term is used in this book.

Vice versa, the same functionality is sometimes named differently in the two types of network and in different technologies, and a generic term is not agreed. In such a case, this book defines a generic name which allows one to reference a particular functionality without reference to a particular technology.

Terms with a book-specific meaning are put in Arial in this Appendix.

3G Network   A 3rd Generation mobile Telecommunication Network satisfying the IMT-2000 requirements, for example UMTS and cdma2000.

4G Network   A 4th Generation mobile Telecommunication Network satisfying the IMT-Advanced requirements.

Access Point   A device connecting a Mobile Station and a Communication Network by means of a wireless interface. In this book, Access Point is used as generic term referring to any Communication Network. The corresponding technology-specific terms are as follows: Base Station for Telecommunication Network; Node B for UMTS; Access Point for Computer Network.

Access Network (AN)   A set of network elements that allow for the attachment of a terminal, e.g. a Mobile Station, in order to access a Core Network and generally other Communication Networks. It always contains access-technology specific elements. A fixed Access Network allows for wireline access, whereas a mobile Access Network allows for wireless access.

UMTS Networks and Beyond   Cornelia Kappler
© 2009 John Wiley & Sons, Ltd

A mobile AN consists of one or (more typically) several Access Points, possibly a controller of the Access Points and one or more gateways to the Core Network(s). The exact functionality and scope of an AN is, however, technology dependent. For example, in UMTS up to Rel–7, the AN comprises the Radio Access Network and the PS Domain. In UMTS Rel–8, the AN comprises the Radio Access Network, SGSN, MME and SGW.

In a telecommunication context, an Access Network supports exactly one access technology, e.g. WLAN, 3GPP or WiMAX. In an IETF context, an Access Network can be a multi-technology entity.

An AN always belongs to a single operator.

All-IP Network   This means different things to different communities (therefore this term is avoided in this book). The Computer Networks community understands an All-IP network to be what this book calls a Computer Network (for a more precise definition see below). The Telecommunication Networks community understands an All-IP Network to be a Communication Network employing the IP protocol (called "IP Network" in this book).

Base Station   The Access Point of a Telecommunication Network.

Business model   While formal definitions exist, in this book we use the term in a somewhat loose sense to denote the following: what products and services does a company offer? What is the competence area of the company? Who are the target customers? How does the company generate revenue?

Communication Network   A network composed of linked Network Elements for transporting user-plane information, e.g. emails, voice, video or web pages. In this book, we use the term Communication Network as the generic term to include both Telecommunication Networks and Computer Networks.

Communication session   A long lasting relationship between two end points of a Communication Network for the purpose of delivering a service, e.g. a data service or a multimedia service.

Computer Network   A Communication Network designed according to the "Computer Network approach" (cf. Chapter 1). The salient features are modularity and a protocol-centric design.

Core Network (CN)   Topologically, the Core Network is located between the (Radio) Access Network and other Core Networks (e.g. the IMS). Its tasks always include transport and routing of information. Further tasks (e.g. mobility control) may also be located in the CN.

Convergence   Two technologies converge when they evolve continuously towards becoming indistinguishable and exchangeable. While a literal interpretation of the term "convergence" would require convergence towards total technical identity, a more pragmatic interpretation understands convergence to mean towards indiscernability and identity *from a user's perspective.*

Data service   A non-real time service offered by a Communication Network, for example file transfer (ftp), email, or web browsing. Data services are delivered typically by packet-switched networks.

Design principle   A guideline according to which a Communication Network is designed. When a particular technical problem allows for several solutions, design principles determine which one is chosen.

Downlink   The direction from network towards Mobile Station, see Figure 5.4.

IETF Protocol   A protocol for an IP network that is developed by the IETF.

IP Network   A packet-switched Communication Network that employs IP.

Mobile network   A Communication Network in which some network elements, e.g. the Mobile Stations, are mobile.

Mobile Station (MS)   A mobile user device, e.g. a mobile phone, a laptop or even a sensor. In this book we use Mobile Station as generic term applicable to any technology. This is in contrast to, e.g. UE (User Equipment), which is a specific type of MS for UMTS, composed of subcomponents such as a User Identity Module. Mobile Terminal (MT) is the ITU-specific term.

Mobile Telecommunication Network: A mobile network that is also a Telecommunication Network.

Network Element   The physical nodes of a Communication Network, e.g. routers, servers, data bases, or controllers, including end-points such as Mobile Stations.

Multimedia service   A service delivered by a Communication Network that includes more than one medium. Examples are streaming video, video conferencing, music on demand, interactive gaming, and joint working on documents while having a voice call. Multimedia services are typically real-time services.

Radio Access Network (RAN)   The Access Points (plus possibly a controller) of a mobile Telecommunication Network. All Access Points support the same radio technology. A RAN is a subset of an Access Network.

Service   In the area of Communication Networks, a service is a somewhat general concept, offered by one abstract entity for consumption by another abstract entity. In this book a service is normally understood to be a user service, i.e. a service offered by the network to the (human) user. An example of such a service is telephony.

Telecommunication Network   A Communication Network designed according to the "telecommunications approach" (cf. Chapter 1), e.g. GSM and UMTS.

Uplink   The direction from Mobile Station towards the network, see Figure 5.4.

Appendix B

The Systematics of 3GPP Specification Numbering

3GPP specifications are identified by a number in the format

$$TS \; ab.xyz \; Vk.m.n$$
or
$$TR \; ab.xyz \; Vk.m.n,$$

for example *TS 23.060 V7.0.0*. While this may look overwhelming in the beginning, these numbers are systematic to some degree, which we shall reveal in this Appendix. This Appendix should be especially useful for those searching documents on the 3GPP web pages.

We begin with an explanation of the individual components of the specification numbering. At the end of this Appendix we include a compass for navigating the 3GPP web pages in order to find a particular specification.

- *TS* stands for **Technical Specification**, and denotes an actual standard. For example, TS 23.228 is the standard for the IMS architecture.

  TR stands for **Technical Report** and is a non-standards document. TRs contain non-binding feasibility studies, concepts and scenario descriptions. An example is TR 43.901 containing a feasibility study for GAN evolution.
- The digits *ab* encode the **specification series**, in other words the general topic of the document. Table B.1 provides an overview of the series which are defined today. The first column of the table contains the subject of the series, and the second column the series number of 3GPP Systems, GPRS and where applicable GSM, starting with R99. The third column contains the series number of the specifications which apply to GSM only, in particular GSM with an EDGE air interface.

  A particular high-level 3GPP component is normally described in specifications of several series. Let us illustrate this with the IMS: when a component—IMS in this case—is newly conceived it is described in the form of a *service* which 3GPP would deliver. This task is performed by the SA1 Working Group of 3GPP (cf. Chapter 3, Section 3.2.1), resulting in a

UMTS Networks and Beyond   Cornelia Kappler
© 2009 John Wiley & Sons, Ltd

Table B.1 Subjects of 3GPP specification series

Subject of specification series	3G System and GSM (R99 and later)	GSM only (Rel-4 and later)
Requirements	21 series	41 series
Service Aspects ("Stage 1")	22 series	42 series
Technical Realization ("stage 2")	23 series	43 series
Signalling Protocols ("stage 3") – UE to network	24 series	44 series
Radio Aspects	25 series	45 series
CODECs	26 series	46 series
Data	27 series	–
Signalling Protocols ("stage 3") RAN – CN)	28 series	48 series
Signalling Protocols ("stage 3") intra-fixed-network)	29 series	49 series
Programme Management	30 series	50 series
SIM/USIM, IC Cards. Test Specifications	31 series	51 series
Operation, Accounting, Management, Provisioning and Charging	32 series	52 series
Security Aspects	33 series	(*)
UE and (U)SIM test Specifications	34 series	(*)
Security Algorithms	35 series	55 series
Evolved UMTS Air Interface Aspects	36 series	–

(*) The specifications on this subject are spread throughout several series

specification of the 22 series. These specifications are called **Stage 1** descriptions. They never contain a technical realization of the service proposed. For example, TS 22.228 is the high-level service description of the IMS (cf. Chapter 10).

The next step consists of an architecture and high-level technical solution. This work is performed by the SA2 Working Group, and results in a **Stage 2** document of the 23 series. In the case of IMS, this is TS 23.228. The reader will note that for ease of reference the last three digits stay identical, although this convention is not always followed. The other 3GPP Working Groups now take over and work out the technical details called **Stage 3**, documented in other series. In the case of IMS, we find that TS 24.228 contains the signalling between UE and the IMS, in other words it documents how SIP and SDP are used (cf. Chapter 15). TS 29.228 is about intra-IMS signalling, in particular it covers the usage of the Diameter protocol (cf. Chapter 13).

In the course of this book we have also encountered documents from a GSM only series, namely TS 43.318 on GAN (cf. Chapter 18, Section 18.2) and TR 43.901 on GAN evolution. The reader will recall that GAN currently only applies to the A/Gb interface for EDGE, it does not apply to a 3G System.

- The digits xyz identify a document in a particular series. To some extent, they carry extra meaning. The reader should be warned, however, that this numbering space is depleted in some zones and that therefore *these conventions cannot always be followed*. Hence, the xyz digits can only hint at the content of a particular specification.
  - $0yz$ applies to 3G Systems, GPRS and GSM, whereas $1yz$ and $2yz$ apply to 3G Systems only. For example, TS 23.060 is the basic PS Domain specification that also applies to GPRS (cf. Chapter 6).
  - $8xy$ and $9xy$ are reserved for TRs.

- *Vk.m.n* identifies the version of a particular specification—most specifications are being worked on continuously and therefore their version number keeps increasing. Thereby *k* denotes the release: V3.m.n is a R99 document, V4.m.n is a Rel-4 document, and the numbering thus continues up to currently V8.m.n for Rel-8.

  A new specification starts with version number V0.0.0. When it reaches a stable state it progresses to V1.0.0. When the responsible Working Group considers the specification to be finalized it goes through a formal approval process and obtains a version number denoting to the release to which the feature belongs. When a new release is issued, the version number of the document is updated accordingly.

  The second digit *m* is augmented whenever—within the same release—the document undergoes a significant update. An amendment of the third digit *n* denotes editorial changes.

  The reader will notice that this book does not reference version numbers unless the item being referenced is indeed described only in a particular version.

3GPP specifications are available from the 3GPP web page [3GPP] by following the link **Specifications**. From there follow another link (*"Numbering page"*) to reach a table containing the desired series number. By clicking the series number, another table will open which lists all of the specifications for the series. By clicking on the desired specification one reaches a final table featuring links to all of the versions in which the specification exists.

Part II Epilogue—Convergence Revisited

This book was about mobile Telecommunication Networks, in particular 3GPP Networks. The focus was on a detailed technical description of UMTS and its evolution towards a 4G Network.

The book, however, also had a second, concurrent focus: how today's Communication Networks are in fact designed based on fundamentally different principles, while delivering comparable functionality: Telecommunication Networks are designed based on the "telecommunications approach" which we characterized as cathedral-style, operator-controlled and architecture-focussed; whereas Computer Networks are designed based on the "Computer Networks approach"—which we characterized as bazaar-style, user-controlled and protocol-focussed. Ultimately, these different design principles are a consequence of the—originally— different business models of the operators: While operators of Telecommunication Networks set out to sell a complete *end-to-end solution*, the operators of Computer Networks traditionally sell *connectivity*. Despite these differences, however, the network technologies are said to converge.

We investigated what convergence means exactly. How, and to what extent, do the different technologies converge in the course of the ongoing network evolution? As a working hypothesis we defined convergence as an evolution towards technical identity and indistinguishability.

In the Epilogue for Part I, we found that in 3G, Telecommunication Networks and Computer Networks have converged only on a superficial level, regarding their functionality and the services they can provide. They have, however, not converged on a deeper technical level. Operators of Telecommunication Networks and Computer Networks follow different business models and, consequently, continue to build their network based on different design principles.

We now revisit the same question for pre-4G Networks, with the example of 3GPP's EPS and Mobile WiMAX—of course the conclusions we can draw are only preliminary since the specifications have not yet been finalized.

The first thing to observe is that business models are becoming more diverse and to some extent they are converging. As discussed in Chapter 20, Section 20.2.2, the operators of Telecommunication Networks have not exactly abandoned the idea of being end-to-end solution providers; however they have opened up to a collaboration with other parties which represent other technologies and other business ideas, see for example the seamless integration

UMTS Networks and Beyond   Cornelia Kappler
© 2009 John Wiley & Sons, Ltd

of heterogeneous Access Networks. They have also opened up to services developed and offered by third parties outside the telecommunications world. At the same time, the ownership of the actual subscriptions "identity provisioning" could emerge as the centre piece of their business model.

What about the business model of operators of Computer Networks? The most interesting example in this respect is Mobile WiMAX. While the aim of Mobile WiMAX operators is indeed selling connectivity rather than services, we may also observe a broadening of focus. Unlike traditional Computer Networks, Mobile WiMAX actually sells "controlled" connectivity, e.g. including mobility support and QoS. Mobile WiMAX also supports explicitly collaboration with Telecommunication Networks.

How is this shift—and convergence—in business models reflected on a technical level?

- The cathedral and the bazaar
  The 3GPP System of Rel-8, EPS, continues to be designed cathedral-style as a complete system with carefully organized interworking between control functions. However, the system is more modular than in previous releases: Access Networks are exchangeable to some degree, and different protocol options exist, simplifying interworking.

  Interestingly, the WiMAX design adopts some "cathedral ideas" in that a high-level, overall architecture is specified; on the other hand, Mobile WiMAX is a modular technology not only on an architectural but also on a functional level: for example, an operator whose business model is selling wireless broadband access can save costs by not deploying mobility support (e.g. Home Agents) and QoS.
- Operator control and user control
  Operator-exerted control is very much en vogue for EPS. We saw how the new IETF mobility protocol which allows mobility control by the network, Proxy MIP, was embraced immediately. Control is exerted preferably by an omniscient entity—a role now played by the PCRF. Mobile WiMAX also locates control with the operator, particularly the operator of the Core Network. For example, user traffic—even local traffic—always travels via the Core Network in order to allow the operator analysis and intervention.
- "In the beginning is the architecture" and "In the beginning is the protocol"
  EPS is architecture-centric just as with previous releases: the architecture is designed first. Almost all Core Network protocols are one-protocol-per-function IETF Protocols, as this allows for better interworking.
  Mobile WiMAX lives in both worlds. A comparatively high-level architecture has been defined in order to allow interworking with Telecommunication Networks. Mobile WiMAX does not have any stake in protocol design; therefore IETF Protocols are adopted conveniently.

Overall, we may observe a slight softening of design principles in 4G Networks, and consequently a technical convergence on a deeper level than for 3G Networks. However, a Telecommunication Network and a Computer Network are still quite simple to tell apart.

It is also worthwhile reconsidering our original hypothesis of the meaning of convergence in the light of 4G. A distinguishing characteristic of 4G is convergence of Access Networks. Convergence is thus understood as simply indiscernability from a user's perspective. In the light of this pragmatic interpretation, 4G Networks are indeed likely to be converged networks. With these observations the book closes and the author wishes the reader a good time following the exciting future evolution of mobile Communication Networks.

Index

References

[3G Americas 2007] 'Defining 4G: Understanding the ITU Process for the Next Generation of Wireless Technology', 3G Americas LLC, June 2007.

[3GPP] http://www.3gpp.org.

[3GPP2] http://www.3gpp2.org.

[3GToday] http://www.3gtoday.com.

[3GPP 22.228] 3GPP TS 22.228, 'Service requirements for the Internet Protocol (IP) multimedia core network subsystem'.

[3GPP 22.259] 3GPP TS 22.259, 'Service requirements for Personal Network Management (PNM)'.

[3GPP 22.278] 3GPP TS 22.278, 'Service requirements for the Evolved Packet System (EPS)'.

[3GPP 22.912] 3GPP TR 22.912, 'Study into network selection requirements for non-3GPP Access'.

[3GPP 22.934] 3GPP TR 22.934, 'Feasibility study on 3GPP system to Wireless Local Area Network (WLAN) interworking'.

[3GPP 22.978] 3GPP TR 22.978, 'All-IP Network (AIPN) feasibility study'.

[3GPP 22.980] 3GPP TR 22.980, 'Network composition feasibility study'.

[3GPP 23.002] 3GPP TS 23.002, 'Network architecture'.

[3GPP 23.060] 3GPP TS 23.060, 'General Packet Radio Service (GPRS)'.

[3GPP 23.107] 3GPP TS 23.107, 'Quality of Service (QoS) concept and architecture'.

[3GPP 23.125] 3GPP TS 23.125, 'Overall high-level functionality and architecture impacts of flow-based charging'.

[3GPP 23.141] 3GPP TS 23.141 'Presence Service'.

[3GPP 23.174] 3GPP TS 23.174 'Push Service'.

[3GPP 23.203] 3GPP TS 23.203, 'Policy and charging control architecture'.

[3GPP 23.205] 3GPP TS 23.205, 'Bearer-independent circuit-switched core network'.

[3GPP 23.207] 3GPP TS 23.207, 'End-to-end Quality of Service (QoS) concept and architecture'.

[3GPP 23.228] 3GPP TS 23.228, 'IP Multimedia Subsystem (IMS)'.

[3GPP 23.327] 3GPP TS 23.327, 'Mobility between 3GPP-Wireless Local Area Network (WLAN) Interworking and 3GPP Systems'.

[3GPP 25.301] 3GPP TS 25.301, 'Radio Interface Protocol Architecture'.

[3GPP 25.303] 3GPP TS 25.303, 'Interlayer procedures in Connected Mode'.

[3GPP 25.304] 3GPP TS 25.304, 'User Equipment (UE) procedures in idle mode and procedures for cell reselection in connected mode'.

[3GPP 25.308] 3GPP TS 25.308 'High Speed Downlink Packet Access (HSDPA)'.

[3GPP 25.309] 3GPP TS 25.309 'FDD enhanced uplink'.

[3GPP 25.331] 3GPP TS 25.331, 'Radio Resource Control (RRC); Protocol Specification'.

[3GPP 25.401] 3GPP TS 25.401, 'UTRAN overall description'.

[3GPP 25.820] 3GPP TR 25.820, '3G Home NodeB Study Item Technical Report'.

[3GPP 25.913] 3GPP TR 25.913, 'Requirements for Evolved UTRA (E-UTRA) and Evolved UTRAN (E-UTRAN)'.

[3GPP 27.001] 3GPP TS 23.001, 'General on Terminal Adaptation Functions (TAF) for Mobile Stations (MS)'.

[3GPP 29.213] 3GPP TS 29.213, 'Policy and Charging Control signalling flows and QoS parameter mapping'.

[3GPP 31.102] 3GPP TS 31.102, 'Characteristics of the Universal Subscriber Identity Module (USIM) application'.

[3GPP 31.103] 3GPP TS 31.103, 'Characteristics of the IP Multimedia Services Identity Module (ISIM) application'.

[3GPP 32.240] 3GPP TS 32.240 'Charging management. Charging Architecture and Principles'.

[3GPP 32.251] 3GPP TS 32.251 'Charging management. Packet Switched (PS) domain charging'.

[3GPP 32.252] 3GPP TS 32.252 'Charging management. Wireless Local Area Network (WLAN) charging'.

[3GPP 32.260] 3GPP TS 32.260 'Charging management. IP Multimedia Subsystem (IMS) charging'.

[3GPP 32.821] 3GPP TR 32.821 'Telecommunication management; Study of Self-Organising Networks (SON) related OAM Interfaces for Home NodeB'.

[3GPP 33.102] 3GPP TS 33.102 'Security architecture'.

[3GPP 33.202] 3GPP TS 33.203 'Access security for IP-based services'.

[3GPP 33.210] 3GPP TS 33.210 'Network domain security; IP network layer security'.

[3GPP 33.220] 3GPP TS 33.220 'Generic Authentication Architecture (GAA); Generic bootstrapping architecture'.

[3GPP 33.401] 3GPP TS 33.401 '3GPP System Architecture Evolution (SAE): Security Architecture'.

[3GPP 33.402] 3GPP TS 33.402 '3GPP System Architecture Evolution (SAE): Security aspects of non-3GPP accesses'.

[3GPP 36.300] 3GPP TS 36.300 'Evolved Universal Terrestrial Radio Access (E-UTRA) and Evolved Universal Terrestrial Radio Access Network (E-UTRAN)'.

[3GPP 36.902] 3GPP TR 36.902 'Evolved Universal Terrestrial Radio Access Network (E-UTRAN); Self-configuring and self-optimizing network use cases and solutions'.

[3GPP 36.913] 3GPP TR 36.913 'Requirements for Further Advancements for E-UTRA (LTE-Advanced)'.

[3GPP 43.318] 3GPP TS 43.318 'Generic access to the A/Gb interface'.

[3GPP 43.901] 3GPP TR 43.901 'Feasibility study on generic access to A/Gb interface'.

[3GPP 43.902] 3GPP TR 43.902 'Enhanced Generic Access Networks Study (EGAN)'.

[3GPP2 C.S0084] 3GPP2 C.S0084-000 'Overview for Ultra Mobile Broadband (UMB) Air Interface Specification', Sept. 2007.

[3GPP2 X.S0011] 3GPP2 X.S0011-001-C 'cdma2000 Wireless IP Network Standard: Introduction' v3.0, Nov. 2006.

[3GPP2 X.S0013] 3GPP2 3GPP2 X.S0013-000-0 'All-IP Core Network Multimedia Domain—Overview', Aug. 2005.

[3GPP2 X.S0054] 3GPP2 S0054-000-0 'CAN Wireless IP Network Overview and List of Parts' v1.0, Dec. 2007.

[Agrawal 2008] P. Agrawal, J. -H. Yeh, J. -C. Chen, T. Zhang, 'IP Multimedia Subsystem in 3GPP2 and 3GPP2: Overview and Scalability Issues', *IEEE Communications Magazine*, Jan. 2008.

[Ahmavaara 2003] K. Ahmavaara, H. Haverinen, R. Pichna, 'Interworking architecture between 3GPP and WLAN systems', *IEEE Communication Magazine*, Nov. 2003.

[Ambient Networks] http://www.ambient.networks.org.

[Bluetooth] http://www.bluetooth.com/bluetooth/.

[Boman 2002] K. Boman, G. Horn, P. Howard, and V. Niemi, 'UMTS Security', *IEE Electronic and Communication Engineering Journal*, Oct. 2002.

[Braden 2002] R. Braden, T. Faber, M. Handley, 'From Protocol Stack to Protocol Heap—Role-Based Architecture'. *HotNets-I*, Princeton, NJ, Oct. 2002.

[Camarillo 2005] G. Camarillo, M. A García-Martín, *The 3G IP Multimedia Subsystem* 2nd ed., John Wiley & Sons, Ltd, Chichester, 2005.

[Chalmers 1999] D. Chalmers, M. Sloman, 'A Survey of Quality of Service in Mobile Computing Environments', *IEEE Communication Surveys*, Feb. 1999.

[Clark 2003a] D. Clark, K. Sollins, J. Wroclawski, D. Katabi, J. Kulik, X. Yang, R. Braden, T. Faber, A. Falk, V. Pingali, M. Handley, N. Chiappa, 'New Arch: Future Generation Internet Architecture', *Final Technical Report*, Dec. 2003.

[Clark 2003b] D. Clark, R. Braden, A. Falk, V. Pingali, 'FARA: Reorganizing the Addressing Architecture'. *ACM SIGCOMM 2003 FDNA Workshop*, Karlsruhe, Germany, Aug. 2003.

[Clark 2003c] D. Clark, C. Partridge, J. C Ramming, J. T Wroclawski, 'A Knowledge Plane for the Internet', *ACM SIGCOMM 2003 FDNA Workshop*, Karlsruhe, Germany, Aug. 2003.

[digitalstrom] http://www.digitalstrom.org.

[Dovrolis 2008] Constantine Dovrolis, 'What would Darwin Think about Clean-Slate Architectures?' *ACM SIGCOMM Computer Communications Review 31.1*, Jan. 2008.

[Emmelmann 2007] M. Emmelmann, S. Wiethoelter, A. Köpsel, C. Kappler, A. Wolisz, 'Moving towards Seamless Mobility, state of the art and emerging aspects in standardization bodies', *Wireless Personal Communications*, April 2007.

[Etoh 2005] M. Etoh (Ed.), *Next Generation Mobile Systems; 3G and Beyond*, John Wiley & Sons, Ltd, Chichester, 2005.

[ETSI 180 001] ETSI ES 180 001 'Telecommunications and Internet converged Services and Protocols for Advanced Networking (TISPAN); NGN Release 1; Release definition', 2006.

[ETSI 282 001] ETSI ES 282 001 'Telecommunications and Internet converged Services and Protocols for Advanced Networking (TISPAN); NGN Functional Architecture Release 1', 2006.

[Femto Forum] http://www.femtoforum.org.

[Gast 2005] M. S. Gast, '*802. 11 Wireless Networks: The Definite Guide*' 2nd ed., O'Reilly, Sebastopol, CA, 2005

[GSMA] http://www.gsmworld.com.

[GSMA IR.33] GSM Association Permanent Reference Document IR.33 'GPRS Roaming Guidelines'.

[GSMA IR.34] GSM Association Permanent Reference Document IR.34 'Inter-Service Provider IP Backbone Guidelines'.

[Guerin 1999] R. Guérin, V. Peris, 'Quality-of-service in packet networks: basic mechanisms and directions', *Computer Networks*, **31**. 1999.

[Holma 2007] H. Holma, A. Toskala, *WCDMA for UMTS: HSPA Evolution and LTE*, John Wiley & Sons, Ltd, Chichester, 2007.

[Hoßfeld 2006] T. Hoßfeld, A. Binzenhöfer, M. Fiedler, K. Tutschku, 'Measurement and Analysis of Skype VoIP Traffic in 3G UMTS Systems', *4th Int. Workshop IPS-MoMe 2006, Salzburg, Austria*, Feb. 2006.

[ID DSMIPv6] Hesham Soliman (Ed.), 'Mobile IPv6 support for dual stack Hosts and Routers (DSMIPv6)', draft-ietf-mext-nemo-v4traversal-v5, July 2008.

[ID pds NSIS] R. Hancock, C. Kappler, J. Quittek, M. Stiemerling, 'Problem Statement for Path-Decoupled Signalling in NSIS', draft-hancock-nsis-pds-problem-04, Oct. 2006.

[ID PMIPv4] K. Leung, G. Dommety, P. Yegani, K. Chowdhury, 'WiMAX Forum/3GPP2 Proxy Mobile IPv4', Internet Draft (work in progress), draft-leung-mip4-proxy-mode-09, July 2008.

[ID RADIUS ext] A. Lior, P. Yegani, K. Chowdhury, H. Tschofenig, A. Pashalidis, 'PrePaid Extensions to Remote Authentication Dial-In User Service (RADIUS)' Internet Draft (work in progress), draft-lior-radius-prepaid-extensions-14, July 2008.

[ID STUN] J. Rosenberg, R. Mahy, P. Matthews, D. Wing, 'Session Traversal Utilities for NAT (STUN)', Internet Draft (work in progress), draft-ietf-behave-rfc3489bis-18.txt, July 2008.

[ID QoS NSLP] J. Manner, G. Karagiannis, A. McDonald, 'NSLP for Quality-of-Service Signaling', Internet Draft (work in progress), draft-ietf-nsis-qos-nslp-16.txt, Feb. 2008.

[IEEE 802] http://standards.ieee.org/getieee802/.

[IEEE 802.1x] 802.1X-2001, 'Port Based Network Access Control', 2001.

[IEEE 802.11] IEEE 802.11-1999, 'IEEE Standard for Information technology—Telecommunication and information exchange between systems—Local and metropolitan area networks—Specific requirements—part 11: Wireless LAN Medium Access Control (MAC) and Physical Layer (PHY) specifications', 1999.

[IEEE 802.11e] 802.11e-2005, 'IEEE Standard for Information technology—Telecommunication and information exchange between systems—Local and metropolitan area networks—specific requirements Part 11: Wireless LAN Medium Access Control (MAC) and Physical Layer (PHY) specifications: Amendment 8: Medium Access Control (MAC) Quality of Service Enhancements', 2005.

[IEEE 802.11f] 802.11f-2003 (withdrawn), 'IEEE Trial-Use Recommended Practice for Multi-Vendor Access Point Interoperability via an Inter-Access Point Protocol Across Distribution Systems Supporting IEEE 802.11 Operation', 2003.

[IEEE 802.11i] 802.11i-2004 Amendment to IEEE Std 802.11, 1999 Edition (Reaff. 2003). IEEE Standard for Information technology—Medium Access Control (MAC) Security Enhancements, 2004.

[IEEE 802.15.4] 802.15.4-2003, IEEE Standard for Information technology—Telecommunications and information exchange between systems—Local and metropolitan area networks specific requirements part 15.4: wireless medium access control (MAC) and physical layer (PHY) specifications for low-rate wireless personal area networks (LR-WPANs), 2003.

[IEEE 802.16] http://ieee802.org/16/.

[IEEE 802.16-2004] 802.16-2004, 'IEEE Standard for local and metropolitan area networks; Air Interface for Fixed Broadband Wireless Access Systems', 2004.

[IEEE 802.16e] 802.16e, 'IEEE Standard for local and metropolitan area networks; Air Interface for Fixed Broadband Wireless Access Systems, Amendment 2: Physical and Medium Access Control Layers for Combined Fixed and Mobile Operation in Licensed Bands and Corrigendum 1', 2006.

[IEEE 802.16h] P802.16h—Amendment to IEEE Standard for Local and Metropolitan Area Networks—Part 16: Air Interface for Fixed Broadband Wireless Access Systems—Improved Coexistence Mechanisms forLicense-Exempt Operation, Draft.

[IEEE 802.21] 802.21, 'Draft IEEE Standard for local and metropolitan area networks: Media Independent Handover Services'.

[IETF] http://www.ietf.org.

[IETF MANET] http://www.ietf.org/html.charters/manet-charter.html.

[ISTAG AmI 2001] K. Ducatel, M. Bogdanowicz, F. Scapolo, J. Leijten, J.-C. Burgelman,'ISTAG Scenarios for Ambient Intelligence in 2010', http://www.cordis.lu/ist/istag, Feb. 2001.

[ITU-D] http://www.itu.int/ITU-D/ict/statistics/.

[ITU Report 2005] ITU Internet Reports 2005, 'The Internet of Things', Nov. 2005.

[ITU H.248] 'Gateway control protocol: Version 3', H.248.1, ITU-T, Sept. 2005.

[ITU I.322] 'Generic protocol reference model for telecommunication networks' I.433, ITU-T, Feb. 1999.

[ITU M.1645] 'Framework and overall objectives of the future development of IMT-2000 and systems beyond IMT-2000', M.1645, June 2003.

[ITU Q.1701] 'Framework for IMT-2000 Networks' Q.1701, ITU-T, March 1999.

[ITU Q.1702] 'Long-term vision of network aspects for systems beyond IMT-2000' Q.1702, June 2002.

[ITU Q.1703] 'Service and network capabilities framework of network aspects for systems beyond IMT-2000', Q.1703, May 2004.

[ITU Y.2001] 'General Overview of NGN', Y.2001, Dec. 2004.

[Kappler 2005] C. Kappler, N. Akhtar, R. Campos, P. Pöyhönen,'Network Composition using existing and new technologies', *IST Summit 2005*, Dresden, June 2005.

[Kasera 2005] S. Kasera, N. Narang, *3G Mobile Networks*, McGraw-Hill, New York, 2005.

[Kempf 2003] J. Kempf, P. Mutaf,'IP Paging Considered Unnecessary: Mobile IPv6 and IP Paging for Dormant Mode Location Update in Macrocellular and Hotspot Networks', *WCNC 2003*, New Orleans, March 2003.

[Kephart 2003] J. O Kephart, D. M Chess, 'The Vision of Autonomic Computing', *IEEE Computer Magazine*, Jan. 2003

[Kühne 2007] R. Kühne, G. Görmer, M. Schläger, G. Carle, 'Charging in the IP Multimedia Subsystem (IMS)—A Tutorial', *IEEE Communications Magazine*, July 2007.

[Lampropoulos 2008] G. Lampropoulos, A. Salkintzis, N. Passas, 'Media-Independent Handover for Seamless Service Provision in Heterogeneous Networks', *IEEE Communication Magazine*, Jan. 2008.

[Lescuyer 2004] P. Lescuyer, *UMTS Its Origins, Architecture and the Standard*, Springer, London, 2004.

[OHA] http://www.openhandsetalliance.com.

[OMA] http://www.openmobilealliance.org.

[PKT FW 2.0 2007] 'PacketCable 2.0 Architecture Framework Technical Report', PKT-TR-ARCH-FRM-V04-071106, Cable Television Laboratories, Nov. 2007.

[Raymond 2000] http://www.catb.org/~esr/writings/cathedral-bazaar.

[RFC 826] D. C Plummer,'An Ethernet Address Resolution Protocol', RFC 826, Nov. 1982.

[RFC 1633] R. Braden, D. Clark, S. Shenker,'Integrated Services in the Internet Architecture: an Overview', RFC 1633, June 1994.

[RFC 2205] R. Braden, Ed., L. Zhang, S. Berson, S. Herzog, S. Jamin,'Resource ReSerVation Protocol (RSVP)—Version 1 Functional Specification', RFC 2205, Sept. 1997.

[RFC 2327] M. Handley, V. Jacobson,'SDP: Session Description Protocol', RFC 2327, April 1998.

[RFC 2396] T. Berners-Lee, R. Fielding, U. C Irvine, L. Masinter,'Uniform Resource Identifiers (URI): Generic Syntax', RFC 2396, Aug. 1998.

[RFC 2409] D. Harkins, D. Carrel,'The Internet Key Exchange (IKE)' RFC 2409, November 1998.

[RFC 2462] S. Thomson, T. Narten,'IPv6 Stateless Address Autoconfiguration,' RFC 2462, December 1998.

[RFC 2475] S. Blake, D. Black, M. Carlson, E. Davies, Z. Wang, W. Weiss,'An Architecture for Differentiated Services', RFC 2475, December 1998.

[RFC 2486] B. Aboba, M. Beadles,'The Network Access Identifier', RFC 2486, January 1999.

[RFC 2570] J. Case, R. Mundy, D. Partain, B. Steward,'Introduction to Version 3 of the Internet-standard Network Management Framework', RFC 2570, April 1999.

[RFC 2607] B. Aboba, J. Vollbrecht, 'Proxy Chaining and Policy Implementation in Roaming', RFC 2607, June 1999.

[RFC 2663] P. Srisuresh, M. Holdrege, 'IP Network Address Translator (NAT) Terminology and Considerations', RFC 2663, Aug. 1999.

[RFC 2722] N. Brownlee, C. Mills, G. Ruth, 'Traffic Flow Measurement: Architecture', RFC 2722, Oct. 1999.

[RFC 2748] J. Boyle, R. Cohen, D. Durham, S. Herzog, R. Raja, A. Sastry, 'The COPS (Common Open Policy Service) Protocol', RFC 2748, Jan. 2000.

[RFC 2753] R. Yavatkar, D. Pendarakis, R. Guerin, 'A framework for policy-based admission control', RFC 2753, Jan. 2000.

[RFC 2784] D. Farinacci, T. Li, S. Hanks, D. Meyer, P. Traina, 'Generic Routing Encapsulation (GRE)', RFC 2784, March 2000.

[RFC 2806] A. Vaha-Sipila, 'URLs for Telephone Calls', RFC 2806, April 2000.

[RFC 2865] C. Rigney, S. Willens, A. Rubens, W. Simpson, 'Remote Authentication Dial In User Service (RADIUS)', RFC 2865, June 2000.

[RFC 2975] B. Aboba, J. Arkko, D. Harrington, 'Introduction to Accounting Management', RFC 2975, Oct. 2000.

[RFC 2996] Y. Bernet, 'Format of the RSVP DCLASS Object', RFC 2996, Nov. 2000.

[RFC 3015] F. Cuervo, N. Greene, A. Rayhan, C. Huitema, B. Rosen, J. Segers, 'Megaco Protocol Version 1.0', RFC 3015, Nov. 2000.

[RFC 3084] K. Chan, J. Seligson, D. Durham, S. Gai, K. McCloghrie, S. Herzog, F. Reichmeyer, R. Yavatkar, A. Smith, 'COPS Usage for Policy Provisioning (COPS-PR)', RFC 3084, March 2001.

[RFC 3132] J. Kempf, 'Dormant Mode Host Alerting ('IP Paging') Problem Statement', RFC 3121, June 2001.

[RFC 3209] D. Awduche, L. Berger, D. Gan, T. Li, V. Srinivasan, G. Swallow, 'RSVP-TE: Extensions to RSVP for LSP Tunnels', RFC 3209, Dec. 2001.

[RFC 3261] J. Rosenberg, H. Schulzrinne, G. Camarillo, A. Johnston, J. Peterson, R. Sparks, M. Handley, E. Schooler, ' SIP: Session Initiation Protocol', June 2002.

[RFC 3264] J. Rosenberg, H. Schulzrinne, 'An Offer/Answer Model with the Session Description Protocol (SDP), RFC 3264, June 2002.

[RFC 3031] E. Rosen, A. Viswanathan, R. Callon, 'Multiprotocol Label Switching Architecture', RFC 3031, Jan. 2001.

[RFC 3312] G. Camarillo, Ed., W. Marshall, Ed., J. Rosenberg, 'Integration of Resource Management and Session Initiation Protocol (SIP)', RFC 3312, October 2002.

[RFC 3313] W. Marshall, Ed., 'Private Session Initiation Protocol (SIP) Extensions for Media Authorization', RFC 3313, January 2003.

[RFC 3315] R. Droms, Ed., J. Bound, B. Volz, T. Lemon, C. Perkins, M. Carney, 'Dynamic Host Configuration Protocol for IPv6 (DHCPv6),' RFC 3315, July 2003.

[RFC 3344] C. Perkins, 'IP Mobility Support for IPv4', RFC 3344, Aug. 2002.

[RFC 3455] M. García-Martín, E. Henrikson, D. Mills, 'Private Header (P-Header) Extensions to the Session Initiation Protocol (SIP) for the 3rd-Generation Partnership Project (3GPP)', RFC 3455, Jan. 2003.

[RFC 3550] H. Schulzrinne, S. Casner, R. Frederick, V. Jacobson, 'RTP: A Transport Protocol for Real-Time Applications', RFC 3550, July 2003.

[RFC 3588] P. Calhoun, J. Loughney, E. Guttman, G. Zorn, J. Arkko, 'Diameter Base Protocol', RFC 3588, Sept. 2003.

[RFC 3746] L. Yang, R. Dantu, T. Anderson, R. Gopal, 'Forwarding and Control Element Separation (ForCES) Framework', RFC 3746, April 2004.

[RFC 3748] B. Aboba, L. Blunk, J. Vollbrecht, J. Carlson, H. Levkowetz (Ed.), 'Extensible Authentication Protocol (EAP)', RFC 3748, June 2004.

[RFC 3775] D. Johnson, C. Perkins, J. Arkko, 'Mobility Support in IPv6', RFC 3775, June 2004.

[RFC 3935] H. Alvestrand, 'A Mission Statement for the IETF', RFC 3935 Oct. 2004.

[RFC 3954] B. Claise (Ed.), 'Cisco Systems NetFlow Services Export Version 9', RFC 3954, Oct. 2004.

[RFC 3955] S. Leinen, 'Evaluation of Candidate Protocols for IP Flow Information Export (IPFIX)', RFC 3955, Oct. 2004.

[RFC 3963] V. Devarapalli, R. Wakikawa, A. Petrescu, P. Thubert, 'Network Mobility (NEMO) Basic Support Protocol', RFC 3963, Jan. 2005.

[RFC 4005] P. Calhoun, G. Zorn, D. Spence, D. Mitton, 'Diameter Network Acess Server Application', RFC 4005, Aug. 2005.

[RFC 4006] H. Hakala, L. Mattila, J. -P. Koskinen, M. Stura, L. Loughney, 'Diameter Credit-Control Application', RFC 4006, August 2005.

[RFC 4066] M. Liebsch, Ed. A. Singh (Ed.), H. Chaskar, D. Funato, E. Shim,'Candidate Access Router Discovery (CARD)', RFC 4066, July 2005.

[RFC 4067] J. Loughney, Ed., M. Nakhjiri, C. Perkins, R. Koodli,'Context Transfer Protocol (CXTP)', RFC 4067, July 2005.

[RFC 4068] R. Koodli, Ed. 'Fast Handovers for Mobile IPv6', RFC 4068, July 2005.

[RFC 4140] H. Soliman, C. Castelluccia, K. El Malki, L. Bellier,'Hierarchical Mobile IPv6 Mobility Management (HMIPv6)', RFC 4140, Aug. 2005.

[RFC 4301] S. Kent, K. Seo,'Security Architecture for the Internet Protocol', RFC 4301, Dec. 2005.

[RFC 4302] S. Kent,'IP Authentication Header', RFC 4302, Dec. 2005.

[RFC 4303] S. Kent,'IP Encapsulating Security Payload (ESP)', RFC 4303, Dec. 2005.

[RFC 4306] C. Kaufman (Ed.), 'Internet Key Exchange (IKEv2) Protocol', RFC 4306, Dec. 2005.

[RFC 4346] T. Dierks, E. Rescorla,'The Transport Layer Security (TLS) Protocol Version 1.1', April 2006.

[RFC 4423] R. Moskowitz, P. Nikander,'Host Identity Protocol (HIP) Architecture', RFC 4423, May 2006.

[RFC 4555] P. Eronen, Ed., 'IKEv2 Mobility and Multihoming Protocol (MOBIKE)', RFC 4555, June 2006.

[RFC 4640] A. Patel, Ed., G. Giaretta, Ed. 'Problem Statement for Bootstrapping Mobile IPv6 (MIPv6)', RFC 4640, Sept. 2006.

[RFC 4830] J. Kempf, Ed., 'Problem Statement for Network-Based Localized Mobility Management (NETLMM)', RFC 4830, April 2007.

[RFC 4944] G. Montenegro, N. Kushalnagar, J. Hui, D. Culler,' Transmission of IPv6 Packets over IEEE 802.15.4 Networks', RFC 4944, Sept. 2007.

[RFC 5113] J. Arkko, B. Aboba, J. Korhonen, Ed., F. Bari,'Network Discovery and Selection Problem', RFC 5113, Jan. 2008.

[RFC 5193] P. Jayaraman, R. Lopez, Y. Ohba (Ed.), M. Parthasarathy, A. Yegin,'Protocol for Carrying Authentication for Network Access (PANA) Framework', RFC 5193, May 2008.

[RFC 5205] P. Nikander, J. Laganier,'Host Identity Protocol (HIP) Domain Name System (DNS) Extension', RFC 5205, April 2008.

[RFC 5213] S. Gundavelli (Ed.), K. Leung, V. Devarapalli, K. Chowdhury, B. Patil, 'Proxy Mobile IPv6', RFC 5213, August 2008.

[Saunders 2007] S. Saunders, 'Three Ages of Future Wireless Communication' in W. Webb (Ed.), *Wireless Communications; the Future*, John Wiley & Sons, Ltd, Chichester, 2007.

[Tafazolli 2006] R. Tafazolli, *Technologies for the Wireless Future: Wireless World Research Forum (WWRF)*. Vol. **2**, John Wiley & Sons, Ltd, Chichester, 2006.

[Tanenbaum 2002] A. Tanenbaum, *Computer Networks*, Prentice Hall, NJ, 4th Edition 2002.

[Thaler 2007] D. Thaler, B. Aboba, 'What Makes For a Successful Protocol?', *IETF Journal 3.3*, Dec. 2007.

[UMTS Forum 2001] UMTS Forum, Report number 13, April 2001.

[UWB] http://uwbforum.org/index.php.

[Walke 2003] B. Walke, P. Seidenbert, M. P. Althoff, *UMTS; The Fundamentals*, John Wiley & Sons, Ltd, Chichester, 2003.

[Webb 2007] W. Webb (Ed.), *Wireless Communications; the Future*, John Wiley & Sons, Ltd, Chichester, 2007.

[Weiser 1991] M. Weiser, 'The Computer for the 21st Century', *Scientific American*, Sept. 1991.

[WiMAX Forum] http://www.wimaxforum.org/home/.

[WF Arch] WiMAX Forum Network Architecture (Stage 2: Architecture Tenets, Reference Model and Reference Points), Release 1.2, 2008.

[WF Overview 2006] 'Mobile WiMAX—Part I: A Technical Overview and Performance Evaluation', WiMAX Forum White Paper, 2006.

[ZigBee 2006] 'ZigBee-2006 Specification', ZigBee Standardization Organization, 2006.

[Zimmermann 2005] K. Zimmermann, S. Felis, S. Schmid, L. Eggert, M. Brunner,'Autonomic wireless network management', *WAC 2005*, Athens, Greece, Oct. 2005.

Printed and bound in the UK by
CPI Antony Rowe, Eastbourne

Printed and bound by CPI Group (UK) Ltd, Croydon, CR0 4YY

27/10/2024

14580168-0002